Retrofitting for Optimal Energy Performance

Adrian Tantau
Bucharest University of Economic Studies, Romania

A volume in the Advances in Environmental Engineering and Green Technologies (AEEGT) Book Series

IGI Global
DISSEMINATOR OF KNOWLEDGE

Published in the United States of America by
IGI Global
Engineering Science Reference (an imprint of IGI Global)
701 E. Chocolate Avenue
Hershey PA, USA 17033
Tel: 717-533-8845
Fax: 717-533-8661
E-mail: cust@igi-global.com
Web site: http://www.igi-global.com

Library of Congress Cataloging-in-Publication Data

Names: Tantau, Adrian Dumitru, editor.
Title: Retrofitting for optimal energy performance / Adrian Tantau, editor.
Description: Hershey, PA : Engineering Science Reference, [2019]
Identifiers: LCCN 2018057857| ISBN 9781522591047 (h/c) | ISBN 9781522591061 (ebook) | ISBN 9781522591054 (s/c)
Subjects: LCSH: Buildings--Energy conservation. | Sustainable construction--Costs. | Buildings--Repair and reconstruction--Environmental aspects.
Classification: LCC TJ163.5.B84 R465 2019 | DDC 696--dc23 LC record available at https://lccn.loc.gov/2018057857

This book is published in the IGI Global book series Advances in Environmental Engineering and Green Technologies (AEEGT) (ISSN: 2326-9162; eISSN: 2326-9170)

British Cataloguing in Publication Data
A Cataloguing in Publication record for this book is available from the British Library.

All work contributed to this book is new, previously-unpublished material. The views expressed in this book are those of the authors, but not necessarily of the publisher.

For electronic access to this publication, please contact: eresources@igi-global.com.

Advances in Environmental Engineering and Green Technologies (AEEGT) Book Series

Sang-Bing Tsai
University of Electronic Science and Technology of China Zhongshan Institute, China
Ming-Lang Tseng
Lunghwa University of Science and Technology, Taiwan
Yuchi Wang
University of Electronic Science and Technology of China Zhongshan Institute, China

ISSN:2326-9162
EISSN:2326-9170

Mission

Growing awareness and an increased focus on environmental issues such as climate change, energy use, and loss of non-renewable resources have brought about a greater need for research that provides potential solutions to these problems. Research in environmental science and engineering continues to play a vital role in uncovering new opportunities for a "green" future.

The **Advances in Environmental Engineering and Green Technologies (AEEGT)** book series is a mouthpiece for research in all aspects of environmental science, earth science, and green initiatives. This series supports the ongoing research in this field through publishing books that discuss topics within environmental engineering or that deal with the interdisciplinary field of green technologies.

Coverage

- Policies Involving Green Technologies and Environmental Engineering
- Pollution Management
- Green Technology
- Green Transportation
- Waste Management
- Air Quality
- Renewable Energy
- Industrial Waste Management and Minimization
- Contaminated Site Remediation
- Electric Vehicles

The Advances in Environmental Engineering and Green Technologies (AEEGT) Book Series (ISSN 2326-9162) is published by IGI Global, 701 E. Chocolate Avenue, Hershey, PA 17033-1240, USA, www.igi-global.com. This series is composed of titles available for purchase individually; each title is edited to be contextually exclusive from any other title within the series. For pricing and ordering information please visit http://www.igi-global.com/book-series/advances-environmental-engineering-green-technologies/73679. Postmaster: Send all address changes to above address.

Titles in this Series

For a list of additional titles in this series, please visit: www.igi-global.com/book-series

Amelioration Technology for Soil Sustainability
Ashok K. Rathoure (Biohm Consultare Pvt Ltd, India)
Engineering Science Reference • ©2019 • 280pp • H/C (ISBN: 9781522579403) • US $185.00

Advanced Agro-Engineering Technologies for Rural Business Development
Valeriy Kharchenko (Federal Scientific Agroengineering Center VIM, Russia) and Pandian Vasant (Universiti Teknologi PETRONAS, Malaysia)
Engineering Science Reference • ©2019 • 484pp • H/C (ISBN: 9781522575733) • US $195.00

Spatial Planning in the Big Data Revolution
Angioletta Voghera (Politecnico di Torino, Italy) and Luigi La Riccia (Politecnico di Torino, Italy)
Engineering Science Reference • ©2019 • 359pp • H/C (ISBN: 9781522579274) • US $195.00

Global Initiatives for Waste Reduction and Cutting Food Loss
Aparna B. Gunjal (Asian Agri Food Consultancy Services Ltd, India) Meghmala S. Waghmode (Annasaheb Magar Mahavidyalaya, India) Neha N. Patil (Annasaheb Magar Mahavidyalaya, India) and Pankaj Bhatt (Dolphin (P.G) College of Biomedical and Natural Sciences Dehradun, India)
Engineering Science Reference • ©2019 • 328pp • H/C (ISBN: 9781522577065) • US $195.00

Green Public Procurement Strategies for Environmental Sustainability
Rajesh Kumar Shakya (The World Bank, USA)
Engineering Science Reference • ©2019 • 228pp • H/C (ISBN: 9781522570837) • US $185.00

Climate Change and Its Impact on Ecosystem Services and Biodiversity in Arid and Semi-Arid Zones
Ahmed Karmaoui (Southern Center for Culture & Sciences (SCCS), Morocco)
Engineering Science Reference • ©2019 • 408pp • H/C (ISBN: 9781522573876) • US $225.00

Bioenergy and the Advanced Application of Bio-Products and Microfluidic Devices
Mohammad Reza Rahimpour (Shiraz University, Iran) Reza Kamali (Shiraz University, Iran) Mohammad Amin Makarem (Shiraz University, Iran) and Mohammad Karim Dehghan Manshadi (Shiraz University, Iran)
Engineering Science Reference • ©2019 • 325pp • H/C (ISBN: 9781522575344) • US $215.00

IGI Global
DISSEMINATOR OF KNOWLEDGE

701 East Chocolate Avenue, Hershey, PA 17033, USA
Tel: 717-533-8845 x100 • Fax: 717-533-8661
E-Mail: cust@igi-global.com • www.igi-global.com

Table of Contents

Section 2
Social Aspects of Green Retrofitting

Section 3
Technical and Ecological Aspects of Green Retrofitting

Detailed Table of Contents

The main financial barrier of large scale implementation of green building retrofitting investments is due to the relatively large investment volume needed, compared to the future flow of yearly energy savings or/and yearly estimated differences of incomes collected, if the building is a commercial building (commercial center, office building, hotel or even residential rental building). The uncertainty implicitly involved in this estimation, both for the future savings and for the yearly differences of incomes, which are usually not very large, both make these investments apparently not so attractive for private investors, especially for owners of residential buildings, with limited self-financing power. Nevertheless, from the society point of view, the benefits created by saving the energy and consequently reducing the carbon foot print, can be very attractive. That is why the public support is often used as an "impulse solution" for implementation of these investments. The Cost Benefit Analysis methodology, particularized for these investments, is presented in this chapter.

Green retrofitting and housing energy efficiency are some of the most debated topics nowadays. They contribute to lower costs for the involved households and more efficient sources of energies, which lead to a cleaner environment. The number of methods and devices for increasing the energy efficiency of housing ranges from very simple methods to complex ones. In spite of the fact that in the research there is high interest for the development of energy-efficient housing systems, it is also important to inform the consumer about the existence of such devices and to convince them to use these types of systems more frequently. The objective of the chapter is to give an overview of the popularity and the interest of the consumers towards different forms of housing energy efficiency systems and green retrofitting methods.

All around the world, taking stock of buildings older than 1970, when the first energy efficiency arose, is an important one. Today, all urban agglomerations confront themselves with environmental problems generated by air pollution, cars, and buildings, vie for the first place as the highest polluters. In the future, cities will become increasingly populated, and as a result these problems will more and more affect the quality of life. A way to prevent this scenario is a gradual transition to smart cities. In this approach, smart houses will become indispensable. In order to maintain unaltered cities' specificity, the only acceptable solution is to retrofit old buildings, especially of those classified as historical monuments or as iconic buildings. By retrofitting these buildings, they could be prepared for integration in future cities.

With all the new technology and scientific progress, the human population is significantly increasing, and the planet will hardly support the overpopulation. The increase of the carbon footprint and natural resources decrease with each human. Society must find solutions to use resources in a smart way, reducing the ecological impact and stopping the cities from sprawling. Instead of extending the manmade environment and using more resources, we can act on what we have, upgrading our buildings by using sustainable and green retrofitting and making them energy efficient and environmentally friendly, while covering our cultural, social and psychological needs. Overpopulation, as well as depopulation/migration, greatly affects architecture. This study aims at finding retrofit solutions and strategies in an attempt to improve the existing urban environment quality and buildings' energy efficiency, provide interior comfort and contribute to aesthetic urban image, as well as doing so in a cheap and efficient way.

Currently, buildings are considered to be a continuously evolving organism that over time has to be treated, rehabilitated, and upgraded to meet the requirements set by the user at a certain stage. Buildings are a central element of the EU member states' energy efficiency policies, accounting for about 40% of final energy consumption, and 36% of greenhouse gas emissions, and about 75% of buildings are not energy efficient. Recent applications and studies establish that green retrofitting has maintained older existing buildings to increase energy efficiency, optimize building performance, increase occupants' satisfaction, and boost economic return while decreasing greenhouse gas emissions. In this regard, this chapter aims to address the main factors that negatively affect the performance of residential buildings and presents the common green retrofitting measures that can be taken to ensure the state of human well-being in residential buildings.

In the era of strong IT development, superior processing of information is a desideratum in any business. Used in modeling and simulation, information has better capitalization, having a cognitive role; introduced in various algorithms and models can generate knowledge and wisdom. The processing of information is realized within the information systems (IS). A sub-domain of the energy sector that has developed in recent years is energy retrofitting, seen as a border subdomain, including specific elements from building construction and the energy field, for building analysis and construction process to achieve much larger energy savings. The development of retrofitting can be organized, through using BEM (Building Energy Model) and Building Information Modeling (BIM) models.

Section 3
Technical and Ecological Aspects of Green Retrofitting

According to ISO 50001, energy management systems are increasingly being implemented in companies, with the continuous measurement of energy consumption playing an important role. At the same time, energy meters are increasingly becoming cyber-physical systems (CPS) in the IoT age. LPWAN radio networks and IoT clouds open up new possibilities for retrofitting meters and industrial components. By using these technologies, the measurement infrastructure of an energy monitoring system can be cost-effectively expanded. This section presents the technologies relevant for energy management and shows their application using selected use cases for LoRaWAN, OPC UA, and the combination of optical character recognition (OCR) and IoT Cloud.

The chapter deals with the green energetic consideration of today's building envelopes for residential and non-residential buildings. It investigates the energetic effects the envelopes have on energy efficiency, energy consumption, material use, sustainable use of resources, lifetime considerations, economic and ecological impact. Today's it is estimated that approximately 30% of the annual primary energy demand for residential and non-residential buildings is needed. Energy resources for heat, electricity, air conditioning and cooling purposes, fossil fuels in form of gas and liquid are predominantly used.

Renovating buildings is more useful than ever. Due to future rising energy prices, energy costs in poorly insulated buildings are an important component of operating costs. Another important point is the rapidly growing emissions from the combustion of fossil energy sources. Good insulation in buildings reduces the amount of primary energy and thus, less greenhouse gases are emitted. The renovation potential is high. A large part of the properties consumes more energy than would actually be necessary. Common construction without thermal insulation is responsible for this. It is advisable to invest in renovation, also, in thermal insulation. This will benefit you in two ways. The ancillary (additional) costs will be reduced massively, the living comfort increases and by today's state subsidies in many countries they will make a contribution to the investment costs.

Within the current context of energy, there are several ways to meet the challenges of durable development. Efficiency in energy use, considered to be the fifth energy source, as well as the use of sustainable energy sources represent critical objectives. Nowadays, almost 50% of the total energy consumption in Europe is consumed by building heating and cooling. The current heat demand is mainly covered by conventional energy—fossil fuels. Consequently, there is a significant growth potential for the use of renewable energy sources (RES) in order to produce heat. One can expect in the near future that the energy systems would include a larger percentage of renewable sources, so the increase of the RES share is one of the main objectives of the thermoenergetic field. This chapter approaches heating system typology, the performance indicators used to asses the hybrid heating systems, and at the same time synthetising the assumptions of ensuring the optimum operating conditions.

Preface

The continuous development of technology influences our expectations, our life style, and also our environment. The high dynamic of the economic, technical and social environment is also reflected in the building field due to the fact that more and more peoples are living in big cities, which are transformed in mega cities and therefore are confronted with new problems because of the limited space for constructions and the depreciation of old buildings. The real estate stock in the most cities in Europe is relatively old and doesn't correspond to modern standards and requirements in terms of energy efficiency. New retrofitting measures and strategies are a response to these factors that bring new challenges in terms of adopting new standards regarding conform and energy efficiency and also a solution to conserve the traditional buildings heritage. In parallel our society is confronted every day with the climate change problem, with the increase of emissions of harmful greenhouse gases and smog in form of fine dust due to the fossil fuels-based life and economy. EU studies indicate that the building sector is responsible for high energy usage, high emissions of greenhouse gases and waste generation. Buildings have therefore an significant impact on the environment and this is also one of the reasons that the building sector was mentioned in the Paris Agreement on climate change from 2015, that stipulates that the global warming has to be limited below 2⁰C and even further to 1.5⁰C, as a main contributor to the emissions of greenhouse gases and to climate change. In order to contribute to this limitations the UE set high targets for reduction of emissions for 2030 and 2050, that are also presented in this book in correlation with main solutions based on retrofitting measures for existing buildings and infrastructure.

The aim of this book is to make clear and to increase the people awareness for the problem, that there is a need to think and act in changing our economy, society and building environment to a more sustainable one. Important contributions for this process have to be implemented in all major fields that have a major impact on the environment and could reduce the climate change impact.

This challenge requires for all actors, managers and stakeholders working in the retrofitting field a rethink about new building structures, building and energy innovations, future energy building visions, new energy concepts for the building environment, new technologies and new energy economic models for a successful energy transition.

This book is about the present and the future of retrofitting with the focus on energy performance and the tendency in applying new life standards in old buildings.

In a traditional approach, retrofitting represents the process of improving something after it has been manufactured, constructed, or assembled. These systems integrate new technologies, new functions, and new services that increase the energy performance in existing private, public, and commercial buildings. This definition can be extended to the concept of green retrofitting. In this perspective green retrofitting also represents a solution for reducing the greenhouse emissions due to a new design, construction and maintenance of buildings based on the sustainability criteria.

This book is dedicated to the retrofitting in the context of increasing the efforts to reduce the climate change impact on the life of future generations. Therefore retrofitting, in its green retrofitting form will improve in a sustainable way the future of the buildings, environment and also the life our next generations. To underline this sustainable approach for retrofitting measures and strategies, the concept of green retrofitting was promoted even in the title of this book to reflect its future environmental orientation.

It is a research and also applied book about business opportunities for retrofitting in the buildings sector that explain in a comprehensive way the main concepts and trends on a sustainable approach with its three main dimensions: economic, social and ecologic.

The text also surveys different approaches to the study of retrofitting: the business and financial approach, the information approach, the social approach, the energy performance approach, the technical and ecological approach. It includes quite a bit of information about the main retrofitting envelopes and their characteristics and about alternative heating systems solutions.

All chapters are practical oriented and even are also explaining theoretical concepts. These are presented and detailed in special case studies in order to be applied into practice. We can affirm that Green retrofitting for optimal energy performance is a book by practitioners for practitioners. It is more operational rather than theoretical in tone, and it is based primarily on the articulated experience of the authors. The authors try to keep the scientific literature in the background and to concentrate the attention to themes that emerge from their collective experience.

This book grew from wider efforts at IGI Global and a group of authors with an important experience in the multiple retrofitting fields of specialization, which have complementary competency, that enable the integration of the business and economic mentality with the engineering one in the green retrofitting sector. Because the authors were selected before either topics, the book present diverse and some personal perspectives in different fields of specialization related to retrofitting of buildings and related infrastructure.

The authors want to find answers in how to design and promote retrofitting solutions that may reduce the energy impact on the environment, which will optimize the energy performance of buildings and increase the health and the comfort of the inhabitants. We selected the themes from the various aspects related to retrofitting solutions in the building sector and invite you to become more familiar to these themes and to create or improve your assumptions about green retrofitting.

Despite this diversity, the chapters are unified by a common philosophy oriented on green retrofitting under the assumption that the retrofitting solutions have to be sustainable on an economic, social and ecological basis and some complementary aspects that assure a large spectrum of applications in the retrofitting field.

Retrofitting for Optimal Energy Performance is a comprehensive reference source that examines environmentally conscious technologies and their applications in advancing retrofitting practices. Providing the latest empirical research findings in the area, but also relevant theoretical frameworks, it highlights an array of topics such as: green buildings, zero emission building, zero carbon building, nearly zero energy building, passive house, smart home, sick building syndrome, well- being building, green retrofitting, sustainable business models, ESCO business models, financial feasibility of retrofitting projects, climate change, energy management, hybrid heating systems and optimization modeling, cyber physical systems and IoT, building energy ratings, heat transfer coefficient for envelopes, life cycle assessment for buildings and is essential for engineers, architects, entrepreneurs, managers, policymakers, building owners but also for academicians, students and researchers. This book, we trust, provides some inspiration for both researchers and readers of the book, which will have the following advantages: understanding the retrofitting as a concept and its evolution, understanding the specificity of a green retrofitting, identifying and analyzing the main factors that characterize the old buildings in Europe, identifying the tendency in the development of retrofitting solution for optimal energy performance, will be connected with financial incentives that are needed for implementing new standards in building field, identifying solutions for retrofitting, defining indicators for retrofit project validation, better understanding for different approaches used for investment valuation in the field of building retrofitting, understanding the importance of social media for promoting green retrofitting projects, understanding the benefits of energy management in buildings, analyzing possible retrofitting models with the energy balance method, identify the main IoT related technologies that make new retrofitting solutions possible.

As a matter of fact, the book can help researchers, people active in the business sector and government officials and not only for a better understanding of the importance of retrofitting models for buildings. The book could help researchers to be closer to the tendency in the green retrofitting field and give them the latest information about this.

The main chapters faced with new retrofitting models close to the newest customers demand and the innovations and technology development in the building sector.

The book is organized in three parts that are integrated in a sustainable perspective that include the economic, the social and the ecologic dimensions as follows. Section 1 which is *economic and business* oriented present the main economic and business aspects of retrofitting and the transition to green retrofitting (Chapter 1 to Chapter 4).

In Chapter 1, "Business Models for Green Retrofitting," there are explored and reviewed the main approaches regarding definitions and perspectives that enable the development of a business models for green retrofitting. In this chapter green retrofitting is considered a strategy for increasing the sustainability of existing buildings. Analyzing the pillars of the business models and understanding the added dimension of sustainability for a business model, the chapter enables the readers to create their own business model based on Canvas structure and the triple bottom line, which illustrates in an original way how an organization creates multiple types of values: economic, social and environmental. Therefore the chapter introduce a more holistic approach that enable to think and integrate the economic, the social and the environmental perspective in a business model.

Chapter 2, "Financial Considerations in Green Retrofitting for Optimal Energy Performance," explains and exemplifies the methods that can be used in order to assess the financial feasibility of a retrofitting project or of its alternative solutions. It is a particular way of applying the general methodology of assessing the financial feasibility of an investment project in the case of retrofitting. This chapter also presents the basic financial concepts necessary in retrofitting project investment decisions and explains their advantages and limits and their functionality. Therefore it provides the readers with the opportunity of understanding how the capital budgeting valuation methods can be used in practice and their results correctly interpreted.

Chapter 3, "Financing the Green Building Retrofitting Investments: A Case Study for a Romanian Seaside Hotel," offers a view about the main approaches used for investment valuation in the field of building retrofitting and analyzes the methodology used in financing of green building retrofitting investments. In the applied part of this chapter the Cost-Benefit-Analysis for a project of the green retrofitting investment in a hotel building located on the Black Sea shore is presented. The benefits of a green building retrofitting investment can be very attractive from the society point of view due to saving the energy and consequently reducing the carbon foot print.

Chapter 4, "Comparative Analysis Regarding the Online Interest of Consumers Towards Green Retrofitting: Case Study in Germany and Romania," tries to determine the popularity of several key terms (e.g. green retrofitting, energy efficiency, residential energy savings etc.) related to retrofitting and energy efficiency in the online environment. The number of searches is a first step in determining the interest and preoccupation of users regarding a certain term. The study reflects the results of a comparative study for users from Romania and Germany. This chapter allow the readers to have a better view about best practices for implementation of a webpage, which should provide the necessary information for increasing the awareness of consumers related to the importance of green retrofitting.

Section 2 introduces many of the key concepts and models for analyzing green retrofitting from a *social and human being approach*. Its four chapters are dedicated to subjects that reflect the development of the standards of living and technologies for retrofitting purposes of existing buildings with a focus to green retrofitting.

Chapter 5, "Smart Homes as a Solution for Sustainable and More Inclusive Retrofitting of Existing Buildings," presents and analyzes the concept of smart home in the context of the transition to smart cities. The problem of a harmoniously retrofitting of existing buildings in the future's smart cities is analyzed in a specific case study that integrates the buildings management system and the intelligent urban management system with the traffic management system.

Chapter 6, "Phenomena Implied by Sustainable and Green Retrofitting: A Quantitative Approach," enables the readers to understand the benefits of a sustainable and green retrofitting and how it may contribute to more sustainable buildings and cities. A specific case study for East Europe explains why in this region, which has a lot of problems that are opposite to green buildings, retrofitting and energy performance of buildings has to play a major role in the house estate. This chapter aims at finding retrofit solutions and strategies in order to improve the existing urban environment and buildings energy efficiency, provide interior comfort and contribute to the aesthetic urban image.

Chapter 7, "Considerations Regarding the Green Retrofitting of Residential Buildings From Human Wellbeing Perspectives," is focused on factors and standards that influence the human comfort and well-being in the building sector. The specificity of this chapter is determined by the fact that people in developed countries spend around 90% of their time inside their workplace or in their own home, so that retrofitting solutions are very important for healthy and comfortable buildings and they should reduce or even eliminate the impact of the sick building syndrome. There are also identified and analyzed the main factors that negatively affect the performance of residential buildings and indicate main green retrofitting measures that can improve the state of human well-being in residential buildings.

Chapter 8, "Building Information Modeling for Energy Retrofitting," presents the importance of information and how it is processed in energy systems for building retrofitting. Taking into account the very large data volume, their different sources and their complexity, the information has to be treated within the information systems specific to the building sector and for retrofitting purposes. From this perspective, the chapter present the framework for developing a Building Information Modeling (BIM) through the creation of models for existing buildings and the analysis of retrofit measures. Supported by specialized software, these models can be used in all stages of the building's evolution, from the conception phase until the end of its existence.

Section 3 covers specific themes related to the technical and the ecological dimension of retrofitting. Here the point of view is applied rather than conceptual.

Chapter 9, "Retrofitting for Energy Management: Using IoT Technologies," makes the connection between energy monitoring and IoT. The chapter presents application examples for energy management based on IoT technologies, that can be used for energy monitoring in energy audits, or for the implementation of an energy management system according to ISO 50001. The retrofitting solutions in this chapter are focused on modernizing the existing equipment (e.g. energy meters). The chapter also presents several case studies in which simple energy meters are retrofitted to cyber physical systems which allow the devices to be networked and to transmit energy measurements to the energy management system at short intervals.

Chapter 10, "Green Retrofit Energy Efficiency Potential on Existing Building Envelope for Residential and Non-Residential Building," presents green energy consideration of today's building envelopes and their features for residential and non-residential buildings and give a good understanding of possible renovations to building envelopes. In the chapter are presented different components for the building envelope and it investigates the energetic effects the envelopes have on energy efficiency, energy consumption, material use, sustainable use of resources and lifetime considerations. The chapter ends with a case study with a simulation regarding the problem of wall renovation and the calculation of the main energy indicators such as heat transfer coefficient. The simulation also present different aspects of insulation, costs and savings potential.

Chapter 11, "Economic Analysis for Green Residential and Non-Residential Building Envelopes," deals with the calculation methodology for energetic considerations of buildings envelopes and the specific parameters that are included in the calculations. There are also analyzed the ecological aspects of building renovations and the energy flows in buildings over a period of time. Therefore, the readers will have an image on how the energy distribution in buildings can look in the future. A case study for a residential building illustrates economic and technical aspects that are related to green buildings envelopes based on the thermal quality of buildings. A second case study elaborated for a non- residential building also presents the calculation for the payback time and energy savings in the case of a green retrofitting solution. Another important concept that is presented in this chapter is the Life cycle assessment for buildings with a special utility for grey energy.

Chapter 12, "Heating Systems: A Comparative Assessment of Alternative Solutions," deals with the thermo energy efficiency of urban thermal consumers and presents and analyzes the conventional heating systems together with modern concepts as the energetic hub that allows the association of one single output – energy demand, to several inputs, related to different energy sources (conventional, renewable or alternative). There the thermo- energy indicators used for assessing the performance of the heating systems at the level of a hybrid energy hub are emphasized. A case study promotes specific indicators that quantify the performance of hybrid heating systems based on a thermo-energetic simulation of a district heating system.

All these transformations, that are presented in this book, will be followed by a transition to new retrofitting models, which should be more focused on new demands and needs on the consumer side.

We realized that there was a need for a book like Green retrofitting for optimal energy performance. The retrofitting field faces with new challenges nowadays due to the continuously technology development and proliferation of information and communication technology and due to the transition from the classical retrofitting solutions to other solutions that have a reduced impact on the climate change. Due to the proliferation of information and communication technology, the energy will be generated, transported and consumed in a more intelligently and efficiently way and it will play a more important role in the green retrofitting solutions and strategies.

Acknowledgment

We began our journey to analyze the main retrofitting models in the building field with the conviction that our book will help other researchers, people active in the business environment and government officials and will not only contribute to the development of measures that will reduce the environmental impact in order to assure a sustainable environment for buildings the next generations.

We are appreciative on the support that we have received from many colleagues in the preparation of this book.

We are grateful to Professor Alina Dima, UNESCO faculty chair for Business Administration at the University of Economic Studies Bucharest who provided assistance in the editorial process with IGI Global.

There are several other people to whom I owe much gratitude, my Phd. Student Ana Maria Santa and my son Nicolas Stefan Tantau, who helped us in the editorial process.

I (Tantau Adrian) am especially grateful to the coauthors for their collaborative spirit, for their orientation and passion on new solutions and approaches that should improve the sustainability awareness in the retrofitting field and also offer a better life and comfort perspectives for the next generation.

On a final note, I am truly grateful to my family (my wife Laura and our two children Carla and Nicolas), to my parents (Rodica and Alexandru) for their love and support through this research.

We wish to our readers success when working with this material.

Section 1

Economic and Business Aspects of Green Retrofitting

Chapter 1
Business Models for Green Retrofitting

Adrian Tantau
The Bucharest University of Economic Studies, Romania

Maria Alexandra Maassen
The Bucharest University of Economic Studies, Romania

ABSTRACT

This chapter is dedicated to business models for green retrofitting on a more holistic approach that enables to think and integrate the economic, social and environmental perspective in a business model. The chapter is a result of research regarding new business models for green retrofitting and presents a framework for developing business models for green retrofitting in the building sector based on the Triple-Layer Business Model Canvas. The business models for green retrofitting could be an important instrument for introducing new green characteristics such as energy efficiency, optimal energy performance, and new comfort standards in the building environment. Green retrofitting is responding to the dynamics of the economic and technological development, and to the new lifestyle of the peoples. The implementation of such a model will be also a catalyst for reducing the emissions of greenhouse gases in the building environment.

INTRODUCTION

The current lifestyle of the society has encountered continuous changes, that can be observed in everything that we are doing, in everything that we are seeing, as well as in all old technologies, that are becoming history and that become quickly obsolete due to the rapid proliferation on new technologies and life standards. The economic development and the changes in the life style of individuals introduced new characteristics in the building environment. These bring new challenges in terms of adapting to new standards regarding comfort and energy efficiency. The environmental protection has to be a criterion for the new building development in order to contribute to the reduction of greenhouse gases and climate change dynamics.

DOI: 10.4018/978-1-5225-9104-7.ch001

Therefore, there is a need for the building sector to transform and evolve from the current buildings that are characterized by high energy usage, high emissions of greenhouse gases and waste generation to new modern ones that respond to the actual climate change problems.

For instance, in Europe the existing building environment is one of the main contributors to the greenhouse gases emissions (Nejat et al., 2015), it is a main energy user and its energy efficiency low.

These factors that characterize the energy performance of buildings are influenced by the age of the building, the type of the materials that were used for their construction, the reduced energy efficiency of the supply system and by the absence of planned maintenance (Filippi, 2015). If for an old building there is an annual energy need for heating of more than 200 kWh/m^2, for new designed buildings the annual energy need for heating is about 50 kWh/m^2 and the energy performance criteria for the annual heating of future buildings will achieve 20 kWh/m^2 or even less energy.

Therefore, from an environmental perspective the buildings are mentioned in the Paris Agreement on climate change from 2015 that stipulate that the global warming has to be limited below 2 °C and even further to 1.5 °C. The limitation of the impact of global warming of 1.5 °C above pre-industrial levels represent the main subject of a special report that was elaborated by the

Intergovernmental Panel on Climate Change in 2018. This report mentioned that this limitation is a must and has to be followed by the implementation of new technologies that are more efficient but also by a change in our behavior that has to be consistent with adapting to the limiting of global worming to 1.5 °C. Therefore, there is a need for education, information and new community approaches regarding climate change impact and how to reduce this impact (IPCC, 2018).

In order to achieve these desired states introduced challenged targets for reducing the emissions. For example, not only EU but also UK target an 80% emissions reduction for 2050. Therefore, the retrofitting of existing buildings and infrastructure represent a main solution in UK for achieving this desiderate (Dixon et al., 2018).

All these considerations contribute to the assumption that in Europe there is an important potential for retrofitting in the building sector and there is also a need to implement the green retrofitting concept. The aim of this paper is to identify the main concepts that are related to the new challenges in the building environment and to analyze them in order to understand the importance of new business models for retrofitting in the building sector from a sustainable development perspective.

In this approach "green building" and "green retrofitting" represent an important response to decreasing greenhouse gases due to a new design, construction and maintenance of buildings based on sustainability criteria. Therefore, in this chapter is designed a generic business model for green retrofitting, based on an own research, that has a triple bottom line structure and illustrate the impact of the main business factors, from an economic, social and environmental perspective.

GREEN BUILDINGS AND GREEN RETROFITTING

Green buildings represent an integrative concept similarly to an umbrella that encompasses closed terms, such as: zero emission building, zero energy building, zero carbon building, passive house, zero net energy or carbon neutral (Riedy et al., 2011).

All these concepts have a common aim to reduce significantly the energy consumption and also carbon emissions. From an EU perspective a nearly zero emission building is defined as a building with a very high energy performance and with a very significant renewable energy input, that is produced onsite or nearby (EU, 2010). The zero-carbon performance of buildings is influenced by the building type and the climate type (Grade & Donn, 2014; Zhao et al., 2018). As an example, in the UK a zero-carbon building has to register a net zero carbon emissions more than one year in order to be classified in this category (Zhao et al., 2018).

In general, green buildings are defined through an integration of specific properties: design that minimizes energy use, maximizes daylight, conserves water, use recycled materials and assure a high comfort and air quality (Jagarajana et al., 2017).

Based on a specific design a green building is defined as the result of a design, that concentrates on resource efficiency increase, while at the same time is decreasing the building effects on human health and the environment through a few elements, such as: better design, construction, operation, maintenance, removal and other (Frej & Browning, 2005). From a general perspective, green buildings can be supported by four pillars:

...minimizing environmental impact, improving the comfort of the occupants, estimate the life cycle characteristics from the planning and development stages and the return on investment to developers and local community (Robichaud & Anantatmula, 2010).

All these definitions of green buildings are focused on four pillars, namely: environmental sustainability, health, community impact and the lifecycle approach (Zuo & Zhao, 2014, Jagarajana et al., 2017).

In this chapter we consider that a green building is a building that in all its development phases from project design up to its use phase has a low impact on the environment, is using recyclable materials that are health for its occupants, is implementing a high level of energy efficiency, is looking for reducing the water consumption, is using renewable energy sources and is assuring an appropriate comfort.

This chapter integrate the green building concept with retrofitting.

Retrofitting is defined in the Oxford English Dictionary as "adding (a component or accessory) to something that did not have it when manufactured". In the building sector retrofitting deals with the improving of the building performance. From a practical perspective retrofitting include renovation or refurbishment. For current buildings, retrofitting can be represented by simple forms, for example, replacing windows, replacing the incandescent lighting with light emitting diode (LED) lighting, replacing the exterior wall isolation or by other changes that include complex activities like replacing the electrical distribution and the security systems, upgrading or replacing the heating system, or improving the heating system controls. Renovation include all these changes if there are interchangeably and contribute to the new needs of the occupants.

Refurbishment is improving the characteristics of buildings by new aesthetical design but also can be used in order to upgrade the building's mechanical system (Eriksson et al., 2015).

The main difference between renovation and refurbishment consist in the fact that renovation is dedicated to buildings, while refurbishment is used also in other sectors.

Building based retrofitting can add new functions to the existing buildings through new technologies and new services but also through rearrangements or replacements of parts of the building. Retrofitting can be used to any part of a building or to the whole building.

Retrofitting is seen as an alternative of demolition and rebuild which reduce the construction waste and use less material resources (Wilkinson, 2012).

From an environmental point of view existing buildings have two main negative impacts. First, most of these buildings are not aligned with the new environmental criteria for buildings and have a low energy efficiency. Second, if these buildings will be replaced by new one, the demolition waste will pollute landfills. The low efficiency of the buildings can be improved by retrofitting the building envelope and/or by retrofit of the building system. By retrofitting the building envelope, the heating demand for the building is decreased. By retrofitting of the building system due to new technologies for upgrading the heating system and the electrical and lighting system has registered a significant reduction of the primary energy demand (Wang et al., 2018). These measures of energy retrofits can reduce also the CO_2 emissions and have also other benefits for the occupants, e.g. reducing bill and maintenance costs (Xu et al., 2011; Moschetti & Brattebo, 2016).

A special concept related to retrofitting is green retrofitting. In this approach energy retrofitting is only a part of green retrofitting.

A recognized definition of green retrofitting is promoted by the United State Green Building Council (USGBC) in form of "any type of upgrade at an existing building that is wholly or partially occupied to improve energy and environmental performance, reduce water use, improve comfort and quality of space in terms of natural lighting, air quality and noise, all done in a way that it is financially beneficial to the owner" (Jagarajana et al., 2017).

This chapter consider green retrofitting as a strategy to increase the sustainability of existing buildings. The measures that can be used in order to implement such a strategy are divided in two major groups: measures that increase the energy efficiency and use renewable energy sources and other measures focused on smart technologies for improving the air quality, reduce the water consumption and control the energy uses. The measures that increase the energy efficiency consist in reducing the energy demand by optimizing the energy flows in the building and/or by implementing new technologies.

Filippi (2015) presented the main measures that can be part of a green retrofitting strategy: increase the energy performances of the building envelope, improve the air conditioned system and the lighting system, create conditions for optimizing the natural ventilation and the passive cooling, introduce automations systems that are monitoring the indoor environmental quality, remove the energy systems with low efficiency and high environmental impact with others based on sustainable technologies and renewable energy sources, introduce measures for reducing the water consumption and for recovering and reusing of water and use recycled or recyclable materials.

The green retrofitting potential of a building can be quantified with indicators such as the Green retrofitting index (GRI). This indicator takes into account the factors that promote greening or act as a barrier against it. If a building is green the GRI indicator has low values while the building has low environmental performance the GRI values are high (Ayyad & Fekry, 2016). Main analysis that contribute to the calculation of green retrofitting potential of buildings are: the cost analysis of the retrofitting alternatives, the building stock analysis, the estimation of the energy savings, and the quantification of other benefits (BPIE, 2013).

Another indicator for quantifying the retrofitting potential of a building is the Deep Retrofit Value (DRV), which is promoted by the Rocky Mountain Institute (2015). The factors that are taken into account in order to calculate DRV are: retrofitting costs, costs and taxes for development, savings in terms of energy costs, tenant-based revenues and other such indicators.

In order to calculate the GRI there is need to take into account additional data regarding the lifecycle of the building: building age/lifetime, building performance, occupancy, retrofitting cost and built-up area, building market value (Ayyad & Fekry, 2016). Furthermore, building performance includes the following factors: building envelope performance, energy consumption rate and water consumption rate.

The implementation on a large scale of green retrofitting indicators, as key performance indicators, will improve the energy and environment performance of green buildings and will contribute to the promotion of new design standards and processes for sustainable buildings.

BUSINESS MODELS (BM) AND SUSTAINABLE BM IN THE BUILDING SECTOR

Business Models

The high dynamics of our life and of our life standards introduce new considerations for developing business models. The new business models have to be more closed and specific to the customers current and future needs.

In the scientific literature business models are analyzed from different points of view and the specialists have defined this concept from a general perspective, but also for specific fields.

In the beginning, around 1960, the business model was seen as a construction plan, an image of a real enterprise environment with its processes, tasks and communication relations, consequently a basis for business processes and data models. Meanwhile many specialists present their specific definition regarding business models.

For a general understanding, we consider a business model as a method of the enterprise to describe key structural and operational characteristics, the way it earns revenue and delivers profit in the current business environment, starting from the value of the business idea (Figure 1). The business model explains the framework on how a company plans to add value, who are the main clients, and which are the final benefits.

Main perspectives that are used for defining and explaining the business model concept are: entrepreneurship perspective, the resourced based perspective, the transaction perspective and the narrative perspective.

Chesbrough defines the business models based on an entrepreneurship perspective where innovation is the main dimension of the business model. In his theory the business model is a tool for analyzing new business ventures and a way to capture value from technical innovations (Chesbrought & Rosenbloom, 2002). The resource-based perspective promotes the business model as an articulation between diverse activities based on financial, human and organizational resources in order to generate value (Demil & Lecoq, 2010). In the transactional structure perspective, the business model could create new markets or innovative transactions (Zott & Amit, 2007). The more general perspective for business models is promoted by Magretta in the narrative perspective as "a structure of a story and its play mode" (Magretta, 2002).

Figure 1. Business model based on own approach
(Source: Tantau & Staiger, 2017; Tantau & Fratila, 2018)

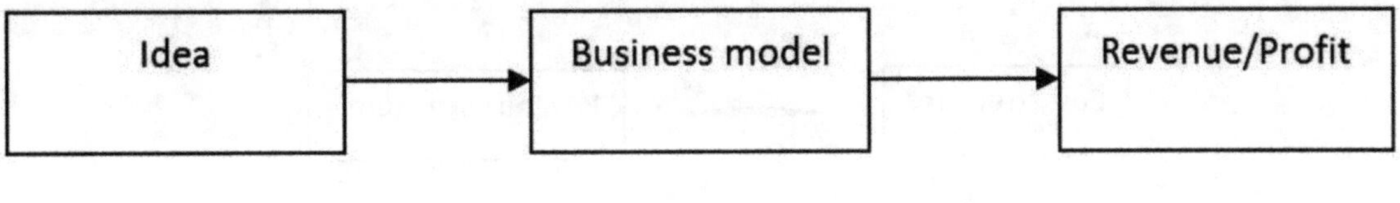

In our analyze for the green retrofitting field we are following the structure of the business model proposed by Osterwalder (2004) that considers a business model as a "conceptual tool containing a set of elements and their relationships", that enables to reflect the logic of the company's earning. In this concept the business model presents the structure of the enterprise and the relationship to the main stakeholders in order to create and deliver value that would contribute to a profitable and sustainable revenue model.

The traditional Osterwalder business model template has nine interrelated building blocks (see Figure 2).

This model was developed by Ostervalder and Pigneur (2010) as a tool for structuring the business models and is also known as Canvas business model. The model has the advantage to be easy understandable due to its visualization form that promoted it as a visual tool for structuring the elements of a business model and their interconnections. The elements that are presented in the canvas structure are: customer value proposition, customer segments, customer relationships, channels, key resources, key activities, partners, costs and revenues. Each element or field of the business model structure introduced some specific characteristics. In the field of distributed photovoltaic generation, customer relationship introduced a pro-active approach to convince customers that utilities, new products and services add value. For the revenues streams module there is a need to be developed in a way to avoid direct competition with established technologies in terms of costs (Richter, 2013, p. 464).

In general, the elements of the business model must be both compatible and complementary in order to be able to create the competitive advantage.

In practice, business models are not static, and their dynamic is caused by the continuous changes in the environment as well as the possibility of imitation by competition.

The essence of business models is to deliver value to customers and to think how to create profit. As a concept the business model was analyzed by many specialists which consider that business models are describing the architecture of the value creation, profit formula, key resources and key processes (Zott et al., 2010; Osterwalder et al., 2010). The business model concept is highly linked to the concept of value creation regarding stakeholders and practically refers to the activities of the organization and its stakeholders in order to deliver such value (Zott & Amit, 2013).

Figure 2. The structure of a business model by Ostervalder and Pigneur (2010)

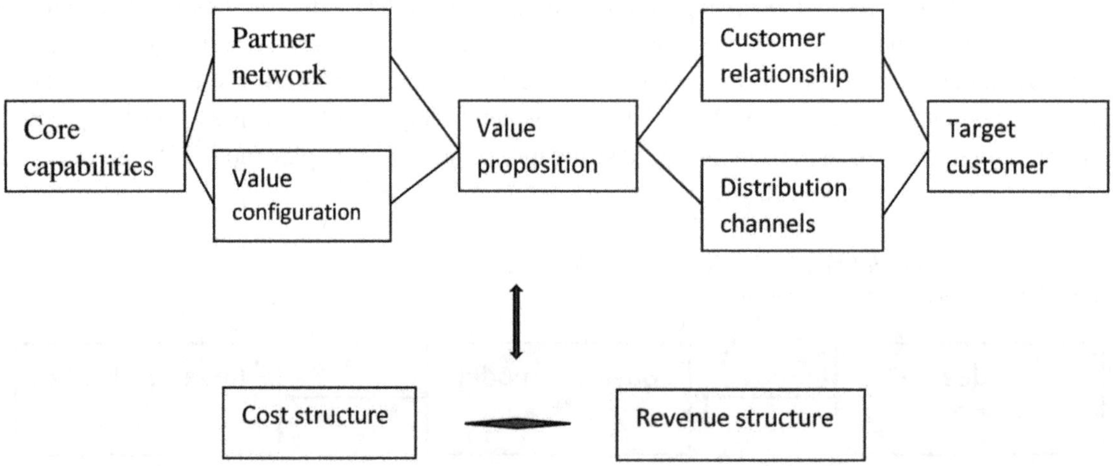

The model developed by Osterwalder and Pigneur has its roots in a four pillars model: customer interface, infrastructure management, product, and financial aspects. All these are illustrated in Figure 3 (Osterwalder, 2004).

In the product pillar, the analysis indicate what kinds of products, services and value proposition could be create and deliver to the main clients. It could state that how and what it's distinction and supremacy from other competitors' organizations.

All customer related aspects, such as customer segments, channels, communication methods, and finally customer relationship management are analyzed in the customer interface pillar.

In the infrastructure management pillar, the focus is about that how an organization efficiently performs infrastructural or logistical issues with whom and what kind of enterprises such as supply chain partners and business partners for the organization's fundamental activities.

Revenue model and cost structure are analyzed in the financial pillar. In fact, this pillar contains all items related to financial issues that any organization should consider for its business. Therefore, a specific pillar assigned to this subject to cover costs and revenues of the firm's business.

By evaluating business model definitions from the year 2000 to the year 2017 by various scholars, Khorshidi (2018) came to the result that financial aspects are a priority for most of the definitions (46%), followed by the product characteristics (33.1%) and infrastructure management (17.0%). These results

Figure 3. Osterwalder four-pillar model
(Source: Osterwalder, 2004)

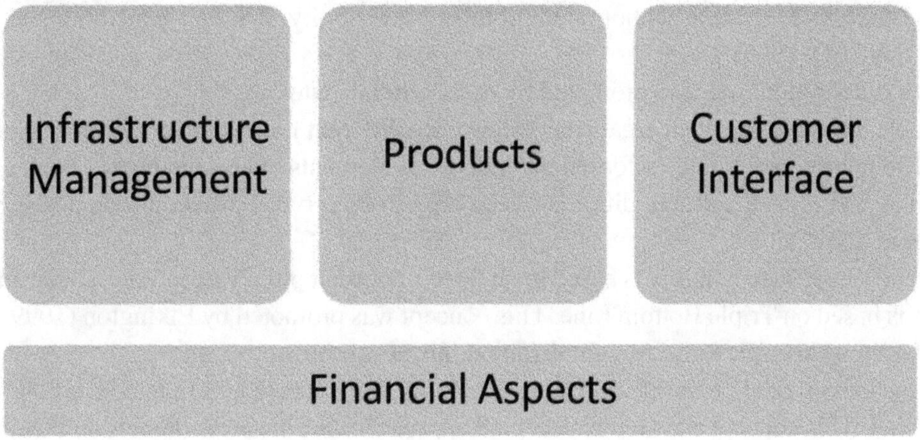

Table 1. Business model pillars priorities

Category	Priority	Rank
Financial Aspects	46.5%	1
Products	33.1%	2
Infrastructure Management	17.0%	3
Customer Interface	3.4%	4

Source: Khorshidi, 2018

Table 2. Triple Layer Business Model Canvas

<table>
<tr><th colspan="2">Infrastructure Management</th><th colspan="2">Product</th><th colspan="2">Customer Interface</th></tr>
<tr><td>Key Partners</td><td>Key Resources</td><td colspan="2" rowspan="2">Value Propositions</td><td>Customer Segments</td><td>Customer Relationships</td></tr>
<tr><td colspan="2">Key activities</td><td colspan="2">Channels</td></tr>
<tr><td colspan="6">Financial Aspects</td></tr>
<tr><td colspan="3">Cost Structure</td><td colspan="3">Revenue Stream</td></tr>
<tr><td colspan="3">Social & Environmental costs</td><td colspan="3">Social & Environmental benefits</td></tr>
</table>

Source: Osterwalder et al., 2014

indicate that for the analyzed definitions, the classic business model is concentrated on the financial aspect and less on the customer interface.

The business model as a concept is nowadays relevant in the sustainability context where the organizations have to assume a more responsible role for achieving the environment targets (e.g. EU environment targets), that may reduce the climate change impact. There is a need to create the framework for developing sustainable business models.

Applying sustainable business models could contribute to the implementation of sustainability within the business, respectively within business objectives and processes. (Moschetti & Brattebo, 2016)

Sustainability as a concept and its implication were analyzed in the last decades.

The first international accepted definition of sustainability was provided in 1987 after the World Commission on Environment and Development (WCED and refers to "the development that meets the needs of the present generation without compromising the ability of future generations to meet their own needs" (WCED, 1987).

Some other definitions are also promoted by other scholars later.

Sustainable development at an organization level has different meanings for different kinds of organizations. The relative weightings afforded the individual elements of sustainability, such as economic, environmental, social, and cultural, differ between organization types depending on a range of factors (Swanson et al., 2012).

In a general sense, sustainability is based in on three important pillars: economic, social and environment which is based on Triple Bottom Line. This concept was promoted by Elkington (1999) and it was developed based on specific indicators for different business sectors.

From a business model perspective, the sustainability finds its utility in the concept of sustainable business model. This concept adds to the accepted approaches for business models and their economic impact new criteria that a business has to fulfill for preserving the social and environment dimensions.

The sustainable business model integrates the stakeholders with the natural ecosystem (Boons & Lüdeke-Freund, 2013, Moschetti & Brattebo, 2016)

From a sustainable development perspective also Osterwalder developed its model based on the Canvas structure and the triple bottom line (Osterwalder et al., 2014).

This model that developed the standard Canvas structure with the environmental and the social value creation part is known as Triple Layer Business Model Canvas. The Triple Layer Business Model Canvas illustrate in an original way how an organization create multiple types of value: economic, social and environmental.

This tool integrates the environmental and also the social dimensions into a more complex concept with three dimensions. The environmental dimension is based on the life cycle assessment (LCA) with its view on environmental impact. Its main elements are: supplies and outsourcing, materials, production, functional value, use phase, distribution, end of life, environmental impacts and environmental benefits. The social dimension has its roots in a stakeholder management perspective that reflect the social impact of the organization and stakeholders' interests. The proposed elements of this dimension in the Triple Layers Business Model are: communities, governance, social value, social culture, employees, end users, social impacts and social benefits (Joyce & Paquin, 2016). The Triple Layers Business Model was dedicated to extending a business model structure, as canvas business structure, to an integrated structure that enable the understanding of the economic, social and environmental impact of a business model. This structure can be use also to define a sustainable business model that has as objective the improving of the economic, environmental and social effectiveness of the business (Tantau & Staiger, 2017).

In this approach, the main strategic elements of a business-like vision, mission, infrastructure, customer interface, products and services, and financial aspects, are analyzed based on the sustainability's issues and indicators.

In order to have a common understanding on sustainability the International Standard Organization (ISO), published in 2010 ISO 26000, which focused on social responsibility. The standard was developed in order to increase the transparency in organizations and to be addressed on the value of public reporting on different stakeholders like local community, employees, investors and, regulators (GRI & ISO, 2014).

New business models and specific business models for green retrofitting are a solution for increasing the buildings performance and for reducing the energy and carbon intensity. With this consideration business models for green retrofitting are attractive for a large category of stakeholders such as building owners or homeowner associations, retrofitting service providers, utility and networks operators, financing companies and local authorities.

The main stakeholders in the case of green retrofitting of residential buildings are: the private users of buildings, the administration as public promoter and the private promoters. Studies indicate that there is a need to analyze the different objectives of these stakeholders' categories in order to find the proper ways for implementing the retrofitting projects (Lizana et al., 2016).

The private users of buildings are looking on retrofitting projects that have an impact on reducing the energy bills that can balance in a positive way the investment costs. They have an important role in the sustainable innovation process and indirect in the new sustainable business models (Wirtz et al., 2016).

The public promoter is looking to achieve the EU targets for CO_2 emissions reduction. The private promoter has focuses on the economic benefit of the retrofitting project. These considerations indicate that different categories of stakeholders have different objectives regarding the retrofitting process therefore there is a need to find new solutions for implementing the retrofitting projects. There is a need for a better cooperation between these stakeholders in order to develop retrofitting projects. For instance, the private users of buildings can have support from the local administration for obtaining financing by credits while having public funds as a collateral or he can find an ESCO private promoter that will develop the retrofitting project. The local administration may simplify the criteria for obtaining permissions for green retrofitting projects. On the other side the private promoter as retrofitting service provider can find a partnership with other stakeholders, such as financing companies.

Nevertheless, traditional business models require modernizing processes in order to be updated to the energy retrofitting standards of existing buildings. Further studies need to be done to establish these standards, as currently the scientific literature in this field is extreme scarce (Moschetti & Brattebo 2016).

Sustainable Business Models

The term ''sustainable business model'' originates from the concept of business model and corporate sustainability. A sustainable business model is similarly to the concept of business model linked to value creation, namely referring to the organization activities and their connections, which practically create, deliver and capture sustainable value, while cooperating with different interested parties in the organization's activity (Geissdoerfer et al., 2017).

By developing a sustainable business model, a company improved its adaptability in complex environments and gain sustainable competitive advantages.

There are also initiatives which are combining the sustainable development approach with the concept of innovative business models in order to design a business model that could reflect the strategic sustainable development (Franca et al., 2016; Broman & Robert, 2016; Upward & Jones, 2016; Rosca et al., 2016).

In practice, there are no innovations or technological improvement without changing the business model. New customer needs generate new business model based on innovations. The main possibilities to develop a new business model are through creating a new hybrid model by combining the new with the current one or a total adoption of a new model, without any connection with the one from the present.

Bocken promoted the concept of business model innovation as "innovations that create significant positive and/or significantly reduced negative impacts for the environment and/or society, through changes in the way the organization and its value-network create, deliver value and capture value, or change their value definition". He also mentioned increasing positive impacts while decreasing negative impacts of business model to each issue of sustainability. The archetype of the sustainable business model based on triple bottom line concept promoted by Bocken are divided in three groups. The technological group includes: maximize material and energy efficiency, create value from waste and identify substitutes for processes as renewable and natural processes. The social group contributes to: deliver functionality rather than ownership, adopt a stewardship role and encourage sufficiency. The third group analyzed in the research coordinated by Bocken is the organizational one: repurpose for the society and for the environment and the development of scale up solutions.

The archetypes promoted by Bocken contribute to the change process of a business model to a more sustainable one. These represent a proposed solution to innovate and to create new development opportunities. Based on this assumption, the business model innovation can be focused on technological, social and organizational innovations. From a technological point of view the business model innovation is focused on new technological innovations (e.g. new process technologies, new automatized productions lines), that ensure a sustainable environment for the implementation of the business model. Main sustainable indicators that are used for developing such a business model for green retrofitting are energy efficiency, use of new materials, use of recyclable materials, use of renewable energy sources or reduce waste.

Business model innovations based on social innovations are dealing with customer behavior creation or alteration. The social group has three archetypes, which are functionality delivery, observation, and propose an incentive for sufficiency.

Business model innovation based on organizational innovations can increase the firm responsibility. This group has two archetypes such as environmental and society repurpose and development enhancement (Boons et al., 2013).

For the building sector the business model innovation contributes to the achievement of an optimal energy performance in a green retrofitting project.

Table 3. Advantages and disadvantages for business models based on traditional individual solutions

Advantages for small size projects	Disadvantages for complex projects
Reduced bureaucracy	Fragmentation of activities
One contractor	Many contractors
Low costs	Lack of communication
	Different planning deadlines
	Different execution deadlines
	Coordination costs

Business Models for Green Retrofitting

International Energy Agency (IEA) developed in 2014 the Demand Side Management (DSM) Technology Cooperation Program and analyzed new business models for energy services. One research question should answer if more sustainable business models may offer multiple types of values and if these are related also to energy efficiency (Tolkamp et al., 2018).

This chapter analyzed the main business models that are proper to green retrofitting projects and highlights their main characteristics.

Business models for green retrofitting can be delimited in business models based on traditional individual solutions or do-it-yourself and business models based on one-stop-shop concept. Business models based on traditional individual solutions are focused on small size projects for single-family house, where, in general, craftsmen offer as service provider different services, usually renovation work or energy audit.

If there is a need for a complex retrofitting process, this solution will face a lot of troubles due to the fragmentation of the retrofitting activities between many contractors. In the case of having many contractors for a single-family house the clients can be confronted with problems of coordination of activities due to lack of communication, and different planning and executions deadlines. Additionally, the clients are confronted with constructions and workplace regulations which can be very complex and increase the risks associated with the retrofitting measures during but also after the end of the retrofitting process (Moschetti & Brattebo, 2016).

ESCO Business Models

Business models based on one-stop-shop concept concentrate all energy retrofitting activities in the hand of a single contractor. In the case of energy retrofitting, this contractor can be an Energy Service Company (ESCO) that offers the necessary framework in order to increase energy efficiency for properties and also covers a certain amount of financial risk. The main difference between ESCO business models and the traditional individual solution or "do it yourself" consist in the fact that the ESCO guarantee the functional performance and the price and can support the technical and economic risk. Therefore, in our case, ESCO are one-stop-shop providers which can offer financing service and can also guarantee the achievement of energy and cost savings. The ESCO is responsible and guarantee for the minimal savings level of energy. This minimal savings level of energy is established based on the energy usage of the building in the past, before the new business model is developed.

In most of the cases the revenues of the ESCO business models are a result of energy consumption reduction (compared with the past energy usage of the building) and are not directly expressed as energy costs nor either reduction of carbon emissions (Garbuzova & Madlener, 2012). The energy savings that exceeding the minimal savings level will be split between the customer and the ESCO.

These revenues are on one side a result of the achieved energy reductions and performance criteria that have to be measurable and observable and on the other side have to quantify also the investment risks (Pätäri & Sinkkonen, 2014; Paiho et al., 2015).

Based on the ESCO experience there are developed three main business models: Energy Supply Contracting (ESC), Energy Performance Contracting (EPC) and Integrated Energy Contracting (IEC).

Energy Supply Contracting (ESC) is a business model based on the provision of useful energy, such as electricity, heat or steam to a building owner or user, through a long-term contract (e.g. 10 to 15 years). ESC is mainly used in decentralized power supply (local) where, for example, the ESCO can support the client to plan and install the heating equipment or other energy installations, and also to operate and maintain it. ESCO can also design and build a local energy distribution grid.

Energy Performance Contracting (EPC) is a business model that offers final energy services based on energy cost savings related to a mean value of the energy consumption in the last year or another reference value. This contract for green retrofitting introduces an important advantage for the owners of the buildings because they will be confronted with zero or minimal upfront costs. In general, such a contract can be implemented with the assumption that the ESCO company will make an energy audit for the building in order to identify the energy savings potential. Based on the energy audit results the ESCO company will negotiate with the owner of the building the duration of the EPC contract (up to 25 years) and the profit-sharing plans. As far as the contract has a longer term the ESCO company has the possibility to recover the initial Capex green retrofitting investment and to make profit from the saved energy costs. Therefore, the ESCO company has the opportunity to balance the risks related to energy price fluctuation or the fluctuations in the building energy consumptions. There are studies that reflect that the owners of the buildings tend to use more the lighting resources or are not switching off the light when they realized that the savings are high due to a higher lighting efficiency (Fouquet & Pearson, 2011; Lu et al., 2017).

The EPC contract is a turnkey service that may include a broad category of services from which are delimited: energy audit, engineering, construction management, project financing and maintenance.

Integrated Energy-Contracting (IEC) is a business models that integrates the useful energy supply (ESC), especially in the case of green sources with measurements of energy conservation (EPC) in the entire building (Würtenberger et al., 2012). Such a model may eliminate some weaknesses of ESC (e.g. completely supply oriented) and EPC contracts (very complex and expensive). Therefore, the IEC will complete the ESC model by adding demand side measures based on energy performance.

Generally, ESCO business models were implemented in public, commercial and industrial buildings, however they pose a certain issue regarding the possibility to guarantee for energy or cost savings, therefore not being largely implemented in the residential field. While EPC contracts have a high complexity and high transactions costs these are suitable for large buildings where the energy savings potential is high.

The occupant behavior is an independent variable, as it cannot be practically influenced, therefore, incomes may be lower, while the risk quite high, especially in terms of small individual projects (Mahapatra et al., 2011; Marino et al., 2011).

Table 4. Advantages and disadvantages for ESCO business models

	Advantages	Disadvantages
Energy Supply Contracting (ESC)	Security of the supply	Completely supply oriented
	Decentralized power supply (local)	
	The ESCO plan and install the energy equipment	
	Use of most recent safety standards	
Energy Performance Contracting (EPC)	Turnkey service	High complexity
	Zero or minimal upfront costs	Expensive (ESCO costs and profit are included in the project's costs)
	Energy savings	Energy audit for the building is needed
	Guarantee that the savings will cover the costs of the project	Duration (up to 25 years)
		ESCO can be focus on lower cost measures
		The ESCO has a low control of the energy used by the owner
Integrated Energy-Contracting (IEC)	Reduction of energy demand	
	Efficient supply	

In order to make the retrofitting business models more attractive new innovations in the revenue models and new financing models need to be developed (Würtenberger et al., 2012; Moschetti & Brattebo, 2016).

Business models based on new and innovative revenue models can originate from the stimulating environment caused through government incentives, that owners can benefit from in terms of gains, as well as the recovery of the high investment costs for deep retrofitting comparing to traditional measurement measures. For instance, tax discounts within the established percentages can be applied for building owners in the case of certain energy upgrading of existing buildings. If systems for generating renewable energy are also installed, then a feed-in subvention framework can be applied for every unit of energy produced (Moschetti & Brattebo, 2016).

Other income sources can also result from the value creation originating from a delivered product or service of the organization. For example, the use of a building label to confirm improved characteristics of the building could lead to higher rent or sale prices for owners (Moschetti & Brattebo, 2016).

The most of the ESCO Projects are financed by third-party finance companies represented by banks and different financial institutions.

New financing schemes for business models focus on the objective of finding solutions to the issue of high upfront costs. Energy efficiency funds, revolving funds, green bonds or crowdfunding may be a proper solution for financing new green retrofitting projects.

1. **Energy Efficiency Funds:** such as the European Energy Efficiency Fund are focused on projects that promote a sustainable energy market and climate protection. The green retrofitting projects fit with the goals of the European Energy Efficiency Fund (www.eeef.eu).
2. **Revolving Funds:** are innovative financing mechanisms that are reinvesting the savings from investments in green projects and energy efficiency in other projects that would reduce the carbon

emissions and their impact to climate change, such as renewable energy or green retrofitting projects (Gouldson et al., 2018). The specificity of a revolving fund consists in the continuing investments without any fiscal year limitations. Therefore, these funds are used for projects developed by government and non-profit organizations.

3. **Green Bonds:** are special bonds, like a debt instrument, that can be used for financing climate protection projects, such as green retrofitting projects. The issuer of the bonds is obliged to pay the borrowed capital with interest, to the investor back within a fixed term. Bonds can be emitted as Government Bonds or municipalities bonds. The main actor for green bonds in Europe is the European Investment Bank, which is promoting renewable energy and energy efficiency projects.
4. **Crowdfunding:** many small investors are part of a green retrofitting project by accessing a crowdfunding platform. Generally, they invest an amount of money in small green projects. For their money they receive interest and after a fixed number of years they are payed back. The main advantage of crowdfunding consists in the highest grade of independence compared with banks or large financial institutions financing (Tantau & Fratila, 2018).
5. **Energy Lease:** An indirect financing scheme that can be used also for green retrofitting projects is energy lease. It is an option for using energy installations without the necessity of buying them. It is an option to obtain the right to use the energy installation without to be in its possession.

Lease is suitable for renovating of buildings that can be removed after the end of the lease term. In the case of ESCO the company can lease to the building owner a new technology that save energy, including service, maintenance and other services such as consultancy for legal, fiscal or financial problems.

In a lease contract the lease company ("lessor"), in principle the distributor, for whom the equipment (e.g. a photovoltaic system) is their property, established a contract with the customer ("lessee"), in principle the building owner, on the use of the photovoltaic system. The home owner has to pay to lessor a monthly payment to over a period of time for the right to use the photovoltaic system and for maintenance.

However, most of the leased systems are connected to the power grid and didn't have an independent energy storage system. Therefore, the lessee consumes electricity from both the photovoltaic system and the grid. Generally, the lessee has the option to purchase the photovoltaic system at the end of the contracted period at a reduced value.

The main leases options for green buildings are operational leasing and financial leasing.

Operational lease is running for a limited term, which is shorter as the economic life of the energy installation and is not transferring the title, the risks and the rewards. Therefore, the operational lease is close to a rental agreement where the energy installation is transferred back to the lessor at the expiry of the lease term.

Financial lease is known also as capital lease and is established for a period that cover the main part of the economic life of the energy installation, without the transfer of title but with the transfer of risk and rewards. At the expiry of the lease term the lessee has the option to purchase the electrical equipment at a price that is less than its market value.

In comparison to consumer credits leasing on a small scale assumes higher transaction costs and risks for the lender, as well as borrower costs in projects with a low component of physical assets. The majority of the leasing projects are financed from private or equity funds. The rentability for energy lease projects is influenced also by other factors. For example, the feed in tariff or the green certificates for promoting the renewable energy are reducing the risks associated with such a business model.

From the lessees' perspective leasing is efficient as far as the monthly leasing fees plus the cost of electricity consumption from the grid is lower than the cost of electricity in the case that all amount of electricity was taking from the grid.

If the financing for a green retrofit project is supported by the owner of the project, there is the case of the customer-side renewable business model.

In the Customer-side renewable business model the renewable system is located on the property of the customer which can be also the owner of the energy facility (Richter, 2012; Huijben & Verbong, 2013).

For example, in the customer owned photovoltaic business model (host owned business model, Figure 4), the customer (host) which is the owner of the rooftop or land where the photovoltaic system is installed take all derived costs of the business, but also the benefit of all revenues.

For the building sector the renewable typical renewable energy systems are represented by photovoltaic systems, wood pellet systems combined heat and power systems, geothermal and heat pumps (Aslani & Mohaghar, 2013). The power of such a system has a range between some kW up to 1 MW (Richter, 2012).

The main advantages of such a business model consist in minimal discomfort for the final consumer, implementation of new technologies with low maintenance requirements, a single party for all emerging issues, more financing parties, such as banks or government bodies (Bell et al., 2011; Würtenberger et al., 2012). Another issue is the application process, that should be simplified and with less barriers for direct payment of contractors' services that could increase the success rate of the programs. (Johnson et al., 2012). Through cooperation with the utility installers of renewable energy equipment may be also integrated (Würtenberger et al., 2012; Paiho et al., 2015).

In this business model the customer can benefit directly 100% of the produced electricity. This model avoids costs that are related to the grid transport costs. In parallel the customer can benefit of the government supporting scheme facilities if this are active.

Connected disadvantages of this business model consist in high investments and maintenance costs and in the risk of poor performance of the system associated with the EPC company, which has to guarantee the performance of the photovoltaic system.

Figure 4. Customer owned photovoltaic business model
Source: Tanţău et al., 2014

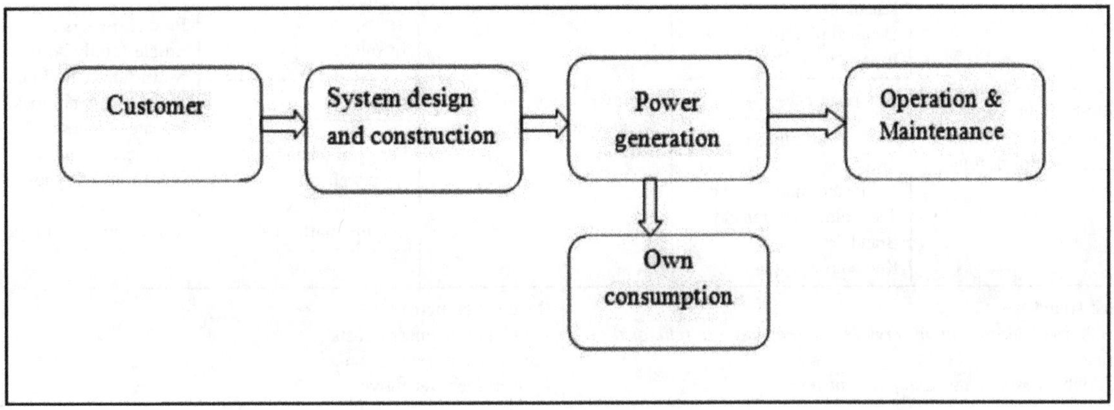

Framework for the Development of a New Business Model for Green Retrofitting in the Building Sector Based on the Triple Layer Business Model Canvas

This case study is based on the traditional Canvas structure for a business model and the theory developed by Osterwalder and Pigneur (2010) and is developed as a triple bottom line model with the economic, environmental and social dimensions. The Triple Layer Business Model Canvas can be adopted in order to develop the traditional business model Canvas to a more sustainable one (Joyce & Paquin, 2016).

The economic part of the model is structured in nine blocks that correspond to the traditional canvas structure: value proposition, customer segment, channels, customer relationship, key resources, key activities, key partners, cost structure and revenues structure (Table 5).

Value Proposition

The value proposition for a green retrofitting business model has to respond to the main needs of a customer and in parallel has to express the essential characteristics of the product or service that is offered to the customers. If the product or the service are innovative then the value proposition has the advantage to be unique for the specific customer segment. The unicity of a product or a service can contribute to the increase of the competitiveness of the organization and has to be clear formulate in the value proposition. Not all products or services that are sold to the clients are innovative. In this case the value proposition has to face with clients' objectives. In such a case the value proposition can be focused on energy savings, comfortable indoor conditions and functional performance for a green retrofitting solution. Energy savings could be quantified in reduction of kWh/year (in %) due to reduction of energy demand for heating or cooling and reduction of primary energy consumption.

Table 5. Economic dimension for the business model for green retrofitting based on the Triple Layer Business Model Canvas

<table>
<tr><th>Key partners</th><th>Key activities</th><th>Value proposition</th><th>Customer relationship</th><th>Customer segment</th></tr>
<tr><td rowspan="3">Organizations from the building industry
Financing institutions
Specialists for design</td><td>Renovations works
Financing
Detailed planning
Installation</td><td rowspan="3">Energy savings
Maintenance
Design</td><td>Mutual trust
Transparency
Loyalty</td><td rowspan="3">Private owners of Single-family houses
Multi-family buildings and Apartment blocks
Private owners of commercial and industrial buildings
Public owners of institutional buildings</td></tr>
<tr><td>Key resources</td><td>Channels</td></tr>
<tr><td>Specialized human resources
Motivated management
Technology equipment,
Brand
Know how</td><td>Personal contacts
Own web site
Specialized Events
On line platforms
E-Mail</td></tr>
<tr><td colspan="3">Cost structure
- Fix costs (salaries, main services, on line platform, rent for the office),
- Variable costs (material input, transport)</td><td colspan="2">Revenue structure
- reduction in energy costs
- revenues for renovation
- government incentives</td></tr>
</table>

The value proposition has not to be limited on the renovated building and has to offer also support service package like maintenance. Other value proposition can include design excellence, energy quality, energy security or structural assurance.

Customer Segment

The customer segment plays a referential role for the business plan because all its blocks have to be developed for a specific customer segment. The target customer is the segment of persons that an organization wants to offer value to.

For a green retrofitting business model, the customer segment is represented by knowledgeable customers, as energy end users, that are willing to introduce new technologies in their buildings such as new ventilation with heat recovery, energy efficient windows and doors, energy efficient lighting and electrical equipment, efficient heating solutions, renewable energy sources or smart metering.

Customers for green retrofitting projects are private owners of single-family houses, multi-family buildings and apartment blocks, private owners of commercial and industrial buildings and also public owners of institutional buildings.

Channels

Main channels to reach clients for green retrofitting are personal contacts, on line platforms, events that are organized for experts and gain attention.

Customer Relationship

Customer relations have to be based on mutual trust. For the green retrofitting business model there is a need for a higher confidence of buildings owners in the sustainability certification of the building (Moschetti & Brattebo, 2016).

Key Resources

Key resources are main resources, such as specialized human resources, technology equipment, brand and know how, that an organization use in order to offer the value proposition.

Green retrofitting projects are often labor intensive and are implementing new technologies or are increasing the buildings standards. As a result, there is a need for specialists also for project management. These projects can be capital intensive and the return on investment could be estimated even on more than 15 years, therefore financial specialists are also very important.

Key Activities

Key activities for green retrofitting are mainly represented by renovations works, but these could be oriented also on services before sale (e.g. financing, detailed planning, obtaining approvals from local administration) and also after sale (e.g. maintenance).

Key Partners

The generic partners for retrofitting projects are other organizations from the building industry, local authorities and universities.

Financial institutions represent a main partner for capital intensive green retrofitting solutions which are capital intensive.

Cost Structure

The cost structure of a green retrofitting project is divided in fix costs and variable costs. The fix costs are represented by salaries, or website costs. The cost structure is dominated by the investment costs for different green retrofitting measures that has to be analyzed and also by the interest rate for the financing.

The variable costs for a green retrofitting project consist mainly in material costs, transportation costs.

Especially, for a green retrofitting project there is a need to take into account also the sustainability certification costs.

Revenue Structure

The revenue structure explains the origins of the main revenue streams. For the green retrofitting solution, the revenues are a result of the reduction in energy costs, but for the next period there is a need to take more into account the energy usage or carbon emissions. Other revenues could be represented by government incentives (e.g. green certificates, feed in tariff). In the case of renovation, the revenues are completed with the owner payments for it.

The second dimension that is analyzed for the business model for green retrofitting based on the Triple Layer Business Model Canvas is the environmental one. This will be analyzed also on a nine-block structure, which is mapping the environmental benefits and the main factors that are contributing to this objective (table 6). The nine blocks of this structure are: functional value, use phase, distribution, end of life, key materials, green retrofitting, supplies and outsourcing, environmental impacts and environmental benefits.

The environment dimension evaluates the environmental impact of different factors that are defining the business model. For a green retrofitting business model are defined also specific indicators that are analyzed: CO_2 emissions, water use or material waste.

Functional Value

Value proposition is translated for the environmental dimension in the functional value, which describe the output of the green retrofitting business as a service. This is represented by the energy reduction, but also by the reduction of CO_2 emissions ($kgCO_2$/year, in %) due to heating or cooling or reduction of water demand.

Table 6. Environmental dimension for the business model for green retrofitting based on the Triple Layer Business Model Canvas

<table>
<tr><th>Suppliers and subcontracting</th><th>Green Retrofitting</th><th>Functional Value</th><th>End of life</th><th>Use phase</th></tr>
<tr><td rowspan="3">Subcontractors
Suppliers of materials</td><td>Energy Performance Contract
Energy Audit</td><td rowspan="3">Energy savings
Reduction of CO_2 emissions</td><td>Increase the building life</td><td rowspan="3">More friendly to the environment
(e.g. Passive House)</td></tr>
<tr><td>Key Materials</td><td>Channels</td></tr>
<tr><td>Material recycling
Materials that are not damaging the environment
Use of local materials</td><td>Transportation of materials</td></tr>
<tr><td colspan="2">Environmental impacts
Potential rebound effect due to new life style standards</td><td colspan="3">Environmental benefits
Reduction of CO_2 emissions
Prevent demolition and avoid material and construction waste
Natural resource savings due to energy saving</td></tr>
</table>

Source: Own contribution for green retrofitting

Use Phase

In the use phase the green house aims to be environmentally friendly (e.g. passive house, zero emissions house). Therefore, there is a need to monitor the energy and water demand and to use smart devices in order to optimize them.

Distribution

Main distribution activities that have an environmental impact for a green retrofitting business models are represented by the transportation of the materials and the logistic part. These have also to be conform to new standards that reduce their environment impact.

End of Life

The green retrofitting increases the building life and from the design phase it has to be taken into account the recycle potential of major parts, equipment and materials at the end of the building life.

Key Materials

For the environmental perspective there is a need to use materials that are not damaging the environment. From the design phase the whole green retrofitting process is looking on recyclable materials and environment friendly materials. In the case of green retrofitting the use of local materials may reduce the embedded energy (Koezjakov et al., 2018). The embodied energy is defined as the total energy that is consuming for extraction, production and transport of building materials to the building site. This is also considered an indicator for the total environment impact of buildings materials and systems.

Green Retrofitting

The main activity that has an impact of the environment is the green retrofitting of the building. This is a very complex activity. Most of the green retrofitting projects are developed with Energy Performance Contracts and are requiring an energy audit.

Suppliers and Subcontracting

A special category of key partners for green retrofitting business models is represented by subcontractors and suppliers of materials. They do not represent the core of the business, but they are a main support for the organization to gain competitive advantages. For a green retrofitting project, the subcontractors and suppliers of materials have also to implement in their organizations standards that contribute to the environmental protection.

Environmental Impacts

Even green retrofitting projects can have an environmental impact. If the Jevons Paradox is admitted, there has to be taken into account the potential rebound effect due to new and improved life style standards. These will generate new needs, which will contribute to an increase demand for energy. Fore a green retrofitting business model the analysis is focused on the energy consumption and CO_2 emissions, but can be extended also for the water use.

Environmental Benefits

Environmental benefits take into account the ecological value that the business model create due to new solutions for reducing the environmental impact of green retrofitting.

The main environmental benefits consist of reduction of CO_2 emissions, preventing demolition and avoiding material and building waste and natural resource savings due to lower demand for energy (the demand of energy per unit of consumption is reduced).

The third dimension that is analyzed for the business model for green retrofitting based on the Triple Layer Business Model Canvas is the social one. This will be analyzed also on a nine block structure that indicates the main social impacts of the green retrofitting project. This block consist of social value, end users, scale of outreach, social culture, employees, governance, communities, social impacts and social benefits (Table 7). The analysis is oriented on the stakeholders' perspective for a green retrofitting business model.

Social Value

Social value illustrates how the business model create value for the main stakeholders. Social value is created by solving material and construction waste and preferably by avoiding their generation. Involving local communities in green solutions or increasing the grade of information about green retrofitting represent other generic social values for a green retrofitting business

Table 7. Social dimension for the business model for green retrofitting based on the Triple Layer Business Model Canvas

<table>
<tr><th>Local communities</th><th>Governance</th><th>Social Value</th><th>Societal culture</th><th>End User</th></tr>
<tr><td rowspan="3">Local authorities
Universities
Social relations with suppliers
Collaboration with other social projects</td><td>Transparent
Create social value</td><td rowspan="3">Solving material and construction waste and preferably avoidance of generating such waste
Involving local communities in green solutions
Increase the grade of information about green retrofitting</td><td>Support active involvement of individuals against construction waste
Militate for reducing the energy consumptions in buildings</td><td rowspan="3">Implying citizens cooperation for reducing the energy consumption</td></tr>
<tr><th>Employees</th><th>Engagement measurement</th></tr>
<tr><td>Promotors of green retrofitting solutions
Trainings for green retrofitting</td><td>Key stakeholders
Best practice</td></tr>
<tr><td colspan="2">Social impacts
Potential rebound effect due to new life style standards</td><td colspan="3">Social benefits
Reduction of CO_2 emissions
Construction waste reduction
Community engagement in reducing the energy demand and in reducing the construction waste
Growth of awareness for implementing solutions for optimal energy performance.</td></tr>
</table>

End User

End users are the owners of private buildings or even commercial, industrial or public institutions which are willing to improve the energy performance of the buildings and want to implement green retrofitting solutions. Such a green retrofitting business model implies citizen cooperation for reducing the energy consumption.

Engagement Measurement

Key stakeholders have to be promoters of new green retrofitting solutions. The green retrofitting solutions could be used as best practice to promote the optimal energy performance in buildings.

Societal Culture

The societal culture block estimates the social impact of the green retrofitting business model on the society. Green retrofitting projects are supporting involvement of individuals against construction waste and are militating for reducing the energy consumptions in buildings.

Employees

The employees have to be promotors of green retrofitting solutions. Hence a need to train the employees to implement green retrofitting solutions for optimal energy performance.

Governance

Governance illustrates the internal organizational structure (e.g. type, relations), the decision- making process and main policies in the organization. These have to be transparent and oriented to create also social value.

Communities

A green retrofitting business model include not only the economic business relations, but also the social relations to the community. Local communities and universities could be key partners for promoting the green retrofitting solutions and make them complementary to other social projects.

Social Impacts

Social impacts of a green retrofitting business model express the social costs of such a business model. For example, these can be quantified based on the potential rebound effect due to new life style standards. The new life style introduce new electrical devices that will be controlled by Internet of Things technology and will contribute to an increase of energy demand for buildings.

Social Benefits

Social benefits are quantifying the positive social value that is created in the green retrofitting business model.

Main social benefits of a green retrofitting projects are the reduction of CO_2 emissions, the reduction of the construction waste, community engagement in reducing the energy demand and in reducing the construction waste and reach growth of popularity for implementing solutions for optimal energy performance.

The business model for green retrofitting in the building sector based on the Triple Layer Business Model Canvas is still in a development phase and has a high degree of subjectivity. The main benefits of using this instrument for elaborating business models for green retrofitting for optimal energy performance consist in offering a high visibility of different relevant aspects that may influence the development and implementation of such a business solution in practice. This instrument can also be used to find and develop new green retrofitting solutions for optimal energy performance.

CONCLUSION

The tool that is developed in this chapter for the green retrofitting integrates the economic, environmental and also the social dimensions into a more complex concept with three dimensions based on the structure of the Triple Layer Business Model Canvas.

The business models for green retrofitting could be an important instrument that is introducing new green characteristics such as energy efficiency, optimal energy performance, new comfort standards in the building environment, and is responding to the dynamic of the economic and technology development and to the new life style of the peoples. The implementation of such a model will be also a catalyst for reducing the emissions of greenhouse gases in the building environment.

The business model for green retrofitting has to be elaborated after collecting data and analyze all relevant information regarding the specificity of the building, which has to be retrofit in a holistic approach from the project design up to the use phase and the end of life phase.

This analysis is based on an own research on green retrofit projects and on the integrated approach on business models which indicate that these projects have significant environmental and social benefits.

This chapter contributes to the applied research on business models for retrofitting and create a framework for developing new business models for green retrofitting based on the triple layer business model canvas.

REFERENCES

Aslani, A., & Mohaghar, A. (2013). Business structure in renewable energy industry: Key areas. *Renewable & Sustainable Energy Reviews*, *27*, 569–575. doi:10.1016/j.rser.2013.07.021

Ayyad, K., & Fekry, A. (2016). Green Retrofitability Index (GRI) as an Indicator for Decision Makers in Green Retrofitting Projects. *Procedia Environmental Sciences*, *34*, 233–244. doi:10.1016/j.proenv.2016.04.022

Bell, C., Nadel, S., & Hayes, S. (2011, December). *On-bill financing for energy efficiency improvements: A review of current program challenges, opportunities and bets practices.* American Council for an Energy-Efficient Economy (ACEEE). Retrieved from http://www.aceee.org/sites/default/files/publications/researchreports/e118.pdf

Boons, F., & Lüdeke-Freund, F. (2013). Business models for sustainable innovation: State-of-the-art and steps towards a research agenda. *Journal of Cleaner Production*, *45*, 9–19. doi:10.1016/j.jclepro.2012.07.007

BPIE. (2013) A Guide to Developing Strategies for Building Energy Renovation - Delivering Article 4 of the Energy Efficiency Directive. Retrieved from www.bpie.eu/.../BPIE/Developing_Building_Renovation_Strategies.pdf

Broman, G., & Robert, K. (2016). A framework for strategic sustainable development. *Journal of Cleaner Production*, *140*(Part 1), 17–31.

Dixon, T., Lannon, S., & Eames, M. (2018). Reflections on disruptive energy innovation in urban retrofitting: Methodology, practice and policy. *Energy Research & Social Science*, *37*, 255–259. doi:10.1016/j.erss.2017.10.009

Eriksson, E., Nenonen, S., Junghans, A., & et al, . (2015). Nordic campus retrofitting concepts - Scalable practices. *Procedia Economics and Finance*, *21*, 329–336. doi:10.1016/S2212-5671(15)00184-7

Filippi, M. (2015). Remarks on the green retrofitting of historic buildings in Italy. *Energy and Building*, *95*, 15–22. doi:10.1016/j.enbuild.2014.11.001

Fouquet, R., & Pearson, P. (2011). The Long Run Demand for Lighting: Elasticities and Rebound Effects in Different Phases of Economic Development. *Economics of Energy & Environmental Policy*, *1*(1), 83–100.

Franca, C., Broman, G., Robert, K., & et al, . (2016). An approach to business model innovation and design for strategic sustainable development. *Journal of Cleaner Production*, *140*, 155–166. doi:10.1016/j.jclepro.2016.06.124

Frej, A., & Browning, W. (2005). *Green office buildings: a practical guide to development.* Washington, D.C.: Urban Land Institute.

Garbuzova, M., & Madlener, R. (2012). Towards an efficient and low carbon economy post-2012: Opportunities and barriers for foreign companies in the Russian energy market. *Mitigation and Adaptation Strategies for Global Change*, *17*(4), 387–413. doi:10.100711027-011-9332-8

Geissdoerfer, M., Savaget, P., & Evans, S. (2017). The Cambridge Business Model Innovation Process. *Procedia Manufacturing*, *8*, 262–269. doi:10.1016/j.promfg.2017.02.033

Gouldson A.,Kerr N., Millward-Hopkins J., Freeman M., Topi C., & Sullivan R. (2018). Innovative financing models for low carbon transitions: Exploring the case for revolving funds for domestic energy efficiency programmes. *Energy Policy*, *86*, 739-748.

Grade, F., & Donn, M. (2014). Solution sets and net zero energy buildings: A review of 30 Net ZEBs case studies worldwide. IEA Joint SHC Task 40, International Energy Agency.

GRI. (2014). ISO GRI G4 Guidelines and ISO 26000:2010, How to use the GRI G4 and ISO 26000 in conjunction, Global reporting initiative and ISO.

Huijben, J., & Verbong, G. (2013). Breakthrough without subsidies? PV business model experiments in the Netherlands. *Energy Policy*, *56*, 362–370. doi:10.1016/j.enpol.2012.12.073

IPCC. (2018). Special Report on Global Warming of 1.5 C (SR15), Intergovernmental Panel on Climate Change. Retrieved from http://www.ipcc.ch/report/sr15/

Jagarajana, R., Asmonia, M., Mohammeda, A., & et al, . (2017). Green retrofitting – A review of current status, implementations and challenges. *Renewable & Sustainable Energy Reviews*, *67*, 1360–1368. doi:10.1016/j.rser.2016.09.091

Joyce, A., & Paquin, R. (2016). The triple layered business model Canvas: A tool to design more sustainable business models. *Journal of Cleaner Production*, *135*, 1474–1486. doi:10.1016/j.jclepro.2016.06.067

Khorshidi, M. (2018). Research for developing sustainable business models for a national oil company in the crude oil supply surplus situation [PhD. Thesis]. The Bucharest University of Economic Studies, ASE Bucharest.

Koezjakov, A., Urge-Vorsatz, D., Crijns-Graus, W., & van den Broek, M. (2018). The relationship between operational energy demand and embodied energy in Dutch residential buildings. *Energy and Building*, *165*, 233–245. doi:10.1016/j.enbuild.2018.01.036

Lizana, J., Barrios-Padura, A., Molina-Huelva, M., & Chacartegui, R. (2016). Multi-criteria assessment for the effective decision management in residential energy retrofitting. *Energy and Building*, *129*, 284–307. doi:10.1016/j.enbuild.2016.07.043

Mahapatra, K., Gustavsson, L., Haavik, T., & Aabrekk, S. (2011). *Report on possible financing schemes for one-stop-shop service for sustainable renovation of single-family house*. Nordic Innovation Publications.

Marino, A., Bertoldi, P., Rezessy, S., & Boza-Kiss, B. (2011). A snapshot of the European energy service market in 2010 and policy recommendations to foster a further market development. *Energy Policy*, *39*(10), 6190–6198. doi:10.1016/j.enpol.2011.07.019

Moschetti, R., & Brattebø, H. (2016). Sustainable business models for deep energy retrofitting of buildings: State-of-the-art and methodological approach. *Energy Procedia, 96*, 435–445. doi:10.1016/j.egypro.2016.09.174

Nejat, P., Jomehzadeh, F., Taheri, M., Gohari, M., & Majid, M. (2015). A global review of energy consumption, CO2 emissions and policy in the residential sector (with an overview of the top ten CO2 emitting countries). *Renewable & Sustainable Energy Reviews, 43*, 843–862. doi:10.1016/j.rser.2014.11.066

Osterwalder, A. (2004). *The Business Model Ontology: A proposition in a design science approach.* Universite de Lausanne, Ecole des Haute Etudes Commerciales.

Osterwalder, A., & Pigneur, Y. (2010). *Business model generation – A handbook for visionaries, game changers, and challengers.* Hoboken, NJ: John Wiley & Sons.

Paiho, S., Abdurafikov, R., Hoang, H., & Kuusisto, J. (2015). An analysis of different business models for energy efficient renovation of residential districts in Russian cold regions. *Sustainable Cities and Society, 14*, 31–42. doi:10.1016/j.scs.2014.07.008

Pätäri, S., & Sinkkonen, K. (2014). Energy service companies and energy performance contracting: Is there a need to renew the business model? Insights from a Delphi study. *Journal of Cleaner Production, 66*, 264–271. doi:10.1016/j.jclepro.2013.10.017

Richter, M. (2012). Utilities' business models for renewable energy: A review. *Renewable & Sustainable Energy Reviews, 16*(5), 2483–2493. doi:10.1016/j.rser.2012.01.072

Riedy, C., Lederwasch, A., & Ison, N. (2011). *Defining zero emission buildings – review and recommendations: final report. Prepared for sustainability Victoria by the Institute for Sustainable Futures.* Sydney: University of Technology.

Robichaud, L., & Anantatmula, V. (2010). Greening project management practices for sustainable construction. *Journal of Management Engineering, 27*(1), 48–57. doi:10.1061/(ASCE)ME.1943-5479.0000030

Rocky Mountain Institute. (2015) How to Calculate and Present Deep Retrofit Value - A Guide for Investors. Boulder, CO, Rocky Mountain Institute. Retrieved from http://www.rmi.org/retrofit_depot_deepretrofitvalue

Rosca, E., Arnold, M., & Bendul, J. (2016). Business models for sustainable innovation - an empirical analysis of frugal products and services. *Journal of Cleaner Production*, 162, 133-145.

Tantau, A., & Fratila, L. (2018). *Entrepreneurship and Business Development in the Renewable Energy Field.* Hershey, PA: IGI Global. doi:10.4018/978-1-5225-3625-3

Tantau, A., & Staiger, R. (2017). *Business Models for Renewable Energy Initiatives.* Hershey, PA: IGI Global.

Tolkamp, J., Huijben, J., Mouric, R., Verbong, G., & Bouwknegt, R. (2018). User centered sustainable business model design. The case of energy efficiency in the Netherlands. *Journal of Cleaner Production, 182*, 755–764. doi:10.1016/j.jclepro.2018.02.032

Upward, A., & Jones, P. (2016). An ontology for strongly sustainable business models defining an enterprise framework compatible with natural and social science. *Organization & Environment, 29*(1), 97–123. doi:10.1177/1086026615592933

Wilkinson, S. (2012). Analysing sustainable retrofit potential in premium office buildings. *Structural Survey, 30*(5), 398–410. doi:10.1108/02630801211288189

Wirtz, B., Pistoia, A., Ulrich, S., & Gottel, V. (2016). Business Models: Origin, development and future research perspectives. *Long Range Planning, 49*(1), 36–54. doi:10.1016/j.lrp.2015.04.001

Würtenberger, T., Bleyl, J., Menkveld, M., Vethman, P., & Tilburg, X. (2012). Business models for renewable energy in the built environment. IEA. Retrieved from http://iea-retd.org/wp-content/uploads/2012/04/RE-BIZZ-final-report.pdf

Xu, P., Chan, E., & Qian, Q. (2011). Success factors of energy performance contracting (EPC) for sustainable building energy efficiency retrofit (BEER) of hotel buildings in China. *Energy Policy, 39*(11), 7389–7398. doi:10.1016/j.enpol.2011.09.001

Zhao, X., Hwang, B. G., & Lu, Q. (2018). Typology of business model innovations for delivering zero carbon buildings. Journal of cleaner production, 196, 1213-1226.

Zuo, J., & Zhao, Z. (2014). Green building research-current status and future agenda: A review. *Renewable & Sustainable Energy Reviews, 30*, 271–281. doi:10.1016/j.rser.2013.10.021

ADDITIONAL READING

Hannon, M. J., Foxon, T. J., & Gale, W. F. (2015). 'Demand pull' government policies to support Product-Service System activity: The case of Energy Service Companies (ESCos) in the UK. *Journal of Cleaner Production, 108*, 900–915.

Johnson, K., Willoughby, G., Shimoda, W., & Volker, M. (2012). Lessons learned from the field: Key strategies for implementing successful on-the-bill financing programs. *Energy Efficiency, 5*(1), 109–119. doi:10.100712053-011-9109-7

Lu, Y., Zhang, N., & Chen, J. (2017). A behavior based decision making model for energy performance contract in building retrofit. *Energy and Building, 156*, 315–326. doi:10.1016/j.enbuild.2017.09.088

Polzin, F., von Flotow, P., & Nolden, C. (2016). What encourages local authorities to engage with energy performance contracting for retrofitting. Evidence from German municipalities. *Energy Policy, 94*, 317–330. doi:10.1016/j.enpol.2016.03.049

Testa, F., Annunziata, E., Iraldo, F., & Frey, M. (2016). Drawbacks and opportunities of green public procurement: An effective tool for sustainable production. *Journal of Cleaner Production, 112*(3), 1894–1900.

Wang, D., Landolt, J., Mavromatidis, G., Orehounig, K., & Carmeliet, J. (2018). CESAR: A bottom-up building stock modelling tool for Switzerland to address sustainable energy transformation strategies. *Energy and Building, 169*, 9–26. doi:10.1016/j.enbuild.2018.03.020

KEY TERMS AND DEFINITIONS

Business Model: Method of the enterprise to describe key structural and operational characteristics, the way it earns revenue and delivers profit in the current business environment, starting from the value of the business idea.

Energy Performance Contracting (EPC): Business model that offers final energy services based on energy cost savings related to a mean value of the energy consumption in the last year or another reference value.

Energy Supply Contracting (ESC): Business model based on the provision of useful energy, such as electricity, heat or steam to a building owner or user, through a long-term contract (e.g. 10 to 15 years).

Green Building: A building that in all its development phases from project design up to its use phase has a low impact on the environment, is using recyclable materials that are health for its occupants, is implementing a high level of energy efficiency, is looking for reducing the water consumption, is using renewable energy sources and is assuring an appropriate comfort.

Green Retrofitting: Upgrade at an existing building that is wholly or partially occupied to improve energy and environmental performance, reduce water use, improve comfort and quality of space in terms of natural lighting, air quality and noise, all done in a way that it is financially beneficial to the owner

Integrated Energy-Contracting (IEC): Business models that integrates the useful energy supply (ESC), especially in the case of green sources with measurements of energy conservation (EPC) in the entire building

Retrofitting: Adding components or accessories to buildings that did not have it when these where initial constructed in order to improve the building performance

Sustainability: Development that meets the needs of the present generation without compromising the ability of future generations to meet their own needs

Triple Layer Business Model: Tool that integrates the economic, environmental and social dimensions into a more complex concept of a business model

Chapter 2
Financial Considerations in Green Retrofitting for Optimal Energy Performance

Andreea Curmei-Semenescu
The Bucharest University of Economic Studies, Romania

Cătălin Valeriu Curmei
The Bucharest University of Economic Studies, Romania

ABSTRACT

This chapter focuses on how to correctly assess the financial feasibility of a retrofitting project using different indicators and methods. It also emphasizes the advantages and limits of each method offering the reader the ability to choose the most appropriate method(s) for each retrofitting project. In its last part, the chapter explains how the financial feasibility can be integrated in a broader set of criteria considered for retrofitting the capital budgeting decision-making process. The chapter also includes examples of computing the financial indicators for green retrofitting evaluation. It thus provides the readers with the opportunity to understand how the capital budgeting evaluation methods can be used in practice and correctly interpret their results.

INTRODUCTION

Green retrofitting is one of the big challenges of the next few decades, especially in Europe. With the objective of the European Union to reduce significantly energy consumption and by this means the global carbon footprint, green retrofitting became an important issue. Statistics suggest that more than one third of the energy consumption in the European Union is related to insuring space comfort within office and residential buildings (Trotta, 2018; Christersson, Vimpari & Junnila, 2015, Lizana, Barrios-Padura, Molina-Huelva & Chacartegui, 2016). In most important cities in Europe, the real estate stock is relatively old and does not correspond to modern times requirements in terms of energy efficiency. The extent of the work to be done in this sense and the complexity of the stakeholders' behavior regarding this aspect require further investigations.

DOI: 10.4018/978-1-5225-9104-7.ch002

The motivations of implementing green retrofitting projects are different and have a complex nature, ranging from the interest in maximizing the value of the building to the incentive of the users to improve their comfort or to associate their image to an environment-friendly project. The type and state of the building subject to retrofitting, its location, the motivation behind the project, the time horizon for the exploitation of the building, the regulatory framework and even moral and cultural values of the society have a role to play in choosing between different retrofitting solutions. In most cases, the financial objectives are interrelated to non-monetary objectives and the decision between several available technical solutions becomes a very complex one.

Though not the only motivation of a retrofitting project, the financial incentive is a very important one and a correct assessment of the feasibility of the available retrofitting solutions is mandatory for making the good choice between them and ultimately for the success of the project. The purpose of the present chapter is to explain and exemplify the methods that can be used in order to assess the financial feasibility of the retrofitting project or of its alternative solutions. It is concerned with providing the reader with the theoretical background needed to establish the feasibility of a retrofitting project from a financial point of view, but also with the practical expertise throughout the examples presented.

A retrofitting solution must be thoroughly analyzed from the point of view of its technical feasibility and compliance with the regulation in force, before assessing its financial feasibility. However, once this stage accomplished, the financial constraints have to be taken into consideration in making the final decision about the opportunity of a certain solution or even of the project itself.

The objective of the present chapter is to present the basic financial concepts necessary in retrofitting project investment decisions and to put in evidence various methods and techniques used to assess the feasibility of a retrofitting project, explaining their advantages and limits and their functioning. The reader will thus be familiarized with the theoretical concepts pertaining to the financial analysis and valuation of a retrofitting project, but also with the stages of its application in practice. The information included will also allow the reader to make the best decision regarding the specific indicators to be used in a particular analysis according to the specifics of the project and to the type of stakeholders benefitting from the results of the analysis. The chapter is equally useful for engineers and other persons interested in acquiring the basic concepts and methods used in assessing the financial feasibility of the project in order to better respond to different stakeholders regarding financial matters related to a retrofitting project, but also to economists who will find within its pages the particularities of the application of a financial feasibility analysis on the particular case of a retrofitting project.

The rest of the chapter will be conducted as follows. The first sub-chapter presents the main theoretical principles and concepts regarding the financial feasibility of retrofitting projects. The second sub-chapter is concerned with explaining the indicators and methods used to assess the feasibility of retrofitting projects, as well as the manner to correctly forecast the financial results of such investments. Also, it will briefly present how the monetary and non-monetary objectives of retrofitting projects can be taken into consideration jointly in the choice among alternative solutions. A practical analysis is provided in guise of example in the following subchapter. The future research directions and the main conclusions are presented in the end of the chapter.

FINANCIAL FEASIBILITY OF PROJECTS

The analysis of the financial feasibility of a green retrofitting project aims to provide an answer to the question "Is the retrofitting project proposed worth more than it costs?" in monetary terms. Basically, the financial feasibility of a project can be defined as its ability to generate sufficient monetary benefits for its initiators to meet or overpass their rational expectations.

When it comes to green retrofitting, people usually associate the concept of sustainability. In general, sustainability means development without harming next generations. In financial terms, the sustainability of a retrofitting project may be defined as its ability to generate during each period of its projected lifetime sufficient cash inflows to cover the necessary payments. Please note that the financial sustainability is different from the global concept of sustainability as it only focusses on the financial issues associated with a retrofitting project, not considering technical and environmental issues. Also, it is important to understand that the financial sustainability does not equal financial feasibility of a retrofitting project. The main reason is that a project may be financially sustainable in the sense of generating enough cash inflows to cover the needed payments, without providing the expected return for the investor. This is the reason why the rest of the chapter will be focused on the financial feasibility of the retrofitting projects.

In order to realize a correct financial analysis of a retrofitting project, the analyst must understand in detail its nature and what monetary costs and benefits it provides for the specific category of stakeholders requiring the assessment.

The stakeholders represent all categories of persons who have an interest into a project or an activity. The choice of using the term in the context of analyzing the financial feasibility of a retrofitting project is motivated by the fact that such project can be undertaken by different persons – landlords, tenants, developers, public authorities etc.- who may have different incentives, different goals, different time horizons and different expectations.

In a very simplistic approach, a retrofitting project is worth undertaking if it provides to the person incurring its initial cost sufficient cash flows during its use as to allow a correct reward for the investment.

The cash flow is defined as the difference between cash inflows and cash outflows generated by the project within a specified timeframe (usually, one year). The investor will be interested to undertake the project if the cash flows generated allow the full coverage of the initial cost and a reasonable return on the investment.

From a practical point of view, the analysis of the financial feasibility of a retrofitting project raises numerous challenges. Firstly, the analyst must always keep in mind who is the stakeholder undertaking the project and what is her/his motivation. This information has a huge impact on the result of the analysis as it influences the nature of the cash inflows and outflows forecasted during the exploitation of the project, the time horizon relevant for the analysis, the expected rate of return on the project. In guise of example, consider as a study case the retrofitting project of a leased one-family residential building. If the landlord is undertaking the investment, the expectation would be of higher rent rates, probably lower maintenance and tax expenses, but also a higher price of the real estate in the future and in some cases even a longer period of exploitation of the project. If the project is undertaken by the tenant, the reasonable expectations would be of lower utilities expenses during the stay and probably lower rent rates if so agreed with the landlord. The time horizon is also different, as the tenant is expected to use the real estate on a period shorter than its lifetime.

Secondly, not all benefits of the retrofitting project are always easily translated in monetary terms and this can have a huge impact on the result of the financial analysis. For example, how can be translated into monetary terms the investor's satisfaction of having used environment-friendly materials? Or else, how can we put in financial terms the fact that the retrofitting project resulted also in solving the mold problem in the house, hence decreasing the illness risk? In a different line of reasoning, how can a company translate into financial terms the image benefits it obtains if having its headquarters in a green building, instead of a conventional one? A possible solution may be to extract information from different transactions on the real estate market, but the task is not always easy because the individual value of such benefits is very subjective and, from a different perspective, prices in recorded transactions cannot always reveal these particular aspects.

There are many cases in the field of retrofitting projects when the prices on the market cannot fully integrate their effects, so that the true beneficiary to be forced to pay the price. In some cases, the reason is the insufficient regulation throughout laws and contracts, in others the subjective view over the results of the projects and the incapacity to correctly assess the risk on a long period. This is the reason why some retrofitting projects are delayed or aborted. The specialty literature refers to this phenomenon as to the landlord-tenant problem, none of them initiating the project as they consider that they would support costs that would eventually result in a net benefit for the other party (sea also Trotta, 2018).

Though the assessment of the financial feasibility of a retrofitting project cannot solve all problems, it is however a necessary step in implementing it successfully. The investors must first verify the technical feasibility of the project, then assess correctly the costs and cash flows generated by it and finally accept it if from a financial perspective the results are satisfactory.

CASH FLOW AND FINANCIAL INDICATORS FOR GREEN ENERGY RETROFITTING

The basic principle of the capital budgeting valuation is to compare the monetary results of an investment to its cost in order to understand if the project is worth being undertaken. Several problems can emerge in practice when applying capital budgeting valuation techniques.

The first one is related to the definition of the investment itself. Ross, Westerfield & Jordan (2002) mentions that an investment can be defined as a more or less certain present payment in exchange of future uncertain cash receipts. The numerous risk factors influencing the results of the investment project including the inability to predict future costs and benefits generated by the project, the changes in the regulatory framework, the unexpected evolution of the real estate market sector etc. make difficult the application of the capital budgeting valuation methodology in practice.

The second problem refers to the estimation of the expected rate of return for such projects that might be difficult in practice. It is influenced by economic factors such as the economic growth rate, the growth rate in the sector, the inflation and interest rate, but also by subjective factors related to investor's willing to improve the comfort and well-being of its family for example.

A third issue is related to the estimated lifetime of the retrofitting project. Although some technical specifications of the materials used, and sometimes different catalogues offer an idea for the analyst to estimate the lifetime of the project, economic and subjective facts should also be considered. For example, the time horizon estimated by the investor for using the real estate.

Last, but not least, sometimes there is a residual result imbedded in the real estate price or value that has to be correctly assessed and accounted for in the estimation of the future results of a retrofitting project. It is called Residual Value (RV) and it represents the cash flow generated by the retrofitting project when it stops being used by the present investor.

When assessing the financial feasibility of a retrofitting project or choosing between several alternative solutions, the analysts must estimate all the inputs listed above and correctly consider the risk embedded in the estimation. The methodology for assessing the financial feasibility of a retrofitting project then consists in aggregating these data into some indicators that are useful tools for making the investment decisions.

Cash Flows Generated by the Retrofitting Project and Other Necessary Input Data

The first step in analyzing the financial feasibility of a retrofitting project is to correctly estimate the cash flows generated by the project and the moments of their occurrence. Please note that in the case of a retrofitting project, the investment refers to improving the building characteristics – mainly in terms of energy efficiency and comfort -, not to the building itself that can be exploited for a given period even without any improvement. Therefore, in order to correctly estimate the cash flows generated by the retrofitting project, the analyst must in fact estimate some so-called incremental cash flows. It means that for each year of the remining lifetime of the construction the analyst should estimate cash flows in two alternative scenarios: if the retrofitting project is not implemented and if it is undertaken and produces results. The cash flows of interest in assessing the financial feasibility of a retrofitting project result as difference between the annual cash flows generated by the building if the retrofitting project was adopted and the corresponding annual cash flows generated by the same building if it was not subject of retrofitting.

It may be useful at this point to remember that the cash flow results from the difference between all cash inflows generated by the building and all payments resulting from operating or using it during a given timeframe, usually one year. It is also useful to mention that in some cases, the cash inflows are not straightforward as they do not correspond to a real operation increasing the amount of money at the investor's disposal, but merely to a cost saving. Think for example of a production space owned by the company which uses it. The result would not be a cash receipt, but a cost saving under the form of the rent that would have to be paid if the space was not owned by the company.

Once the cash flows were estimated under the two possible scenarios, usually on a regular basis – yearly or monthly -, the cash flows resulting from the retrofitting project are computed by the difference between the cash flow for a given period if the project is undertaken and that for the same period if the project is not implemented.

The analyst should pay attention to the type of investors for whom the analysis is made in the sense that the cash flow has to be estimated from the point of view of the cash inflows and outflows incumbent to the specific type of investors. If the investor is the owner and user of the building, then a retrofitting project may affect future utilities payments, probably the property tax if tax reductions are allowed for the specific type of retrofitting project and possibly the future value of the property if it is to be sold in the future. If the landlord undertakes the retrofitting project, future increased rents, small decreases in the utilities cost, possible tax reductions and an increased future value of the property in case of sale may be expected. If the tenant invests in a retrofitting project, the utilities costs are expected to diminish on the time horizon of the rental contract and some rent reductions can be granted by the landlord.

The basis for the estimation of the cash flows may be the technical assessment of the retrofitting project results, usually in terms of reducing energy and water consumption. However, the analyst must be aware that these technical results are only a start point, as they may be integrated to a smaller or full extent in the market prices (for example in the rent rates). Although in some cases the estimation of the monetary value of the utilities cost may be satisfactory, in many cases it isn't, and an estimation of the cash flows based on the market information is needed. For example, it may be useful to observe the difference in rent prices between retrofitted and conventional buildings. Sometimes, a reduction can be observed in the vacancy rates for retrofitted buildings and last, but not least, the project may also influence the lifetime of the building. Due to the lack of market information allowing the analyst to estimate the financial value of all types of effects generated by the project, it is a widespread solution in practice to consider all costs and the main benefits, for example those resulting from the energy or water cost savings. This choice bases on the idea that the remaining benefits are less important and, in any case, if the project is worth considering only a part of its benefits, than it will certainly be worth when considering all benefits. Although underestimating a little bit the retrofitting project, this approach can have significant advantages for the analyst in terms of time and costs needed to conduct the financial assessment. The mandatory condition in this case for the analyst is to be sure that all costs have been properly taken into consideration, otherwise the conclusion being useless.

Another important task for an analyst would be to correctly estimate the economic lifetime of the project. The economic lifetime can be defined as the period when effects of the project in monetary terms are available for the investor. Please notice that there are objective, but also subjective factors that influence the economic lifetime of the project and they depend on the nature of the retrofitting project, but also on the investor's profile. For example, a rooftop thermic insulation has a 15-30 year lifetime and this is the right economic lifetime of the project if undertaken by the owner. However, if it is implemented by a tenant based on a rent contract signed for 10 years from now on, then the analyst must consider that 10 years represent the economic lifetime of the project from the tenant's perspective.

In particular cases, the retrofitting project may even increase the lifetime of the building itself. This may result from a technical expertise conclusion or from regulatory disposition. In this case, the analyst must also consider yearly cash flow increases and a longer exploitation period, when estimating the marginal cash-flows generated by the retrofitting projects.

The feasibility of a retrofitting project cannot be correctly assessed unless an adequate discount rate is estimated. The reason why the discount rate is necessary is that the cash flows generated by the project are obtained at different moments in the future and their comparison to the initial investment cost cannot be made correctly without a discount process. The essence of the discounting procedure is the need to insure the equivalence of amounts obtained at different moments, knowing that the investors associate a bigger value to 100 USD available presently than to 100 USD available at some point in the future. Two main explanations are given for this behavior: the inflation that erodes the purchasing power of the given amount and the possibility for the investor to get a reward for holding the money by investing it. The second explanation is the reason why generally, a discount rate higher than the inflation rate is expected. In order to keep things simple, the discount rate would be the reasonable expected return for investments with similar risk profiles.

A given amount X invested with an annual return rate equal to r would result after n years in an amount $Y=X*(1+r)^n$. Y is called the future value of X. Following the same line of reasoning, the present value of the amount Y to be cashed in after n years equals $X=Y/(1+r)^n$. Hence, in order to make a cash flow obtainable after n years comparable with an initial cost of an investment considered to be undertaken presently, one should use its present value.

One possible benchmark for the discount rate can be the rate of return for investment in similar buildings on the local market. Although widely used in assessing the retrofitting project feasibility, this benchmark can be used only if the analyst considers that the project risk is similar to that of the investment in the local real estate market sector. The local market characteristics are in this case very important. In Great Britain, the expected rate of return in nominal terms in Euro for prime office buildings varies between 4.5% and 5.5%, while in Poland is averages 6.5%-7%. In Romania, it varies between 8% and 9%. Of course, the data provided are indicative and refer to the year 2019. They are subject to changes determined by the economic cycle, the real estate market stage, but also macroeconomic or regulatory decisions.

Another source can be the assessment of the discount rate as a risk-free rate plus risk premia for market and idiosyncratic risks. A basis in this approach can be Damodaran database available at http://pages.stern.nyu.edu/~adamodar/ providing yearly estimates for the country risks premia and volatility coefficients. Of course, based on the market information the analyst should add idiosyncratic risk premia characteristic to the project itself. For instance, a location premium for projects in secondary or tertiary locations, etc. may be needed. If the discount rate is thus estimated, the analysts must also estimate the possible variation in the discount rates in the future.

The discount rates must be estimated in such a way as to adequately consider the investors' expectations corresponding to the type of cash flow estimated. For example, the discount rate must correspond to the monetary unit used to estimate cash flows. Also, it should be in nominal or real terms according to how the cash flows were estimated. Expressing the cash flows (or discount rates) in real terms means not taking into consideration the annual increase of prices due to the inflation. For cash flows, it usually leads to constant or almost constant yearly values. For discount rates, Fisher formula must be applied:

$$(1+Rn)=(1+Rr)*(1+Ri),$$

where Rn represents the nominal rate of return, Rr the rate of return in real terms and Ri the inflation rate.

The annual discount rate used must consider the economic conditions of the specific years. Therefore, numerous authors recommend using different discount rates if the economic reality imposes it (Baker and English, 2011).

According to the specialty literature, the discount rates vary largely between 2% and 8% for retrofitting projects in Europe, with values around 3% for cash flows estimated in Euro in real terms (Rocchi, Kadzinski, Menconi, Grohmann, Miebs, Palotti, & Boggia, 2018).

The residual value of a project is represented by the cash flow obtained when the investor stops exploiting the project. According to the particular features of the project, the residual value may be positive, negative or null. In retrofitting projects, the residual value is usually computed marginally by the difference between the sale price of the building at some point in the future if the retrofitting project is undertaken and the sale price of the building at the same moment if it is not retrofitted. According the specialty literature, the retrofitting project is expected to increase the sale price of the building compared to one where retrofitting was not implemented. Filippi (2015) offers evidence on this effect in case of historical buildings in Italy. The estimation must consider the expected trends in the specific real estate sector, but also the changes in buyers' preferences until the forecasted sale.

A particular case is that of choosing between alternative solutions for retrofitting, especially when specific regulatory costs are imposed. The residual value in this case also considers the cost of the disposal of the materials used for retrofitting which may differ between alternatives. Hence, some alternative solutions may also present negative residual values, if taxes and other costs are imposed for their disposal, that have to be considered in the feasibility assessment.

A general issue in the feasibility analysis of an investment is the estimation of the investment cost, as there are numerous sources of uncertainty in its ex-ante estimation. In the case of retrofitting projects, it is based on offers required by the investors from construction and installation companies, which implies a small risk embedded in the estimation of the initial investment cost. The duration of the retrofitting process is also easy to estimate based on the discussion with specialists. Moreover, the moments when the initial cost is paid is also easy to forecast based on the financial arrangements with the constructors and/or installers.

In the estimation of all input data, the assessment team must in most cases rely on persons with technical expertise who can make adequate estimations of the technical efficiency and costs of the retrofitting projects. For example, it is essential to have a good approximation of the annual quantity of energy or water saved and of efficiency loss during its functioning to estimate the annual cash inflows, but also a clear estimation of the project initial cost, of the quantity of material disposed at the end of the project and its optimal disposal that will influence the residual value, of the lifetime of the project and so on. Based on these data and considering the present and forecasted prices on the local market, the analyst can estimate the input indicators necessary: the initial investment cost, the project lifetime, the cash flows, the residual value and the expected rate of return.

Indicators for Green Retrofit Project Valuation

Based on the cash flows estimated, the analysts can compute financial indicators that provide information on the feasibility of the retrofitting projects. These indicators are based on the financial principle that an investment project should generate sufficient cash flows to provide a required return on investment estimated by the annual discount rate.

The Net Present Value (NPV) is considered by both theoreticians (Berk & DeMarzo, 2017; Dragotă, Obreja-Brasoveanu & Dragotă, 2016 etc.) and practitioners as the most important indicator in capital budgeting valuation. It is born from the need to compare the initial cost of the investment to the present value of the cash flows generated by it, being computed as follows:

$$NPV = -I + \sum_{i=1}^{n} \frac{CFi}{\left(1+r\right)^{i}} + \frac{RV}{\left(1+r\right)^{n}}$$

with I being the initial investment cost, CFi the cash flow generated in the year i, r the expected return rate on the investment and RV the residual value of the project.

Please note that for any positive or null NPV the investment may be undertaken, as the cash flows cover the initial investment cost offering an annual rate of return equal (if NPV=0) or higher than r. The NPV gives an image of the excess amount returned by the project above the required annual rate of return. Therefore, higher the NPV of a project or solution, more desirable it will be for the investor.

However, if the investment cost varies significantly among alternative solutions or projects and there are also important budgetary constraints, as in most of the retrofitting projects, the investor may be interested in choosing the project that brings the higher return for each monetary unit invested. In this case, the profitability index (PI) may be useful.

$$PI = \frac{\sum_{i=1}^{n} \frac{CFi}{\left(1+r\right)^{i}} + \frac{RV}{\left(1+r\right)^{n}}}{I} = \frac{I + NPV}{I}$$

The indicator leads the analysts towards the solution that provides the highest result for each monetary unit invested, but it may also induce too much weight of the financial criterion on the overall investment decision. It may for example point out as the best solution a relatively small investment with sufficient monetary results, but not too much importance in term of overall performance in reducing energy consumption or carbon footprint. Therefore, analysts are called for prudence when using profitability index in their retrofitting project valuation.

Another useful indicator in capital budgeting valuation is the internal rate of return, representing the annual return generated by the project. It would be computed as the discount rate that results from the equation NPV = 0, considering the cash flows equal to that estimated for the analyzed project. The internal rate of return is easily obtainable by using the IRR financial function in an excel spreadsheet or on a financial calculator.

The main advantage of this indicator is that it is easily understood by the investors and the decision condition results naturally: the IRR must equal or overpass the required discount rate. However, a major problem of this indicator is that there are numerous cases when the equation NPV=0 returns several solutions, in which case IRR becomes inadequate as a tool for the investment decision. In the retrofitting projects precisely, if alternative solutions are analyzed, the existence of a disposal cost for the materials at the end of the project may often result in negative residual values and probably in multiple solutions for the internal rate of return. An IRR higher than or equal to the expected rate of return leads to the decision of implementing the retrofitting project. Of course, the project will not be considered if providing a negative IRR or one smaller than the rate of return required by the investor. An analyst will choose the project with the highest internal rate of return. However, basing the investment decision upon this indicator has some limits. For instance, the comparison is adequate if the different alternative projects proposed have similar risk, otherwise the simple comparison of the IRR may mislead the investor as higher IRRs are expected for riskier investments. Also, the internal rate of return has the same drawback as the profitability index, as higher internal rate of returns may be displayed by small investments with not so much environmental impact.

The payback period is probably the most popular indicator analyzed in retrofitting decisions. In simple terms, it gives a measure of the time needed to completely regain the initial investment throughout the cash flows generated by the retrofitting project. Two alternatives of the indicator are used in practice: the static payback period and the dynamic payback period. The first one does not consider an opportunity cost of money (a required rate of return on the investment). It basically estimates the period needed for the sum of the cash flows generated by the investment to equal its initial cost. Although with clear limits resulting from the ignorance of the time value of money principle, the static payback period is a popular indicator as it is simple to compute and to understand by the investors. The dynamic payback period overcomes this drawback by using the discounted cash flows at the moment of the initial investment. It will result in higher values than the static payback period and has the advantage of allowing the comparison between projects with different risk, if the risk is dully incorporated in the estimation of the discount rate.

One possible issue in the application of the payback period for retrofitting projects is the fact that the residual value of the investment project may be negative. In this case, a re-estimation of the initial cost must be made in order to incorporate the present value of the residual value in the initial investment cost before computing the payback period. By this means, the indicator will be suitable for comparison between alternative solutions, some of them having a negative residual value induced by high disposal costs.

Although not commonly used in capital budgeting valuation, the life cycle cost is one of the indicators largely used in analyzing retrofitting projects. The indicator is especially valuable for those types of projects that have as main result a reduction in the operational costs. The indicator is based on computing the annual operating cost during the entire lifetime of the project, and it is expressed as the present value of all costs supported by the investor, including the initial investment cost and the material disposal cost after the end of the project. The investment project will be undertaken if the life cycle cost of the project implementation scenario is lower than the life cycle cost of the not adopting the project scenario. Among several alternative solutions, the preferred one is that with the lowest life cycle cost. Please note that the life cycle cost of a retrofitting project does not provide an indication about the rate of return generated by it. It does not even consider the cash receipts when generated by the project and therefore its use is limited to the retrofitting projects generating only cost savings.

$$LCC = -I + \sum_{i=1}^{n} \frac{\mathrm{Cos}\, t_i}{\left(1+r\right)^i} + \frac{Disposal \; \cos t}{\left(1+r\right)^n}$$

One mention needs to be made regarding the content of the life cycle cost. It is a financial construct, useful to understand the economic impact in terms of cost reduction for retrofitting projects, but it is not to be confounded with the life cycle assessment. The last concept refers to a more complex analysis, focusing on financial, but also on technical and environmental issues and conducted on a larger timeframe basis. For instance, the life cycle assessment considers even the environmental impact of the fabrication process for the material to be used in the retrofitting project, which clearly exceeds the scope of the life cycle cost analysis in terms of period, type of effects considered – social, environmental, financial, technical etc. – and content.

Techniques for Dealing With the Uncertainty in Retrofitting Projects

All the indicators presented above are indicators used in capital budgeting valuation that are suitable for the retrofitting projects analysis in financial terms. The reader must be aware that all of them are highly sensitive to the estimation of cash flows, the discount rate, the residual value and the initial investment cost, as well as to unpredicted changes in the economy or in the users' behavior. Therefore, the analyst is interested in understanding how the indicators above would vary if forecasting errors or unexpected market conditions changes occurred.

The most popular technique used to cope with uncertainty in retrofitting projects valuation is the sensitivity analysis. It is based on the idea of varying only one input within the analysis – let's say, energy price – and re-estimate cash flows and the capital budgeting valuation indicators. Hence, one can easily obtain the sensitivity of the capital budgeting valuation indicator – NPV, for example – considering a one percentage point change in the input factor. The technique is mainly used to understand the impact of a forecasting error on the outcomes of an investment project. If the change in the input factor leads to a huge change of the result, then one can appreciate the risk of the project as significant. The analyst must do a thorough market research to understand how likely a change of the input factor is and how important it can be, based on the present information. According to the conclusion, the risk of the project is appreciated as important or not and the decision to implement the project or not will be taken accordingly. In most capital budgeting analyses, the sensitivity analysis exhibits a major drawback, in the sense that it assumes a change in one input factor, the other remaining constant, which is not a realistic assumption. However, it is widely used in retrofitting projects, because the outcomes in this case are strongly linked to the utilities' prices and the results of the sensitivity analysis have a high informational value.

Less used in retrofitting projects, but important in the sense that it adapts better to the market evolution, the scenario analysis allows the analyst to construe several scenarios consisting in varying several input factors in the same time and to realistically assess a probability of occurrence for every scenario, considering that the indicator has a normal probability distribution. The capital budgeting indicators are then computed for every scenario. Based on the probability of occurrence estimated by the analyst, the average or median capital budgeting indicators, and more important a variation interval of each indicator for a given probability may be assessed. By this means, the risk of the investment project can be studied and dully considered in the capital budgeting decision. Please note that the scenario analysis is a very useful tool whenever there is a good chance for several influence factors of the results of a retrofitting project to change simultaneously (for example, energy price and the cost of the material or energy prices and the disposal cost of the material used and/or the property tax regime).

The Monte Carlo analysis is nothing but a generalization of the scenario technique. It is based on generating a big number of scenarios (10000, for example) in a randomized manner, but keeping the input factors in reasonable limits. All scenarios will be given an equal probability of occurrence and for each of them the capital budgeting valuation indicators are being calculated in order to obtain their average and standard deviation values as basis for the risk assessment.

The decision tree technique also copes with uncertainty, in the sense that it associates different probabilities to different scenarios also following the project behavior in time and introducing periodical decision moments. For each possible combination of subsequent decisions, the capital budgeting indicators and their corresponding probability resulting of the multiplication of the probabilities estimated at each intermediary period are computed. Based on the above information, the average or median capital budgeting indicators and a specific interval of variation of each of them with a given probability may be computed, thus allowing the assessment of the investment decision in uncertainty conditions.

In guise of example, we provide below the graph of the decision tree used to analyze the viability of a green retrofitting project, considered to have two yearly stages each with two possible alternative outcomes, understood as the favorable and respectively the unfavorable scenarios. Based on the alternative chosen, the NPV (in this case) or any other capital budgeting valuation indicator can be computed. For the simple case chosen, four alternative outcomes are expected, each of them obtained with a given probability which results by multiplying the probabilities estimated rigorously for each period and scenario.

Figure 1. The decision-tree for the implementation of a green retrofitting project

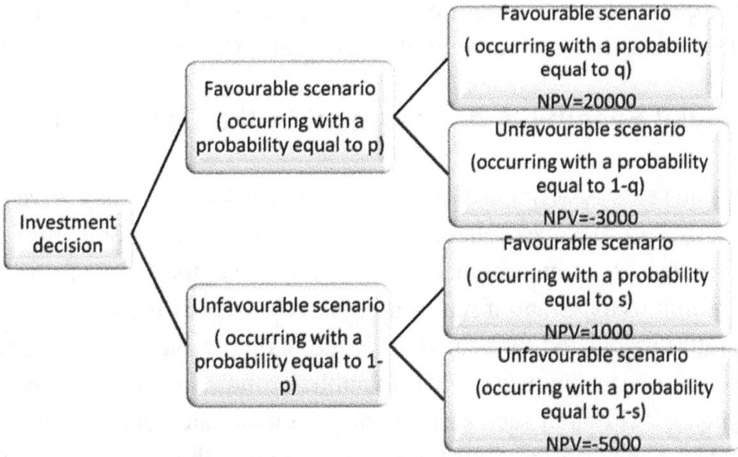

The practical application of this method for the given example consists in estimating the probabilities associated to each scenario in the first period (p for the favorable scenario and 1-p for the unfavorable one) and in the second period (let's assume the probabilities being q and (1-q) for the favorable alternative and s and (1-s) for the unfavorable one). In this case, our investment would have one of four possible estates at the end of the second period, in terms of Net Present Value: 20k€ with the probability p*q, -3k€ with the probability p*(1-q), 1k€ with the probability (1-p)*s and -5k€ with the probability (1-p)*(1-s).

The average net present value to be associated to the retrofitting project is:

E(NPV)=(20k€)*p*q+(-3k)€*p*(1-q)+1k€*(1-p)*s+(-5k€)*(1-p)*(1-s)

The standard deviation of the net present value would then be the square root of the variance indicator, where:

$$Var\left(NPV\right) = p * q * \left(20k€ - E\left(NPV\right)\right)^2 + p * \left(1-q\right) * \left(-3k€ - E\left(NPV\right)\right)^2$$
$$+\left(1-p\right) * s * \left(1k€ - E\left(NPV\right)^2 * + \left(1-p\right) * \left(1-s\right) * \left(-5k€ - E\left(NPV\right)\right)^2$$

$$Stdev\left(NPV\right) = \sqrt{Var\left(NPV\right)}$$

A very interesting perspective from the point of view of the capital budgeting valuation is given by the valuation of a certain option referring to the investment project such as the delay option, the abandonment option and so on. Although very useful in capital budgeting valuation in general, this method is not very often used in retrofitting projects, as they are relatively small projects, with insignificant change in conditions from one year to another. The real option valuation method is based on the valuation techniques for financial options and can give a clue for understanding if it would be better for the investor to delay by a certain period the implementation of the project or to abandon it. It may be very useful in case of cash flows strongly dependent on the regulatory framework, if major changes are envisaged.

Because of the complexity of real option valuation, but also considering the rare use of real options in green retrofitting projects, this chapter does not describe in detail how the method can be implemented in practice. The reader might find useful the book of Berk and DeMarzo (2017) in this regard.

Integrating Financial Feasibility Assessment in the Decision-Making Process

The green retrofitting for energy efficiency has clearly a broader set of objectives than the financial ones, often including social objectives related to improvements of the population' health, prevention of diseases, comfort increases, or environmental objectives. The appraisal of this kind of benefits in monetary terms is difficult to make and if realized, it is in most cases incomplete, as it cannot estimate precisely the externalities associated to them. This is the main reason why an important stream of the literature recommends or uses techniques specific to public economics such as the multi-criteria analysis or the hedonic techniques. Although such benefits may be associated also to conventional retrofitting projects, the non-monetary benefits are clearly more important in the case of green retrofitting projects and the use of multi-criteria analysis is very useful in valuing this kind of investment.

Considering the particularities of the green retrofitting projects in terms of monetary and non-monetary costs and benefits, the analyst focuses on quantifying the results of the project in monetary terms whenever possible – estimating for example the annual utilities cost savings and the increase of the value of the property after a number of years when it is likely to be sold-, but also the non-monetary benefits such as a 3% smaller probability for the inhabitants to develop respiratory diseases, a one year period necessary for the materials to decompose after the end of the project instead of 20 years for conventional material and so on. In guise of example, Rocchi et al., (2018) use in their multi-criteria analysis the following seven criteria: energy saving, non-renewable energy, comfort performance, global warming, ozone layer depletion, respiratory inorganics, net present value.

After that, importance coefficients are assigned to each type of benefit usually resulting from interviews with professionals in the field or with a significant number of end users.

By this means, the result would be ranking the alternative solutions according the preference of the stakeholders, including the entire range of benefits into the analysis. The method is very useful for the analysts when there is not sufficient information available on the market in order to correctly assess the monetary value of all benefits resulting from a green retrofitting project. The difficulty of correctly appraise the non-monetary benefits leads to the use of multi-criteria analysis very often in practice. However, there is a significant drawback of the method which is represented by the subjective nature of the importance coefficients assigned to different types of benefits that may significantly bias the results.

CASE STUDY

In order to give an example of how the financial indicators for capital budgeting valuation can be used in green retrofitting project, this subchapter gives an example of how to compute the financial indicators in a hypothetical case of two alternative solutions of roof insulation, using polystyrene or wool-based insulation. Based on the information on the market the cost of insulating an individual residential building with a footprint area equal to 80 sq m is estimated at 1100 Euro for polystyrene insulation and 1500 Euro for wool-based insulation. The initial investment cost is estimated based on offers obtained from constructors. The annual energy saving costs are estimated at 120 Euro in both cases, computed as the quantity of energy saved, obtained from specialists, multiplied by unitary energy prices considered constant in real terms. The energy costs for an insulated residential house will equal 130 Euro/year decreasing from 250 euro and they are assumed constant over the entire period in constant prices. The disposal cost for the polystyrene at the end of 15 years is estimated at 250 Euro based on the regulatory framework, computed as the quantity disposed multiplied by the unitary disposal cost available at waste management companies, while no disposal cost is necessary in the case of wool-based insulation material, as it copes with the environment protection regulation.

Please have in mind that a 5.5% rate of return is expected from investments in residential buildings, the long-term inflation rate being provided by Eurostat database and equaling on average 2%. This leads to analyst to estimate a real expected rate of return following Fisher's formula as follows:

$$Rr = \frac{\left(1 + 5.5\%\right)}{\left(1 + 2\%\right)} - 1 = 3.4\%$$

The analysts were also able to compute that a 5 Euro discount is obtained yearly on the property tax for the houses using environment-friendly materials, according to the national regulation in force and the advantage is estimated to maintain in the future. It represents 5% of the usual property tax. Also, the market information suggests that tenants are willing to pay on average a supplementary monthly rent of 2.5 Euro for the houses insulated with environmental-friendly materials than for the buildings with conventional insulation systems.

The yearly cash flow, computed as a marginal cost saving over the status quo situation of a non-insulated roof residential home, for the first alternative solution is in this case 120 Euro (the annual energy savings in monetary units), with a negative residual value at the end of the 15th year equal to 250 Euro. The estimated yearly cash flow for the second alternative is also estimated as a marginal cash flow over the status quo situation of a non-insulated roof home. It is computed as the sum of energy and tax savings plus supplementary cash inflows obtained equal to the increase of monthly rent (2.5 Euro) multiplied by 12. It is estimated at 120 Euro plus the 5 euro obtainable from tax reduction plus the yearly incremental rent equal to 30 Euro, resulting in a total of 155 euro. The residual value in this case is considered as equal to zero, as no disposal costs are expected in this case.

The net present value of the first alternative, noted NPV1 is computed using the formula below:

$$NPV = -I + \sum_{i=1}^{n} \frac{CFi}{\left(1 + r\right)^{i}} + \frac{RV}{\left(1 + r\right)^{n}}$$

where:

- I represents the initial cost of the investment, meaning the roof-insulating cost equal to 1100 Euro in the case of polystyrene insulation and 1500 Euro in the case of wool-based insulation
- CF represent the annual cash-flows estimated at 120 Euro per year in the first case and 155 Euro for the second insulation solution

Figure 2. The computation of the internal rate of return in an Excel spreadsheet

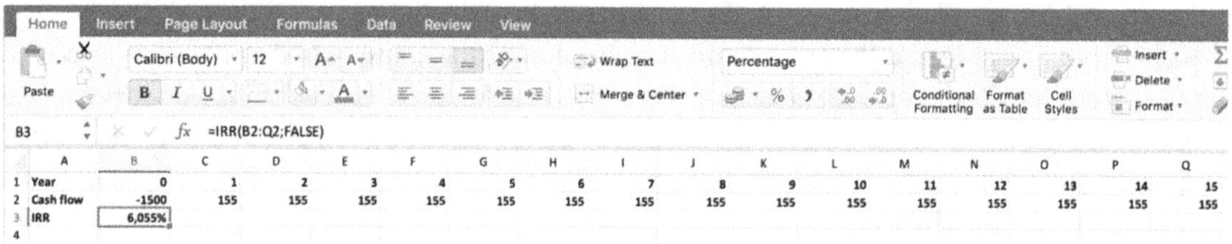

	A	B	C	D	E	F	G	H	I	J	K	L	M	N	O	P	Q
1	Year	0	1	2	3	4	5	6	7	8	9	10	11	12	13	14	15
2	Cash flow	-1500	155	155	155	155	155	155	155	155	155	155	155	155	155	155	155
3	IRR	6,055%															
4																	

- RV represents the residual value estimated as equal to -250 Euro for the polystyrene insulation that needs to cover a disposal cost at the end of the lifetime of the project and 0 for the wool-based insulation
- r is the expected discount rate equal to 3.4% as computed above. The choice for the discount rate in real terms instead of nominal terms is motivated by the fact that cash flows are in real terms.

For the first roof-insulation alternative, the net present value is estimated as follows:

$$NPV1 = -1100 + \sum_{i=1}^{15} \frac{120}{\left(1+3.4\%\right)^{i}} - \frac{250}{\left(1+3.4\%\right)^{15}} = 140.57\,Euro$$

The NPV of the second alternative solution (NPV2) is:

$$NPV2 = -1500 + \sum_{i=1}^{15} \frac{155}{\left(1+3.4\%\right)^{i}} = 297.96\,Euro$$

The net present value favors the second alternative, which brings to the investor a higher present cash flow above the required return than the first one, the green retrofitting alternative being recommended.

However, for a financial constrained project, the investor would be also interested in understanding which is the result provided by the projects for each Euro invested. Therefore, the profitability index is computed for each alternative (PI1 and PI2).

$$PI1 = \frac{1100 + 140.57}{1100} = 1.128$$

$$PI2 = \frac{1500 + 297.96}{1500} = 1.199$$

The analysis of the profitability index also favors the second alternative, the preferable solution being the one that maximizes the profitability index. The second solution provides an overall present result for each Euro invested equal to 1.199 Euro.

Table 1. Static payback period for the retrofitting project consisting of roof insulation with wool-based insulation

Year	Cash flow	Cumulated cash flow
1	155	155
2	155	310
3	155	465
4	155	620
5	155	775
6	155	930
7	155	1085
8	155	1240
9	155	1395
10	155	1550

Source: Own computation

The internal rate of return confirms the result, with a value equal to 6.055% for the second solution, higher than the expected rate of return equal to 3.4%.

However, a relevant internal rate of return cannot be computed for the first solution due to the fact that the total cash flow is negative in the last year leading to multiple IRR solutions.

An interesting perspective will also be to estimate the payback period. In this case, for the ease of understanding, the payback period for the second project will be computed firstly.

Considering that different annual cash flows can be generated for other projects, the payback period will be computed for the general case. As the reader probably remembers, the analyst can compute a static payback period showing the period needed to cover the investment cost throughout its cash flows not considering the time value of money and the dynamic payback period, taking into consideration an expected rate of return.

In order to compute the payback period, one would be interested to sum up every year all passed cash flows until the amount obtained overpasses or equals the initial investment. In the table below the annual cash flows and the cumulated annual cash flows are presented and the payback period is obtained linearly.

The static payback period (SPB2) is computed below. We look at the year 10 because this is the year when the cumulated cash flows become higher that the initial investment cost (1550>1500).

Based on an interpolation procedure, the static payback period is computed as:

$$SPB2 = 9 \ years \ and \frac{1500 - 1395}{155} * 360 \ days = 9 \ years \ and \ 244 \ days$$

The Dynamic payback period (DPB2) is:

$$DPB2 = 11 \ years + \frac{\left(1500 - 1402.9\right)}{103.77} * 360 \ days = 10 \ years \ and \ 337 \ days$$

Table 2. Dynamic payback period for the retrofitting project consisting of roof insulation with wool-based insulation

Year	Cash flow	Discounted cashflows	Cumulated cash flow
1	155	155/(1+3.4%)=149.9	149.90
2	155	$155/(1+3.4\%)^2$=144.97	294.88
3	155	140.21	435.08
4	155	135.60	570.68
5	155	131.14	701.82
6	155	126.83	828.65
7	155	122.66	951.30
8	155	118.62	1069.92
9	155	114.72	1184.65
10	155	110.95	1295.60
11	155	107.30	1402.90
12	155	103.77	1506.67

Source: Own computation

For the first alternative solution, the existence of a negative residual value requires the analyst to re-estimate the investment cost by adding the present value of the disposal cost. The initial investment cost will be considered as below:

$$I = 1100 + \frac{250}{\left(1 + 3.4\%\right)^{15}} = 1251.4$$

Table 3. Static payback period for the retrofitting project consisting of roof insulation with polystyrene

Year	Cash flow	Cumulated cash flow
1	120	120
2	120	240
3	120	360
4	120	480
5	120	600
6	120	720
7	120	840
8	120	960
9	120	1080
10	120	1200
11	120	1320

Source: Own computation

Table 4. Dynamic payback period for the retrofitting project consisting of roof insulation with polystyrene insulation

Year	Cash flow	Discounted cashflows	Cumulated cash flow
1	120.00	116.05	116.05
2	120.00	112.24	228.29
3	120.00	108.55	336.84
4	120.00	104.98	441.82
5	120.00	101.53	543.34
6	120.00	98.19	641.53
7	120.00	94.96	736.49
8	120.00	91.84	828.33
9	120.00	88.82	917.15
10	120.00	85.90	1003.04
11	120.00	83.07	1086.11
12	120.00	80.34	1166.45
13	120.00	77.70	1244.15
14	120.00	75.14	1319.30

Source: Own computation

The static payback period is computed below.
The static payback period (SPB1) equals:

$$SPB1 = 10\ years + \frac{1251.4 - 1200}{120} * 360\ days = 10\ years\ and\ 155\ days$$

The Dynamic payback period (DPB1) is:

$$DPB1 = 13\ years + \frac{\left(1251.4 - 1244.15\right)}{75.14} * 360\ days = 13\ years\ and\ 35\ days.$$

The reader observes that all capital budgeting valuation indicators lead to similar results, favoring the second solution as being the best and recommend its adoption. However, things are not so clear in every case, in many instances the four indicators leading to controversies. The analyst will make the decision considering the limits of each indicator and the particular conditions of each project as explained in the previous subchapter.

In guise of example, the life cycle cost of the project is provided below.

$$LCC = -I + \sum_{i=1}^{n} \frac{\mathrm{Cos}\,t_i}{\left(1+r\right)^i} + \frac{Disposal\ \cos t}{\left(1+r\right)^n}$$

where:

- I represents the initial investment cost
- Cost i represent the operating cost in the year i computed in our case as the cost of utilities (130 Euro) plus the property tax (100 Euro for the first alternative and 95 Euro for the second).
- Disposal cost represents 250 Euro necessary at the end of the project for material disposal.

For the first alternative (LCC1):

$$LCC1 = 1100 + \sum_{i=1}^{15} \frac{130+100}{\left(1+3.4\%\right)^i} + \frac{250}{\left(1+3.4\%\right)^{15}} = 3919.35\,Euro$$

The life cycle cost in the alternative of not implementing the project (LCC0) would be:

$$LCC0 = \sum_{i=1}^{n} \frac{250+100}{\left(1+3.4\%\right)^i} = 4059.91\,Euro$$

The rational investor would choose to implement the polystyrene insulation instead of not doing so, as this alternative reduces the overall cost of the project. However, the life cycle cost does not represent a capital budgeting indicator, as it focuses on costs, not considering incomes in the analysis. Thus, the life cycle cost for the second alternative project (LCC2) computed below suggests that the alternative is inferior to the first one, because this indicator does not consider rent increases due to tenants' preference for environment-friendly materials.

$$LCC2 = 1500 + \sum_{i=1}^{n} \frac{130+95}{\left(1+3.4\%\right)^i} = 4109.9\,Euro$$

In order to have a clue about how the forecast risk influence the results, the analyst conducts a sensitivity analysis considering that the energy price may remain reasonably by 5% below the forecasted price. In both cases, only energy cost savings diminish by 5%. In this case, the net present value for the two alternative solutions (NPV1s and NPV2s) will be as follows:

$$NPV1s = -1100 + \sum_{1=1}^{15} \frac{120 * 0.95}{\left(1+3.4\%\right)^{i}} - \frac{250}{\left(1+3.4\%\right)^{15}} = 70.97\,Euro$$

$$NPV2s = -1500 + \sum_{i=1}^{15} \frac{120 * 0,95 + 5 + 30}{\left(1+3.4\%\right)^{i}} = 228.36\,Euro$$

The analyst may also be interested in computing the sensitivity of the net present value to a change in the energy price. In the case of the first alternative solution, the elasticity (e1) equals:

$$e1 = \frac{\dfrac{70.97 - 140.57}{140.57}}{-5\%} = 9.90$$

For the second alternative the elasticity (e2) is computed below:

$$e2 = \frac{\dfrac{228.36 - 297.96}{297.96}}{-5\%} = 4.67$$

The result shows that not only the second alternative is better for the investor, but it also has the lowest risk in case of energy price changes or forecasting errors regarding it. Therefore, the investor will be advised to undertake the solution of roof insulation with a wool-based material.

FUTURE RESEARCH DIRECTIONS

Important steps have been made lately in green retrofitting capital budgeting valuation, but there are numerous issues still insufficiently developed. One important point will be to find better ways to take into consideration together monetary and non-monetary costs and benefits.

Also, an optimization of the real economic and regulatory system in order to better internalize the costs and benefits associated to each stakeholder would certainly have a positive effect on the implementation of green retrofitting projects.

Moreover, multidisciplinary work is still needed in order to optimize the cash flows estimation for green retrofitting projects.

CONCLUSION

This chapter focusses on introducing the capital budgeting valuation tools available in the analysis of green retrofitting projects for energy efficiency performance. The specialty literature and practice does not provide specific indicators used for assessing the feasibility of retrofitting projects. Therefore, the present chapter focusses on explaining the particularities of the methodology of assessing the financial feasibility of an investment project in the case of a retrofitting one.

It explains what are the input data necessary for assessing the financial feasibility of the green retrofitting projects and how the analyst can obtain them. The chapter then explains the methodology used to assess the financial feasibility of green retrofitting projects for energy efficiency, explaining the most important indicators used in practice and pointing out their economic meaning, their advantages, and their limits.

The chapter also explains the methods that can be used in order to assess the risk associated to a green retrofitting project and their functioning, as well as their advantages and limits when used in the financial assessment of this kind of projects.

The most important particularity in assessing the financial feasibility of the retrofitting projects is to correctly estimate the annual cash flows as incremental cash flows obtained if the projects are undertaken compared to the status quo situation of no retrofitting. The chapter explains in detail that sometimes also a prolongation of the useful lifetime of the building is to be considered in the analysis. Also, a retrofitting project is likely to have a positive influence on the residual value of the building, contributing to the financial return on investment.

The particularities of green retrofitting projects often require the analyst to account for both monetary and non-monetary costs and benefits. The chapter also includes some considerations about how non-monetary effects can be integrated into the analysis.

In order to make things clearer to the reader, the last sub-chapter presents a numerical example of estimation of capital budgeting valuation indicators, along with the necessary comments about their economic meaning.

REFERENCES

Baker, H. K., & English, P. (Eds.). (2011). *Capital Budgeting Valuation. Financial Analyses for Today's Investment Projects*. New Jersey: Wiley. doi:10.1002/9781118258422

Berk, J., & DeMarzo, P. (2017). *Corporate Finance*. Harlow, UK: Pearson Education.

Bouhal, T., Fertahi, S.-D., Agrouaz, Y., El Rhafiki, T., Kousksou, T., Zeraouli, Y., & Jamil, A. (2018). Technical Assessment, Economic Viability and Investment Risk Analysis of Solar Heating/Cooling Systems in Residential Buildings in Morocco. *Solar Energy*, *170*, 1043–1062. doi:10.1016/j.solener.2018.06.032

Christersson, M., Vimpari, J., & Junnila, S. (2015). Assessment of financial potential of real estate energy efficiency investments – A discounted cash flow approach. *Sustainable Cities and Society*, *18*, 66–73. doi:10.1016/j.scs.2015.06.002

Dragotă, V., Obreja-Brasoveanu, L., & Dragotă, I. M. (2016). *Management financiar. Management financiar strategic*. Bucharest: Editura Economică.

Filippi, N. (2015). Remarks on the Green Retrofitting of Historic Buildings in Italy. *Energy and Building*, *95*, 15–22. doi:10.1016/j.enbuild.2014.11.001

Lizana, J., Barrios-Padura, A., Molina-Huelva, M., & Chacartegui, R. (2016). Multi-criteria Assessment for the effective decision management in residential energy retrofitting. *Energy and Building*, *129*, 284–307. doi:10.1016/j.enbuild.2016.07.043

Rocchi, L., Kadzinski, M., Menconi, M. E., Grohmann, D., Miebs, G., Paolotti, L., & Boggia, A. (2018). Sustainability Evaluation of Retrofitting Solutions for Rural Buildings through Life Cycle Approach and Multi-Criteria Analysis. *Energy and Building*, *173*, 281–290. doi:10.1016/j.enbuild.2018.05.032

Ross, S.A., Westerfield, R. W., & Jordan (2002). *Fundamentals of Corporate Finance.* USA: The McGraw-Hill Companies.

Trotta, G. (2005). The determinants of energy efficient retrofit investments in the English residential sector. *Energy Policy*, *120*, 175–182. doi:10.1016/j.enpol.2018.05.024

ADDITIONAL READING

Brealey, R. A., Myers, S. C., & Marcus, A. J. (2001). *Fundamental of corporate finance.* USA: The McGraw-Hill Companies.

Certiner, I., & Edis, E. (2014). An Environmental and Economic Sustainability Assessment Method for Retrofitting of Residential Buildings. *Energy and Building*, *74*, 132–140. doi:10.1016/j.enbuild.2014.01.020

Hagentoft, C.-E. (2017). Reliability of Energy Efficient Building Retrofitting – Probability Assessment of Performance and Cost (Annex 55, RAP-RETRO). *Energy and Building*, *155*, 166–171. doi:10.1016/j.enbuild.2017.09.007

Wang, B., Xia, X., & Zhang, J. (2014). A Multi-Objective Optimization Model for the Life-Cycle Cost Analysis and Retrofitting Planning of Buildings. *Energy and Building*, *77*, 227–235. doi:10.1016/j.enbuild.2014.03.025

KEY TERMS AND DEFINITIONS

Cash Flow: The difference between cash inflows and payments generated by an investment project within a given period, usually a year

Financial Feasibility: The capacity of a project to generate cash incomes superior to the payments needed

Financial Sustainability: The capacity of an investment project to generate the expected return for the investor

Life Cycle Cost: The sum of present annual costs generated by the project during its entire lifetime, including the implementation phase, functioning and the disposal of the project components after the end of its economic lifetime.

Net Present Value: The present value of all cash flows generated by an investment project during its lifetime.

Chapter 3
Financing the Green Building Retrofitting Investments:
A Case Study for a Romanian Seaside Hotel

Eugen Mitrica
The Bucharest University of Economic Studies, Romania

ABSTRACT

The main financial barrier of large scale implementation of green building retrofitting investments is due to the relatively large investment volume needed, compared to the future flow of yearly energy savings or/and yearly estimated differences of incomes collected, if the building is a commercial building (commercial center, office building, hotel or even residential rental building). The uncertainty implicitly involved in this estimation, both for the future savings and for the yearly differences of incomes, which are usually not very large, both make these investments apparently not so attractive for private investors, especially for owners of residential buildings, with limited self-financing power. Nevertheless, from the society point of view, the benefits created by saving the energy and consequently reducing the carbon foot print, can be very attractive. That is why the public support is often used as an "impulse solution" for implementation of these investments. The Cost Benefit Analysis methodology, particularized for these investments, is presented in this chapter.

INTRODUCTION

The green building retrofitting investments are viewed as an optimal solution for improving the net operating cash flows for the buildings' owners, either only by the energy savings incurred for the residential buildings owners, or by the energy savings and by larger revenues for the owners of commercial buildings (i.e. commercial centers, office buildings, hotels) and of the residential rental buildings, but also for providing significant benefits to the society by reducing the pollutant emissions, due to energy savings. From the second point of view, this type of investments is considered very effective in the prospect of the 2020 and 2050 assumed targets of emissions reduction (Torregrossa, 2015), considering that about one third of the greenhouse gases emissions are produced by buildings (He et al., 2019).

DOI: 10.4018/978-1-5225-9104-7.ch003

Typically, the cost of these investments is significantly larger than the expected yearly net cash flow change, even if the additional net cash flow is due both to the energy savings and to the larger revenues or only to the energy savings. In the last case it is possible that prudent financial projections for such investments to show a zero or even negative financial net present value. Besides, especially these energy savings are affected by uncertainty or at least are perceived as uncertain by the buildings' owners, due to the variation of the market prices of the electricity and gas but also to the variation of the future need of energy, at least partly due to the effects of the climate change.

The other benefits of the green building retrofitting, like improvement of the residents' life quality and the corresponding reduction of the public health costs due to the improper quality of housing and infrastructure, as the rise in the value of the retrofitted buildings and even the expected new jobs created and the economic growth incentive effects, are also perceived as uncertain by the buildings' owners and not enough interesting to counterbalance the cash out-flow for the cost of investment. Furthermore, sometimes the total effort of investment is even larger than the investment cost, if the opportunity cost of fully or partly operation interruption of the green retrofitted building during the investment is considered.

From the point of view of financial resources, the residential buildings owners have most usually limited available resources and the private lenders have low experience and interest in this kind of investments, perceived as investments with high initial cost and long payback period.

The above presented barriers are the main reasons that reduce these investments attractiveness for the building owners, as for their relatively slow deployment pace. Therefore, considering the significant social benefits generated by reducing the energy consumption and, correspondingly, by significant reduction of the buildings' carbon footprint, the public support is required for promoting these building retrofitting investments. For this purpose, countries' governments set various public funding mechanisms, from public supported grants for building retrofitting investment, to tax incentives and public warranted loan schemes with corresponding lower interest rate, or even loan schemes for which the interest rate is directly subsidized. For the example, in Germany direct subsidies were initially granted by the German Alliance for Work and Environment, a broad coalition between the trade unions, federal government, industry and non-governmental organizations with the purpose to provide public financed incentives for the green building retrofitting investments, as for the new green building projects and for the domestic market of the necessary materials and services. Later, the subsidies granted by preferential loans and grants were managed by the German development bank (KfW). Whereas in the U.S., at the federal government level, the tax incentives were preferred, being implemented in the form of different income tax incentives for residential and commercial buildings. Nevertheless, after 2012 at the federal level, the tax incentive for residential buildings was eliminated (Neuhoff et al., 2012).

Summarizing, the issues involved by the green building retrofitting investments are: the scarcity of the relatively large initial capital required by the investment, long payback periods for the investment (especially in case of residential buildings) and partly or fully interruption of building operation during the investment.

In dealing with these issues, involved by the green building retrofitting investments, and considering the need of prioritization and assessment of the different buildings' investments for financing purposes, specific approaches were developed in the literature. Following are presented a few of these:

1. Cost-optimal approach is a limited cost-benefit analysis, which ignores the positive environmental effects involved by the green building retrofitting, focusing exclusively on the monetary impacts of the energy savings on the retrofitted building owners. This method was used for example in

assessment of the investment for the renovation of school buildings in the Alps (Stocker et al., 2015), considering only the cost savings related to the heating energy and ignoring any other eco-indicators like reduction of CO_2 emissions. The sensitivity analysis run in this case showed as critical variables the energy price, the price variation in the individual renovation measurements and the interest rate. The Cost-benefit analysis was used for valuation of different investment strategies for envelope retrofitting, including the roof isolation, of a residential building in Israel, under warm Mediterranean temperatures (Friedman et al., 2014). Both direct economic benefits for the owners and the external benefits provided to the society by the electricity savings were quantified and what resulted was an economic unattractive retrofitting investment: direct benefits were only one third of the investment cost and the external benefits were insufficient to compensate the difference. A more complete Cost-benefit analysis methodology was developed for valuation of a building retrofitting investment in China, quantifying the costs and benefits over building life cycle (Liu et al., 2018). This study revealed an overall economic unattractive retrofitting investment, with different cost effectiveness between retrofitting measures taken into account. The sensitivity analysis revealed energy price, investment cost and energy conservation rate, as critical factors.

2. Multi-objective optimization approach uses different solving methods for mathematical programming problems to optimize objectives and constraints which can be opposite, like the most frequent considered objectives, related to the cost of investment, the energy savings and the net present value. For example, a mixed-integer linear programming approach was used to select the right energy efficiency measures among all the available ones to optimize financial or environmental benefits subject to budget constraints for a university campus (Tan et al., 2016), and the multi-objective neighborhood field optimization algorithm was involved to obtain the optimal retrofit strategies in a 1980's office building (Wu et al., 2015).
3. Life cycle cost analysis (LCCA) is other approach used for investment valuation in the field of building retrofitting. This method is focused on the changes in all the facility ownership costs incurred in connection with the green building retrofitting investments. Typically, these costs cover the initial investment, energy costs, operation costs, maintenance, repairs and replacement costs and disposals costs (net of residual value). Other similar approaches (Life cycle analysis - LCA and Whole-life costing) exclusively cover the possible externalities generated for the society (typically environmental externalities). The LCCA approach was used for deriving the economic net present value included between the objective functions in a multi-objective optimization model (Wang et al., 2014); and LCCA and LCA were combined for optimization of the retrofitting process of a traditional rural building, based on a multi-criteria approach, which included between the seven criteria used: energy saving, comfort performance, global warming, and Net Present Value (Rocchi et al., 2018).

Looking from the point of view of financial decision, in analyzing the desirability of a green building, the classical assessment and prioritization tool for direct investments, based on the financial analysis and focused exclusively on the owners' benefits, is not always enough suitable for these investments, especially for those regarding residential buildings. The suitable approach consists of a type of multi-criteria analysis, accounting not only for the owners benefits but also for all the other benefits or costs affecting any of the investment project stakeholders, as for the positive and negative externalities affecting the third parties. This is the Cost-Benefit Analysis, which is so designed to allow for the monetization, quantification and aggregation of all the cost and benefits generated by the investment for the society.

From one point of view the Cost-Benefit Analysis can be practically considered a simplification of the more extended Life-Cycle Cost Analysis (LCCA) approach, actually the Cost-Benefit Analysis being a limited scope LCCA approach (Pombo et al., 2016). The use of the Cost-Benefit Analysis is suitable due to the practically limited scope of the approach in case of green building retrofitting and because of the limited available data and information. On the other side, LCCA is not focused on the potential revenues generated by these investments, in the case of the commercial buildings, and this makes the Cost-Benefit Analysis more suitable for a larger prospect financial approach.

THE METHODOLOGY INVOLVED IN FINANCING OF THE GREEN BUILDING RETROFITTING INVESTMENTS

In line with the above presented ideas the best method to assess and prioritize the financing of a green building retrofitting investment compared to other alternative investment projects, of the same or of different kind, seems to be the Cost-Benefit Analysis.

Generally, there are 9 basic steps involved in the Cost Benefit Analysis (Boardman et al., 2018), assuming as implicit the initial step related to identification of the project base case scenario, which is not included in the following list:

1. The initial step is to identify and consider the alternative investment projects. Which for this specific case of a green building retrofitting can consist in any of the available variants for what can be considered the complete investment, respectively the base case, starting with "no investment" ("do nothing"), or so called BAU (business as usual) scenario, to any less or more extended investment scenarios, if such scenarios exist. Nevertheless, most usually, as in the production investments, the approach taken is based on the incremental analysis of the case "with the project" compared only to the case "without the project" or BAU (Belli et al., 2001). The case study presented in the third subchapter considers only two scenarios: "do nothing" and "full investment" base scenario, because there is no other viable alternative, from the commercial point of view, consisting in a limited investment.

Nevertheless, not for all the investment projects the case of BAU is relevant as a viable alternative to compare with. In terms of green building retrofitting investments, it is possible, as an extreme case, that the current state of the building to impose an immediately ceasing of using it, due to the risks affecting the tenants. Consequently, no BAU scenario is available, and in this case only the alternative scenarios are compared.

2. In the next step all the benefits and cost, including negative externalities, are identified and those which are relevant for the purpose of the CBA are selected to be compared for the alternative scenarios (Boardman et al., 2018).
3. The third step consists in the analysis of the impacts of the alternative projects considered, and in choosing the measures and formulas that properly quantifies them.

For example, the case study presented in the third subchapter uses certain formulas to quantify the energy savings benefits. In the same time, considering the case of a summer vacation hotel, the investment is expected to positively affect the occupancy rates which generate higher rental benefits.

4. In the fourth step, a quantitative forecast is derived for the considered benefits, costs and all the other negative externalities over the lifespan horizon of the investment project, including for the expected residual value. In the example of the summer hotel building presented below, in the third subchapter, the investment time horizon considered was 15 years, being so extended because the business operates only six months during the year (from May to October) and the investment is assumed to be properly conserved.
5. The fifth step consists in the monetization of all the impacts, respectively conversion and quantification in currency units of all the benefits, costs and negative externalities.

In the following example all the direct benefits were expressed in currency units using the formulas of estimated energy savings, or in the case of the additional rental revenue projection, the higher estimated occupancy rate is used. As long as the indirect benefit provided by the reduction of CO_2 is monetized by using the market price of the CO_2 emission per ton.

6. Then, the monetary benefits, the costs and the monetized benefits are discounted depending on the moment when are generated, to be expressed in their present values. The discount rate used is the social discount rate, for which the possible approaches are presented below.
7. In the seventh step, the Net Present Value of all the alternative projects is derived as the sum of the discounted monetary benefits, costs and monetized benefits, for every project. The costs are summed up in the Net Present Value as negative values. Additionally, in the case of green building retrofitting, the cost of investment can be reduced by the saved maintenance costs, if such maintenance operations are necessary in the BAU scenario (Liu et al., 2018)
8. The next step consists in the sensitivity analysis for the risky variables included in the projections of the above derived Net Present Value. Almost all the variables can be uncertain, but the analysis focus results from the analyst's judgment related to the most important variables (Boardman, et al. 2018). Nevertheless, the analyst's choices need to be confirmed by the sensitivity analysis, which can reflect or not the status of "critical variable" for the selected variables. Then, a qualitative risk analysis for the critical variables is run (European Commission, 2015).
9. Finally, based on the criterion of highest Net Present Value, mixed with the sensitivity and risk analysis conclusions, the best alternative project is selected.

In the case of green building retrofitting investments, because, in principle, the risks are similarly affecting all the alternative projects, the decision is eventually made on the criterion of the highest Economic Net Present Value.

The above general steps of the Cost-Benefit Analysis were developed and adjusted in the Cost-Benefit Analysis guide recommended by the European Commission (European Commission, 2015), under three sections: Financial Analysis, Economic Analysis and Risk Analysis. The initial steps of defining the base case scenario and the other alternative scenarios are implicit steps included in the Financial Analysis.

Considering the last presented approach, the one recommended by the European Commission, the steps of the methodology involved are derived from the steps of the Cost-Benefit Analysis, being adapted to the green building retrofitting investments.

1. Financial analysis defines the base case scenario and the other alternative investment scenarios.

Overall, Financial Analysis basically compares the present value of the cost of investment to the sum of the present values of the estimated operating costs savings and of the possible additional incomes expected to be generated, if such revenues exist. The relevant indicators involved in the financial analysis are the Financial Net Present Value and the Financial Internal Rate of Return.

Financial Net Present Value (FNPV) is the sum of the present value of all the investment and operating cash flows, negative and positive, including the residual value:

$$\mathrm{FNPV} = \sum_{i=0}^{n} \frac{NCF_i}{(1+r_d)^i}$$

where: NCF_i = net cash flow from operation and investment estimated for the year i, respectively the cash outflows due to the costs of investment, the net operational cash flow and the residual value, the last being considered only for the last year, if the time horizon of the investment is shorter than the economic life of the investment project;

n = the number of years considered for the investment time horizon, reasonably set at 15 years;
r_d = the discount rate.

The reference period of 15 years corresponds to the life span of the investment projects in "Other sectors", as it is assumed in the European Union (European Commission, 2015).

For the commercial buildings, if after the investment an increase in the occupancy rate is reasonable, the net operating cash flow is estimated based on the incremental analysis approach, compared to the current estimated net operating cash flow, before the deployment of the investment.

The discount rate used (the financial discount rate) reflects the opportunity cost of the funds providers, typically best estimated by the weighted average cost of capital (*wacc*).

On the one hand, in the case of a residential building, the weighted average cost of capital quantifies the proportion of the own contribution of the residents in the total cost of the investment, for which a proper opportunity cost is the available interest rate for deposits in the banking system, and assumes the rest of the necessary financing is assured by loan for which the estimated cost is the average interest rate charged in the banking system for similar projects loans (or the interest rate asked for the loans by a specific bank, especially if an indicative offer exists), without considering the tax shield effect.

$$wacc = i_{dep}\,(Own\ C/I) + i_{loan}\,(L/I)$$

where: i_{dep} = deposit interest rate;

i_{loan} = average interest rate asked for loans in the banking system;

Own C = own contribution of the building owners;
I = total cost of investment;
L = loan financing of the investment (*L= I – Own C*)

On the other hand, in the case of a commercial building, the cost of the capital considered for the own investment contribution is still the investor opportunity cost, expected to be significantly higher than the opportunity cost of the individuals. The cost of the debt is in this case the corporate loans interest rate and this cost is discounted by taking into the account the tax shield.

$$wacc = r_{alt}\,(Own\ C/I) + i_{loan}\,(L/I)(1\text{-}t)$$

where: r_{alt} = return rates available to the investor company;

i_{loan} = average interest rate asked for corporate loans in the banking system;
Own C = own contribution of the building owners;
I = total cost of investment;
L = loan financing of the investment (*L= I – Own C*)
(*1-t*) = the tax shield effect, with *t* = income tax rate

The other key indicator used for assessment and prioritization of the green building retrofitting investments is the Financial Internal Rate of Return (FIRR), which is derived by solving the following equation:

$$\text{FNPV(FIRR)} = 0 <=> \sum_{i=0}^{n} \frac{NCF_i}{(1+FIRR)^i} = 0$$

In this stage, if the resulted FNPV is positive and the FIRR is higher than the discount rate (*wacc*) the project can be financed by private lenders (i.e. banking institutions), and in this case no public assistance and consequently no further economic analysis are necessary, the only additional step is the risk analysis, that identifies the critical variables, quantifies their effects on the results of the investment project and assesses the manageability of the specific risks.

2. The economic analysis is the assessment tool additionally involved in valuation and prioritization of the investment projects for which the initial financial analysis shows negative results. For these investment projects the benefits generated for the society (i.e. the positive externalities) need to be considered also, and this approach reflects if the public support is justified.

For example, the main benefit (positive externality) of a building retrofitting investment for energy savings is the CO_2 emissions reduction, due to the lower need of energy. This benefit is possible to be monetized by estimating the CO_2 emissions quantity reduction (in tons) and using the price per ton of CO_2 on the market. Additionally, in the economic analysis, all the possible negative externalities, if exist, need to be considered, monetized and subtracted from the yearly cash flows.

The monetized externalities are treated similarly as the net cash flows from operation and investment, in the Economic Net Present Value (ENPV) and Economic Rate of Return (ERR) indicators, using as discount rate the social discount rate (SDR). The economic indicators are correspondingly reflected by the following formulas:

$$\text{ENPV} = \sum_{i=0}^{n} \frac{NCF_i + MonExt_i}{(1+SDR)^i}$$

$$\text{ENPV(ERR)} = 0 <=> \sum_{i=0}^{n} \frac{NCF_i + MonExt_i}{(1+EIRR)^i} = 0$$

where *MonExt* = monetized externalities considered.

For deriving this discount rate there are different approaches: from the use of the private discount rate, due to the financial competition between private and public projects, to the use of different discount rates, decreasing in time, due to the higher importance of the public benefits, like CO_2 reduction, for the future generations. Also, in the case of the intergenerational projects, with impacts beyond a period of 50 years, it is recommended the same approach of time-declining social discount rates (Boardman et al., 2018).

Nevertheless, in the EU, most of the public authorities propose a standardized level of SDR, for example the European Commission recommends a 5% SDR for 2014-2020 programming period, for Cohesion countries (European Commission 2015).

The conditions for the public investments to be considered desirable for the society, reflected by the economic indicators, are that the resulted ENPV to be positive and the EIRR to be higher than the SDR. Most often, if these conditions are met the project qualifies to be subsidized by the public authorities, under various subsidies programs, if such suitable programs exist.

In the case of the example provided below, in Romania, the subsidies programs for green building retrofitting are only available for individuals, households and tenants' associations. Companies are not included in, any subsidies program, because the benefits from energy savings, and in some cases, the increase in the revenues are considered to be enough strong incentives for businesses.

Risk Analysis

The final step in the Financial Analysis of the financial effective investment projects, as well as in the Cost-Benefit Analysis for the investments projects which are only socially effective, is the risk analysis. This is a mixed qualitative and quantitative approach, that identifies the variables with potential to negatively influence the valuation results and, based on the sensitivity analysis, which estimates the extent of the influence of these variables, the critical variables are confirmed. These variable are those for which any of elasticity of the valuation indicators is larger than 1 (that means the valuation indicators can be significantly affected by a smaller proportional change in the critical variables).

Examples of the potentially negative influence variables for green building retrofitting investments are the cost of investment, in the case of all types of the projects, and the occupancy rate in the case of commercial buildings.

Finally, the risk analysis considers the possible actions to be taken for these risk managements. The risk assessment conclusion depends on the reliability of these actions.

THE COST-BENEFIT ANALYSIS FOR A GREEN BUILDING RETROFITTING INVESTMENT IN A SUMMER HOTEL

In the following example, the methodology of financing the green building retrofitting investment is applied for the case of a commercial building, respectively a summer hotel. The financial analysis covers also the monetization of the benefits of the investment, in the context of the economic analysis, even though the investment project is developed by a private company, and no public grant or subsidy is considered for co-financing the investment project, consequently the Cost-Benefit Analysis is not necessarily necessary. Generally, the available government subsidies or grants programs for companies in this field are inexistent. The households, individuals and tenants' associations are the main targets of these subsidy programs.

Financial Analysis

The investment considered is implemented for a hotel building, located in a Black Sea resort, and for this type of business there is a significant seasonality: the only operating six months are those between May and October. Besides, out of these six months only two had significant occupancy rates, before the investment. The potential positive Net Present Value for the green building retrofitting investment derived in this case of a hotel which is operated only during the warmer months of the year would additionally strengthen the conclusion about the high desirability of this type of investments.

Consequently, for the investment considered, there are two main net operating cash flow sources estimated: the supplementary cash flows resulted from the estimated increase of the occupancy rate during the months of the tourist season, especially during the colder months of the season: May, June, September and October, and the net operating cash flows resulted from the energy savings. The energy savings expected are due to the lower quantity of natural gas used for heating the building and the domestic water especially during the colder months: May and October; and due to the lower electricity consumption especially with the air conditioning during the hottest months of the season: July and August, but also because of the replacement of all the lamps, light bulbs and rooms appliances with some new and more energy effective. Both of these types of energy savings have also a certain impact in reducing the carbon footprint of the hotel building. Additionally, to these savings, the investment is expected to significantly improve the living standard for the guests by replacing all the fixtures and fittings in the hotel, including the sanitary objects in the bathrooms. In this way the occupancy rates expected for the years after the investment will show an upward evolution.

The approach is based on the incremental analysis: the cost of investment being compared to the present value of the additional net operating cash flows generated by it. The time horizon considered for the investment is 15 years. Any required reinvestments, which are less frequent than those necessary for a regular hotel, which is not affected by the seasonality (actually, the periods after which the reinvestments are necessary are two times longer), are assumed similar as value to the reinvestment necessary before the green building retrofitting investment, and consequently being negligible in the analysis.

The cost of investment, of EUR 445,798, was considered based on the total construction cost and on the average prices offered by suppliers for fixtures and fittings. This cost of investment is resulted from the following cost estimate and it is included in the analysis on the net basis, subtracting the costs for the currently required maintenance and repairs (see Table 1).

Table 1. Investment cost estimate

	Unit cost EUR/ room (or per m2*)	Cost of investment	Maintenance cost**
Cost of replacing windows and doors	258	77,419	3,300
Cost of heating (boiler, pipes, fittings and radiators)*	8.70	65,217	8,100
Cost of external thermal insulation (enveloping of the outside walls and the roof)	505	151,571	
Cost of water pipes, sewage and bathroom sanitary objects	204	61,290	7,200
Replacement of the air conditioner	150	45,000	2,400
Replacement of the TV set	95	28,500	
Replacement of the room's small refrigerator	65	19,500	
Replacement of 7 lamps and light bulbs (including the lights in the common spaces)	49	14,700	
Replacement of the hairdryer	12	3,600	
Cost of investment		**466,798**	**21,000**
Saved maintenance costs		21,000	
Total (net)cost of investment		**445,798**	

**These costs are imposed by the current state of the building

The financial structure assumed to cover this cost is limited to a single source: a banking loan, for which the expected fixed interest rate charged by the bank is 6%. This approach, of financing the investment fully by a banking loan, is possible due to the current low indebted balance sheet of the company.

The income tax is 16%, and correspondingly the discount rate (based on the *wacc* approach) is 5.04%.

$$wacc = 6\% \,(1\text{-}16\%) = 5.04\%$$

The data and quantitative assumptions used in the projection of the net operating cash flows over the investment horizon of 15 years, due to the higher revenues collected and to the energy savings, are presented in the following Table (see Table 2):

Both types of quantitative assumptions related to the expected evolution of the occupancy rate and expected rooms prices after the investment deployment resulted from a market study, which is critical for the financial analysis purposes. The study also revealed a required period of seven years for the hotel to attain a maximum level of the occupancy rates. After this seven-year period, the occupancy rates are expected to remain constant, as the improved quality of the hotel resort will become well known for a large and significant number of customers.

Based on these data and assumptions, the additional revenues expected to be directly generated by the investment are the following (see Table 3):

The other monetary benefits generated by the building retrofitting investment result from the energy savings due the reduction of gas consumption for both the water and the room heating, but for the room heating savings, these are generated only during the first half of May and the last half of October, when the room heating is necessary. For the purpose of the financial analysis, the gas savings are considered on the net basis, taking into the account, in the same time, the compensatory effect of the higher occu-

Table 2. The data and quantitative assumptions used in projection

Operating months	*May*	*Jun*	*Jul*	*Aug*	*Sep*	*Oct*
Average room price before the investment - EUR -	56	65	65	69	59	37
Average room price after the investment - EUR -	69	80	85	92	77	55
Occupancy rate before the investment	28%	37%	68%	78%	52%	21%
Occupancy rate after the investment year 1	37%	45%	73%	82%	67%	27%
Occupancy rate after the investment year 2	45%	50%	78%	85%	70%	36%
Occupancy rate after the investment year 3	55%	57%	82%	88%	73%	45%
Occupancy rate after the investment year 4	65%	67%	84%	90%	76%	55%
Occupancy rate after the investment year 5	66%	68%	85%	91%	77%	66%
Occupancy rate after the investment year 6	67%	69%	86%	92%	78%	67%
Occupancy rate after the investment year 7 -15	68%	70%	88%	95%	80%	68%

Table 3. The projected additional accommodation revenues, based on the available data and assumptions -EUR-

Revenues data and projections	**May**	**Jun**	**Jul**	**Aug**	**Sep**	**Oct**	**Yearly Total**
Monthly accommodation revenue before the investment	4,704	7,215	13,260	16,146	9,204	2,331	52,860
Monthly accommodation revenue after the investment - year 1	7,659	10,800	18,615	22,632	15,477	4,455	79,638
Monthly accommodation revenue after the investment - year 2	9,315	12,000	19,890	23,460	16,170	5,940	86,775
Monthly accommodation revenue after the investment - year 3	11,385	13,680	20,910	24,288	16,863	7,425	94,551
Monthly accommodation revenue after the investment - year 4	13,455	16,080	21,420	24,840	17,556	9,075	102,426
Monthly accommodation revenue after the investment - year 5	13,662	16,320	21,675	25,116	17,787	10,890	105,450
Monthly accommodation revenue after the investment - year 6	13,869	16,560	21,930	25,392	18,018	11,055	106,824
Monthly accommodation revenue after the investment - year 7-15	14,076	16,800	22,440	26,220	18,480	11,220	109,236
Revenues due to the investment - year 1	2,955	3,585	5,355	6,486	6,273	2,124	**26,778**
Revenues due to the investment - year 2	4,611	4,785	6,630	7,314	6,966	3,609	**33,915**
Revenues due to the investment - year 3	6,681	6,465	7,650	8,142	7,659	5,094	**41,691**
Revenues due to the investment - year 4	8,751	8,865	8,160	8,694	8,352	6,744	**49,566**
Revenues due to the investment - year 5	8,958	9,105	8,415	8,970	8,583	8,559	**52,590**
Revenues due to the investment - year 6	9,165	9,345	8,670	9,246	8,814	8,724	**53,964**
Revenues due to the investment - years 7-15	9,372	9,585	9,180	10,074	9,276	8,889	**56,376**

pancy rates expected which determine larger heating needs. The following Table (see Table 4), reflects the aggregated effects of the savings and higher occupancy rates:

The price of the natural gas involved in the projections is the current price, based on statistics data: EUR 0.0258/kWh, excluding the VAT (https://ec.europa.eu/eurostat), as these are made in the current prices (the increase of the room prices assumed in the projections is due only to the higher quality of the accommodation services offered after the investment).

The above Table shows that starting with the 4th year, after the investment, the larger heating energy consumption due to the increase of the occupancy rate exceeds the saving. For the financial analysis, only the net effect is relevant, even if it is negative, reflecting higher heating costs.

The other energy savings come from the replacement of all air conditioners, of the lamps and bulbs (7 per room, including the corridors and reception), of the TV sets, of the small refrigerators in every room and of the bathrooms' hairdryers. The new devices are more energy effective, but as in the case of the energy used for heating, the higher power consumption due to the larger occupancy rates partly compensate the savings, and for the financial analysis the net effect is relevant. In the quantification of the net monetary savings effect (see Table 5), the price used for the electricity is the current price based on statistics: EUR 0.0786/kWh, excluding the VAT (https://ec.europa.eu/eurostat).

Assuming there are not any additional variable operational costs affected by the increase in the occupancy rates, the above presented are the necessary data for the financial assessment of the hotel green building retrofitting investment.

The aggregation of the above resulted projections generated the following key financial indicators, used for the investment assessment (FNPV, FIRR and Simple Payback Period) (see Table 6).

Economic Analysis

In this case, the investment presented seems to be desirable for the investor company, especially because an important share of the benefits directly generated come from the additional revenues, due to the higher expected occupancy rates. Nevertheless, in the case of the residential buildings such benefits are not generated, and the financial analysis is limited to the energy savings benefits. These benefits are most often not so transparent and significant for the owners, but the benefits of reducing the building carbon footprint can be desirable for the society and this is the reason that made the governments to use different subsidies programs as an incentive for these investments.

In our particular case study, besides the lower CO_2 emissions due to the net energy savings, from the point of view of the society the benefits to be considered are larger: the CO_2 emissions reduction to be monetized is resulted as the difference between potential emissions in the absence of the investment, but considering the projected occupancy rates for the years after the investment and the corresponding emissions after the investment deployment, derived based on the same projected occupancy rates. For this purpose, the potential electricity savings are derived by comparing the potential consumptions before the investment, based on the expected increasing occupancy rates for all the years after the investment, to the projected yearly consumptions; and the gas savings are derived as the difference between the potential consumptions before the investment, under the same occupancy rates assumed for the years after the investment, and the projected consumptions.

In the following Table (see Table 7), the potential savings of the investment from the electricity consumption are presented.

Table 4. The net effect of the investment on the gas consumed for heating purposes

Year before the investment	May	Jun	Jul	Aug	Sep	Oct	Yearly savings kWh	Yearly savings EUR
Natural gas - water heating before inv. kWh	38,546	50,936	93,613	107,379	71,586	28,910		
Natural gas - rooms heating before inv. kWh	13,074					16,231		
1st year after inv.								
Natural gas - water heating after inv. kWh	33,957	41,300	66,997	75,257	61,491	24,780		
Natural gas - rooms heating after inv. kWh	11,619					8,479		
Savings 1st year	**6,043**	**9,637**	**26,615**	**32,122**	**10,095**	**11,882**	**96,395**	**2,487**
2nd year								
Natural gas - water heating after inv. kWh	41,300	45,888	71,586	78,010	64,244	33,040		
Natural gas - rooms heating after inv. kWh	14,132					11,305		
Savings 2nd year	**-3,811**	**5,048**	**22,026**	**29,369**	**7,342**	**796**	**60,770**	**1,568**
3rd year								
Natural gas - water heating after inv. kWh	50,477	52,313	75,257	80,764	66,997	41,300		
Natural gas - rooms heating after inv. kWh	17,272					14,132		
Savings 3rd year	**-16,129**	**-1,377**	**18,355**	**26,615**	**4,589**	**-10,290**	**21,763**	**561**
4th year								
Natural gas - water heating after inv. kWh	59,655	61,491	77,093	82,599	69,751	50,477		
Natural gas - rooms heating after inv. kWh	20,412					17,272		
Savings 4th year	**-28,447**	**-10,554**	**16,520**	**24,780**	**1,836**	**-22,608**	**-18,475**	**-477**
5th year								
Natural gas - water heating after inv. kWh	60,573	62,408	78,010	83,517	70,668	60,573		
Natural gas - rooms heating after inv. kWh	20,726					20,726		
Savings 5th year	**-29,679**	**-11,472**	**15,602**	**23,862**	**918**	**-36,158**	**-36,927**	**-953**
6th year								
Natural gas - water heating after inv. kWh	61,491	63,326	78,928	84,435	71,586	61,491		
Natural gas - rooms heating after inv. kWh	21,040					21,040		
Savings 6th year	**-30,911**	**-12,390**	**14,684**	**22,944**	**0**	**-37,390**	**-43,062**	**-1,111**
7th to 15th year								
Natural gas - water heating after inv. kWh	62,408	64,244	80,764	87,188	73,422	62,408		
Natural gas - rooms heating after inv. kWh	21,354					21,354		
Savings 7th to 15th year	**-32,143**	**-13,308**	**12,849**	**20,191**	**-1,836**	**-38,622**	**-52,868**	**-1,364**

Table 5. The net effect of the investment on the electricity consumed

Year before the investment	May	Jun	Jul	Aug	Sep	Oct	Total kWh	Savings kWh/ year	Savings EUR/ year
Electrical power consumed before the inv. kWh	19,608	25,075	47,620	54,623	35,240	14,706	196,872		
1st year									
Electrical power consumed kWh	13,274	15,623	26,189	29,417	23,261	9,686	117,449	79,423	**6,243**
2nd year									
Electrical power consumed kWh	16,144	17,359	27,982	30,494	24,302	12,915	129,195	67,676	**5,319**
3rd year									
Electrical power consumed kWh	19,731	19,789	29,417	31,570	25,344	16,144	141,995	54,877	**4,313**
4th year									
Electrical power consumed kWh	23,319	23,261	30,135	32,287	26,385	19,731	155,118	41,754	**3,282**
5th year									
Electrical power consumed kWh	23,677	23,608	30,494	32,646	26,732	23,677	160,835	36,037	**2,833**
6th year									
Electrical power consumed kWh	24,036	23,955	30,852	33,005	27,080	24,036	162,964	33,908	**2,665**
7th -15th year									
Electrical power consumed kWh	24,395	24,302	31,570	34,081	27,774	24,395	166,517	30,355	**2,386**

Table 6. The financial analysis of the green building retrofitting investment -EUR-

	Investment completion	*Year 1*	*Year 2*	*Year 3*	*Year 4*	*Year 5*	*Year 6*	*Year 7 to 15*
Additional revenues		26,778	33,915	41,691	49,566	52,590	53,964	56,376
Total net savings		8,730	6,887	4,875	2,356	1,712	1,554	1,022
Cost of investment	-445,798							
Net cash flow (investing and operating)	-445,798	35,508	40,802	46,566	51,922	54,302	55,518	57,398
FNPV (@5.04%) 94,818								
FIRR 7.83%								
Simple Payback Period[1] 8 years & 10 months								

Table 7. The potential electricity savings - kWh -

	May	*Jun*	*Jul*	*Aug*	*Sep*	*Oct*	*Total kWh*	**Savings kWh/ year**
Potential electricity cons. 1st year w/o savings provided by the inv.								
Electrical power cons. before the inv. kWh	25,911	31,513	51,121	57,424	46,919	18,908	231,796	
Electrical power cons. after the inv. year 1	13,274	15,623	26,189	29,417	23,261	9,686	117,449	**114,347**
Potential electricity cons. 2nd year w/o savings provided by the inv.								
Electrical power cons. before the inv. kwh	31,513	35,015	54,623	59,525	49,020	25,210	254,906	
Electrical power cons. after the inv. year 2	16,144	17,359	27,982	30,494	24,302	12,915	129,195	**125,710**
Potential electricity cons.3rd year w/o savings provided by the inv.								
Electrical power cons. before the inv. kWh	38,516	39,917	57,424	61,626	51,121	31,513	280,116	
Electrical power cons. after the inv. year 3	19,731	19,789	29,417	31,570	25,344	16,144	141,995	**138,121**
Potential electricity cons. 4th year w/o savings provided by the inv.								
Power cons. before the inv. kWh	45,519	46,919	58,824	63,026	53,222	38,516	306,027	
Electrical power cons. after the inv. year 4	23,319	23,261	30,135	32,287	26,385	19,731	155,118	**150,909**
Potential electricity cons. 5th year w/o savings provided by the inv.								
Power cons. before the inv. kWh	46,219	47,620	59,525	63,726	53,922	46,219	317,231	
Electrical power cons. after the inv. year 5	23,677	23,608	30,494	32,646	26,732	23,677	160,835	**156,397**
Potential electricity cons. 6th year w/o savings provided by the inv.								
Electrical power cons. before the inv. kWh	46,919	48,320	60,225	64,427	54,623	46,919	321,433	
Electrical power cons. after the inv. year 6	24,036	23,955	30,852	33,005	27,080	24,036	162,964	**158,469**
Potential electricity cons. for years 7th to 15th w/o savings provided by the inv.								
Electrical power cons. before the inv. kWh	47,620	49,020	61,626	66,528	56,023	47,620	328,436	
Electrical power cons. after the inv. year 7-15	24,395	24,302	31,570	34,081	27,774	24,395	166,517	**161,919**

The yearly potential savings of electricity are reflected into the lower CO_2 emissions, consequently their social benefit can be quantified by estimating the CO_2 emissions reduction. For this purpose, the average emissions from electricity generation are used (CO_2 intensity of electricity production): 0.282 kg CO_2/kWh – last available data for the summer period (2017), in Romania (http://www.stopco2.ro/).

Based on the estimations related to the CO_2 emissions saved (avoided), and considering the market CO2 price per ton: 18.53 EUR/t (https://markets.businessinsider.com), it is possible to quantify the benefits in monetary units (EUR), by the process which is called "monetization of the benefits" (see Table 8.).

The next benefit considered is related to the reduction of the CO_2 emissions due to the lower consumption of natural gas, compared to the potential consumption before the investment, corresponding to the

Table 8. The monetization of the benefits of electricity savings

	Year 1	*Year 2*	*Year 3*	*Year 4*	*Year 5*	*Year 6*	*Year 7 to 15*
Electricity savings - kWh	114,347	125,710	138,121	150,909	156,397	158,469	161,919
Electricity emissions - t CO_2	32.246	35.450	38.950	42.556	44.104	44.688	45.661
Electricity emissions monetized - EUR	598	657	722	789	817	828	846

projected occupancy rates for the period after the investment. The potential gas savings are presented in the following Table (see Table 9).

The potential natural gas savings estimated are the basis for CO_2 emissions reduction projected in the economic analysis, based on the unitary CO_2 emission per kWh gas consumed, respectively 0.18 kg CO_2 / kWh (https://www.engineeringtoolbox.com). The economic benefits of the investment monetized are estimated based on the avoided CO_2 emission and on the current CO_2 price per ton: 18.53 EUR/t (see Table 10.).

Table 9. The potential natural gas savings - kWh –

Natural gas potential savings	*May*	*Jun*	*Jul*	*Aug*	*Sep*	*Oct*	*Total*
1 st year potential savings							**161,675**
Savings heating water	16,979	20,650	33,499	37,629	30,745	12,390	151,891
Savings rooms heating	5,656					4,128	9,784
2nd year potential savings							**179,417**
Savings heating water	20,650	22,944	35,793	39,005	32,122	16,520	167,034
Savings rooms heating	6,879					5,504	12,383
3rd year potential savings							**198,842**
Savings heating water	25,239	26,156	37,629	40,382	33,499	20,650	183,554
Savings rooms heating	8,408					6,879	15,288
4th year potential savings							**218,878**
Savings heating water	29,828	30,745	38,546	41,300	34,875	25,239	200,533
Savings rooms heating	9,937					8,408	18,345
5th year potential savings							**228,055**
Savings heating water	30,286	31,204	39,005	41,759	35,334	30,286	207,875
Savings rooms heating	10,090					10,090	20,180
6th year potential savings							**231,114**
Savings heating water	30,745	31,663	39,464	42,217	35,793	30,745	210,628
Savings rooms heating	10,243					10,243	20,485
7th to 15th years potential savings							**236,008**
Savings heating water	31,204	32,122	40,382	43,594	36,711	31,204	215,217
Savings rooms heating	10,396					10,396	20,791

Table 10. The monetization of the benefits of natural gas savings

	Year 1	*Year 2*	*Year 3*	*Year 4*	*Year 5*	*Year 6*	*Year 7 to 15*
Gas savings - kWh	161,675	179,417	198,842	218,878	228,055	231,114	236,008
Gas emissions - t CO_2	29.102	32.295	35.791	39.398	41.050	41.600	42.481
Gas emissions monetized - EUR	539	598	663	730	761	771	787

Both benefits monetized are added to the net direct benefits generated for the investor and the resulted net benefits are discounted with the purpose to derive the key economic analysis indicators (ENPV and ERR). The discount rate used is the recommended social discount rate by the European Commission: 5% (see Table 11).

The monetization of the benefits shows a relatively small additional economic value of the reduction of CO_2 footprint of the hotel building. This is mainly due to the seasonality of the activity for the analyzed hotel, as it is closed during the cold season, most of the emissions due to rooms heating, by the use of gas, are avoided because of this, and the green retrofitting investment has no additional emissions reduction effect, because between November and April there is not any gas consumption.

Generally, for residential buildings the economic benefits can be significant, especially during the cold season, but the financial benefits (savings) for the owners can be not so transparent and by their delayed timing can represent a poor incentive for such investments' deployment.

Therefore, as it was already mentioned above, the governments promote different grants and/or subsidies programs, designed to work as an incentive for the owners in accepting this type of investments: from the subsidies for the total cost of investments to the loans granted with partly or totally subsidized interest rates. Nevertheless, as it was already mentioned, in Romania none of these subsidies programs are available for the companies, either regarding residential, commercial or industrial buildings, at least until 2019.

Table 11. The economic analysis of the green building retrofitting investment -EUR-

	Investment completion	**Year 1**	**Year 2**	**Year 3**	**Year 4**	**Year 5**	**Year 6**	**Year 7 to 15**
Additional revenues		26,778	33,915	41,691	49,566	52,590	53,964	56,376
Total net savings		8,730	6,887	4,875	2,356	1,712	1,554	1,022
Benefits from reducing the building CO_2 footprint		1,137	1,255	1,385	1,519	1,578	1,599	1,633
Cost of investment	-445,798							
Net benefits of the investment	-445,798	36,644	42,058	47,951	53,440	55,880	57,117	59,031
ENPV (@5%)	112,141							
ERR	8.27%							

Sensitivity Analysis and Risks Assessment

For completion of the analysis of the above presented investment a risk assessment is necessary. The first step in the risk assessment is the identification, based on a qualitative analysis, of the basic variables involved in the projections, with potential to highly affect the results. Then, by sensitivity analysis, the effects of the changes in these variables on the key valuation indicators of the investment are quantified, using the elasticity of the indicators depending on the changes in each relevant variable. For this purpose, a graphical approach is taken to reveal the critical variables, respectively those for which the results show elasticity (the percent change in the indicators is larger than the percent change in the variable). Critical variables for the investment project are considered those for which the elasticity of the indicators is higher than 1.

The final part of the risk analysis consists in a qualitative analysis of the potential effects of the critical variables, possibly by including extreme case simulations, assuming changes in the variables larger than 10%, and planning response measures, to cover the risks.

For the example presented, the variables considered for the sensitivity analysis were the occupancy rates and the cost of investment, both being subject to the uncertainties. Generally, the price considered in the revenue projections can be also a very probable risky variable for the results, but in the example used the price is correlated to the occupancy rates, and ultimately both has the same impact on the results.

The following are the simulations of the effect of the changes in the selected variables: Table 12 and Figure 1 for the occupancy rate (*OR*), and Table 13 and Figure 2 for the investment cost (IC).

The analysis showed a high sensitivity of the results to the changes in the occupancy rate, more obvious in the negative scenarios, because in case of the positive scenarios the occupancy rate is limited to 100%. Nevertheless, even if in the negative scenarios the (negative) change of indicators significantly exceeds the change in the occupancy rate, and this reaction reveals the fact that the occupancy rate is a critical variable, the risk is still manageable. Only in the extreme case of a 15% lower occupancy rate, negative FNPV and ENPV result, but the negative values are not very significant. Considering the limited increases assumed for the occupancy rates in the base case scenario, and the resulted occupancy rates not very high, especially for the spring and autumn months, the risk of a deviation of 15% in the overall occupancy rate can be viewed as extreme, and low probable.

Table 12. The sensitivity analysis of the investment valuation indicators depending on the occupancy rate

Scenarios	*Δ OR%*	*FNPV%*	*FRR%*	*ENPV%*	*ERR%*	*FNPV (EUR)*	*FRR*	*ENPV (EUR)*	*ERR*
S -15%	-15%	-129%	-47%	-111%	-44%	-27,488	4.17%	-12,876	4.60%
S -10%	-10%	-86%	-30%	-74%	-29%	13,281	5.45%	28,796	5.88%
S -5%	-5%	-43%	-15%	-37%	-14%	54,050	6.66%	70,468	7.10%
Baseline	0%	0%	0%	0%	0%	94,818	7.83%	112,141	8.27%
S +5%	5%	43%	14%	37%	14%	135,587	8.96%	153,813	9.40%
S +10%	10%	80%	27%	69%	25%	170,837	9.92%	189,862	10.37%
S +15%	15%	113%	38%	98%	30%	202,279	10.77%	222,033	10.77%

Figure 1. Effects of changes in the occupancy rate

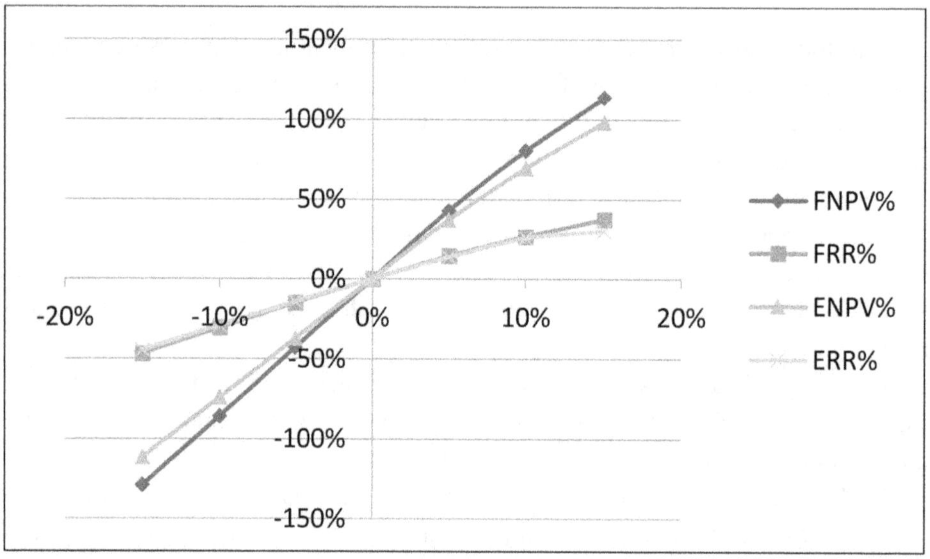

Figure 2. Effects of changes in the investment cost

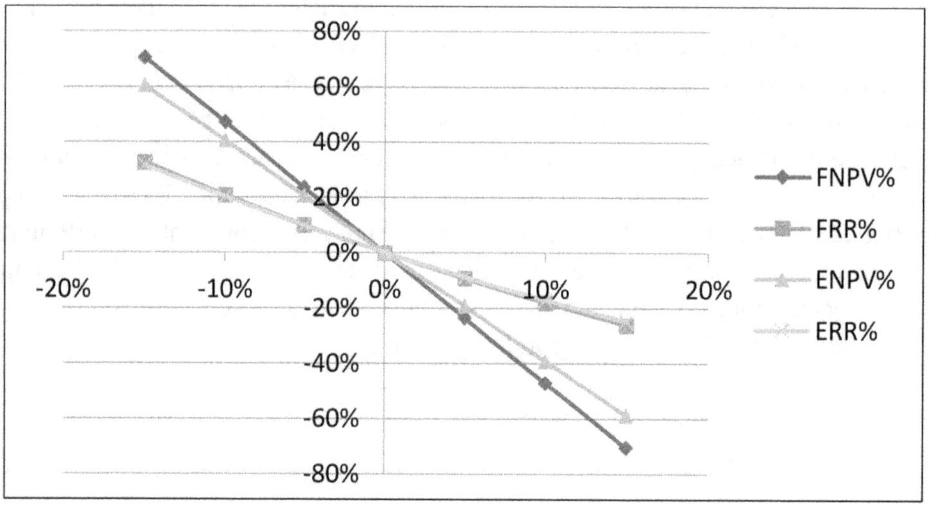

The next part of the sensitivity analysis is focused on the second variable considered, respectively the cost of investment (*CI*). In this case the risk is opposite as direction of variation, respectively, the investment cost overrun. The sensitivity of the indicators is not so high, like in the case of the occupancy rate.

The proportional change in the indicators is still significantly higher than the change in the cost of investment, and this makes the cost of investment a critical variable. But this risk seems to be also manageable, as the extreme case of an investment cost overrun of 15% still generates a non-negligible positive financial net present value.

Table 13. The sensitivity analysis of the investment valuation indicators depending on the cost of investment

Scenarios	*Δ CI%*	*FNPV%*	*FRR%*	*ENPV%*	*ERR%*	*FNPV (EUR)*	*FRR*	*ENPV (EUR)*	*ERR*
S -15%	-15%	71%	33%	60%	32%	161,688	10.39%	179,759	10.88%
S -10%	-10%	47%	21%	40%	20%	139,398	9.47%	157,469	9.94%
S -5%	-5%	24%	10%	21%	10%	117,108	8.62%	135,179	9.08%
Baseline	0%	0%	0%	0%	0%	94,818	7.83%	112,141	8.27%
S +5%	5%	-24%	-9%	-19%	-9%	72,529	7.10%	90,599	7.55%
S +10%	10%	-47%	-18%	-39%	-17%	50,239	6.42%	68,309	6.86%
S +15%	15%	-71%	-26%	-59%	-25%	27,949	5.78%	46,019	6.21%

Testing the sensitivity of the valuation indicators can be done for other variables included in the projections, also, like the energy (electricity and gas) price or the rooms' price for example. But, in the example provided, it was assumed that there are not any other relevant variables estimated to negatively impact the projections.

CONCLUSION OF THE COST-BENEFIT ANALYSIS

The example of the Cost-Benefit Analysis ran for the project of the green retrofitting investment in a hotel building located on the Black Sea shore, highly affected by the seasonality of the activity, showed that the expected monetary benefits for the owner company, respectively the additional accommodation revenues and the net energy savings, in real terms, are higher than the cost of investment, as the Financial Net Present Value is positive, even if the building considered is just a summer hotel. There are some relatively small monetized benefits due to the reduction of the building CO_2 footprint, because the hotel operating activity is highly affected by the seasonality. Nevertheless, the benefits generated by such an investment can be significantly larger for buildings which are operated all the year and especially during the colder months, between November and April, when the rooms heating is necessary and consequently, the gas and the CO_2 emissions saved through the green building retrofitting investment become significant. But, on the other hand, generally for residential buildings there are neither estimated revenues, nor additional revenues, and in the absence of these revenues most probable the financial analysis indicators shows an undesirable investment, even though the energy savings are more significant. In this case, considering the social benefits generated, the public support is the only "impulse solution" for implementation of such type of investments.

In the above presented case study, the risk assessment revealed manageable risks related to the revenues and the investment cost. Actually, under these risks the project is still desirable for the investor and society, in the same time. In the case of a residential building the risk related to the revenues is not present, because generally there are not any revenues projected, the only risk remained to be taken into the account is related to the overrun of the investment cost. This is why, in accepting the green building retrofitting investment for residential buildings from the financial point of view, the most important supporting tool for decision making process is an accurate and complete investment cost estimate.

From a larger prospect, the analysis involved in financing the green building retrofitting investments can be extended, more than the above presented example. As it was mentioned before, the presented approach was limited to the case of a seaside hotel, opened only six months a year. Consequently, the case of a building used all the year and especially during the colder months will probably generate significantly higher energy savings, being even more attractive for owners. Besides, in such cases an extended version of the green retrofitting investment, including more retrofit measures, is more suitable. The measures could be able even to convert the building to a (net) zero-energy one, possibly by installing solar panels on the roof of the building and feeding-in the grid with the electricity excess.

Finally, it is necessary to mention that the investment attractiveness for the hotel building operated only during the warmer months of the year in Romania is not convergent with the unattractiveness revealed by the study of a residential green building retrofitting investment in Israel, operated all the year, under the warm temperature conditions (Friedman et al., 2014). The positive difference of the results in Romanian case is due to the expected increase of the hotel's revenues, both because of the occupancy rate and the rooms' prices. Both of these two variables are reasonable expected to increase, because the green building retrofitting investment makes the hotel more attractive for the tourists.

REFERENCES

Belli, P., Anderson, J.R., Barnum, H.N., Dixon, J.A., Tan, J-P, 2001. Economic analysis of investment operations: analytical tools and practical applications. The World Bank.

Boardman, N. E., Greenberg, D. H., Vining, A. R., & Weimer, D. L. (2018). *Cost-benefit Analysis: Concepts and Practice* (4th ed.). Cambridge University Press. doi:10.1017/9781108235594

Engineering Toolbox. (n.d.). Combustion of Fuels - Carbon Dioxide Emission. Retrieved from https://www.engineeringtoolbox.com/co2-emission-fuels-d_1085.html

European Commission. (2015). *Guide to Cost-Benefit Analysis of Investment Projects - Economic appraisal tool for Cohesion Policy 2014-2020*. European Union.

Eurostat. (n.d.). Retrieved from https://ec.europa.eu/eurostat

Friedman, C., Becker, N., & Erell, E. (2014). Energy retrofit of residential building envelopes in Israel: A cost-benefit analysis. *Energy*, *77*, 183–193. doi:10.1016/j.energy.2014.06.019

He, Y., Liao, N., Bi, J., & Guo, L. (2019). Investment decision-making optimization of energy efficiency retrofit measures in multiple buildings under financing budgetary restraint. *Journal of Cleaner Production*, *215*, 1078–1094. doi:10.1016/j.jclepro.2019.01.119

Liu, Y., Liu, T., Ye, S., & Liu, Y. (2018). Cost-benefit analysis for Energy Efficiency Retrofit of existing buildings: A case study in China. *Journal of Cleaner Production*, *177*, 493–506. doi:10.1016/j.jclepro.2017.12.225

Markets Insider. (n.d.). CO2 European emission allowances. Retrieved form https://markets.businessinsider.com/commodities/co2-emissionsrechte

Neuhoff, K., Stelmakh, K., Amecke, H., Novikova, A., Deason, J., & Hobbs, A. (2012). Financial incentives for energy efficiency retrofits in buildings. In ACEEE Summer Study on Energy Efficiency in Buildings (pp. 236–248).

Pombo, O., Rivela, B., & Neila, J. (2016). The challenge of sustainable building renovation: Assessment of current criteria and future outlook. *Journal of Cleaner Production*, *123*, 88–100. doi:10.1016/j.jclepro.2015.06.137

Rocchi, L., Kadzinski, M., Menconi, M. E., Grohmann, D., Miebs, G., Paolotti, L., & Boggia, A. (2018). Sustainability evaluation of retrofitting solutions for rural buildings through life cycle approach and multi-criteria analysis. *Energy and Building*, *173*, 281–290. doi:10.1016/j.enbuild.2018.05.032

Stocker, E., Tschurtschenthaler, M., & Schrott, L. (2015). Cost-optimal renovation and energy performance: Evidence from existing school buildings in the Alps. *Energy and Building*, *100*, 20–26. doi:10.1016/j.enbuild.2015.04.005

Stop CO2. (n.d.). Nivelul de CO2 pentru producerea unui kwh de electricitate, in scadere. Retrieved from http://www.stopco2.ro/2017/10/23/nivelul-de-co2-pentru-producerea-unui-kwh-de-electricitate-in-scadere/

Tan, B., Yavuz, Y., Otay, E. N., & Çamlibel, E. (2016). Optimal selection of energy efficiency measures for energy sustainability of existing buildings. *Computers & Operations Research*, *66*, 258–271. doi:10.1016/j.cor.2015.01.013

Torregrossa, M. (2015). *Energy-efficiency investment with special regard to the retrofitting of buildings in Europe. In B. Galgoczi (Ed.), Europe's energy transformation in the austerity trap. European Trade Union Institute*. Brussels: ETUI.

Wang, B., Xia, X., & Zhang, J. (2014). A multi-objective optimization model for the life-cycle cost analysis and retrofitting planning of buildings. *Energy and Building*, *77*, 227–235. doi:10.1016/j.enbuild.2014.03.025

Wu, Z., Xia, X., & Wang, B. (2015). Improving building energy efficiency by multiobjective neighborhood field optimization. *Energy and Building*, *87*, 45–56. doi:10.1016/j.enbuild.2014.10.079

ADDITIONAL READINGS

Abdullah, A. (2016). Sustainable and Energy Efficient Commercial Retrofit: Case Study of Perkins+ Will Atlanta Office. *Perkins+ Will Research Journal, 8*(1), 48-66.

Pedinotti-Castelle, M., Astudillo, M. F., Pineau, P. O., & Amor, B. (2018). Is the environmental opportunity of retrofitting the residential sector worth the life cycle cost? A consequential assessment of a typical house in Quebec. *Renewable & Sustainable Energy Reviews*, *101*, 428–439. doi:10.1016/j.rser.2018.11.021

KEY TERMS

Business as Usual Scenario (BAU) or do Nothing Scenario: Scenario in which the investment is not implemented

Cost Benefit Analysis: Assessment and prioritization tool used for investments which generate significant externalities, typically social benefits, additional to financial results.

Economic Analysis: Main part of the Cost Benefit Analysis, focused on monetization of externalities, including their present values comparison to the cost of investment.

Incremental Analysis: Approach used for quantification of the additional revenues, benefits and operating costs of a certain investment scenario, compared to BAU scenario.

Monetization of Externalities: Process under which the externalities, positive and negative, are quantified and converted in currency units.

Sensitivity Analysis: Analysis under which the critical variables of investment are confirmed or not, by observing the elasticity of the valuation indicators depending on the selected variables.

Social Discount Rate: Discount rate used for deriving the present values of social benefits and costs.

ENDNOTES

1. The Simple Payback Period is the period after which the investment cost is recovered by the revenues and savings generated, in nominal terms.

Chapter 4
Comparative Analysis Regarding the Online Interest of Consumers Towards Green Retrofitting:
Case Study in Germany and Romania

Corina Pelau
The Bucharest University of Economic Studies, Romania

Mihai-Ionut Pop
The Bucharest University of Economic Studies, Romania

Andra Pop
The Bucharest University of Economic Studies, Romania

ABSTRACT

Green retrofitting and housing energy efficiency are some of the most debated topics nowadays. They contribute to lower costs for the involved households and more efficient sources of energies, which lead to a cleaner environment. The number of methods and devices for increasing the energy efficiency of housing ranges from very simple methods to complex ones. In spite of the fact that in the research there is high interest for the development of energy-efficient housing systems, it is also important to inform the consumer about the existence of such devices and to convince them to use these types of systems more frequently. The objective of the chapter is to give an overview of the popularity and the interest of the consumers towards different forms of housing energy efficiency systems and green retrofitting methods.

DOI: 10.4018/978-1-5225-9104-7.ch004

INTRODUCTION

One of the goals of the sustainability strategy within the energy policy refers to an increasing domestic efficiency for assuring the welfare of its tenants and home owners (Gooding & Gul, 2016). Taking into consideration the fact that the comfort needs of the population represent a big amount of the private household energy consumption, it is important to identify means of reducing the waste of energy. In this sense housing energy efficiency plays an important role for the wellbeing of the population, as it reduces the energy costs of households, by offering optimal thermic solutions. The preoccupation for better living conditions from an energetic point of view has led to a big range of methods and devices for increasing the energy efficiency of housing. Simple methods include the changing of lighting bulbs with ecological ones or the replacement of old electronic equipment with newer ones, with lower energy consumption. In the category of complex energy efficiency methods, several applications or smart devices which are able to optimize the energy consumption in the household as well as intelligent devices with an independent programming for energy optimization can be included. Green retrofitting and the modernization of houses is one of the methods to increase energy consumption efficiency. In spite of the fact that in the research area, there is a high interest for the development of energy efficient housing systems, it is also important to inform the consumer about the existence of such devices and to convince them, to use these types of systems more frequently. The lack of information, a certain inertia in the consumption or the limited financial resources are factors that slow down the process towards an increased use of housing energy efficiency systems.

The main objective of our chapter is to determine the popularity of several key terms related to green retrofitting and energy efficiency in the online environment. With the help of Google Keyword Planner, we have determined the number of searches for the defined key terms, such as green retrofitting, energy efficiency, residential energy saving, building retrofitting, modernization, refurbishment, solar panel for the year 2017. The research has been done as comparison between the interest of German and Romanian users. The number of searches is a first step in determining the interest and preoccupation of users regarding a certain term. In the final part of the paper, we have also presented a best practice example for the implementation of a webpage, which should provide the necessary information for increasing the awareness of consumers related to the importance of energy efficient housing systems and green retrofitting. Only by knowing the advantages of such systems, will the attitude and behavior of consumers' switch towards a clean energy usage.

LITERATURE REVIEW REGARDING THE EFFECT OF INFORMATION SOURCES ON THE IMPLEMENTATION OF HOUSING ENERGY EFFICIENCY SYSTEMS

Factors Influencing the Pro-Environmental Behavior

Different researches have tried to determine the factors that affect the worldwide energy consumption. The standard of living and the development of technology are assumed to be factors that increase the energy consumption of households, but there are also researches which contradict this assumption (Stoppok, et al., 2018). On one hand in developed countries in a household there are definitely more domestic appliances that use energy, but on the other hand technological innovations have led to more modern appliances with a better energy performance. The use of energy resources with an inefficient technology can use more energy than a large number of running electric devices (Stoppok, et al., 2018).

In spite of the existing technological innovations, the willingness of consumers to use more efficient energy resources depends to a great extent to their willingness to adopt a pro-environmental behavior and to develop energy efficient consumption patterns. Different demographic factors (Volland, 2016; Mukherjee & Chakraborty, 2013), such as: the income (Cayla et al., 2015), the education (Sardianou & Genoudi, 2013) or the cultural background (Beunder & Groot, 2015; Pelau & Pop, 2018) play an important role in the willingness of consumers to adopt an energy efficient behavior. The price (Hast Alimohammadisagvand & Syri 2015; Yoo & Kwak, 2009) and the easiness of implementing more efficient resources (Lauren et al., 2016) are other important incentives for the development of an energy efficient consumption pattern.

Despite these demographic factors that favor the pro-environmental behavior, it is important for consumers to know about the existence of energy efficient devices (Iwata, et al., 2015). They should be also informed in order to adopt a behaviour that favors the reduction of the energy consumption. It is important to increase the consumer's awareness regarding the negative effects associated with energy waste on our planet (Soon & Ahmad, 2015) and to develop an environmentally friendly attitude (Claudy et al., 2013; Tantau & Staiger, 2018). Therefore, it is important to create different information and communication sources in the online and offline environment in order to determine the consumers to use energy efficient technologies. Knowing the demographic and the other factors that affect the pro-environmental behavior can determine decision-makers to adopt the proper communication strategies in order to change their beliefs and behaviors. For instance, researches have shown that more educated people have a higher willingness to recycle products. Consequently, in order to increase the recycling rate in a society, decision-makers should focus on one hand on the educated people to intensify their pro-environmental behavior. On the other hand, the increase of education, with focus on pro-environmenta education, will probably have an impact on the recycling rate (Pelau & Chinie, 2018).

There are different typologies of pro-environmental behavior. There are people who recycle products, there are people who eat and use bio-products, while other people are trying to reduce the energy consumption. For the energy consumption there are also different typologies of behavior. One may have an energy-friendly behavior by trying to reduce the energy consumption of different household devices. Other people may use different technologies and devices in order to reduce the energy waste. One of the most debated topics in this sense is the increased energy efficiency of housing and the different means by which it can be achieved. An increased energy efficiency of housing has not only a positive impact on the environment, but at the same time it also contributes to reduced energy costs for its consumer and consequently it leads to its wellbeing (Jagarajan et al., 2017). Therefore, it is convenient for the consumers to invest in different means for increasing the energy efficiency of their housing.

Green Retrofitting and Housing Energy Efficiency

One of the methods of increasing energy efficiency of existing housing is green retrofitting. It refers to different means to upgrade the characteristics of existing buildings with the purpose of increasing its sustainability (Jagarajan et al., 2017). For the new built buildings and homes, it is easy to develop energy efficiency schemes, as these can be planned and implemented from the beginning. For existing older buildings several modernization and refurbishment schemes have been developed in order to increase their sustainability and to offer energy efficient and adequate costs for the wellbeing of their inhabitants.

According to some authors a new industry of Energy Efficiency Retrofit Services (EERS) has emerged (Gooding & Gul, 2016) for offering upgrade solutions for older buildings.

The new industry of Energy Efficiency Retrofit Services (EERS) includes several activities and schemes which refer to the whole building or to just some parts of it. On one hand, there are different services which replace old parts of a building with newer, innovative and more efficient ones. On the other hand, there are different means of creating solar shadings and green roofs for increasing the thermal comfort in buildings. According to Filippi (2015) among the elements that contribute to an increased energy performance of housing, there are the enveloping of buildings, the substitution of old inefficient parts in the air conditioning and lightning system, the replacement of air conditioning with natural ventilation or the reduction of water consumption and the use of recycled materials. Gooding and Gul (2016) confirm these schemes by associating green retrofitting with refurbishment, installation and maintenance of existing equipment and materials (Gooding & Gul, 2016). Window retrofitting (Litti et al., 2018) plays a key role in the process of increasing the thermal comfort of a building and it is important to determine which parts of the window should be replaced. Especially in historic buildings, the retrofitting of windows may influence the market value of the property and its marketability (Litti et al., 2018). Besides the classic schemes, there are also disruptive technologies for building retrofitting such as heat capture and storage materials, high performance thin installation, controllable optical films for windows, modular constructions and phase change materials (Dixon et al., 2018).

Thermal comfort of buildings doesn't refer only to keeping high temperatures in the cold winter times, but also to the cooling of buildings in the warm summer time. There are opinions that affirm that in warm climate regions, more energy is being used for cooling the buildings in comparison to heating them. Therefore, high energy efficiency refers also to different means of reducing the energy cost for cooling the buildings. Solar shadings (El-Darwish & Gomaa, 2017), green roofs and walls with rain water retention or with vegetation layers (Cascone et al., 2018) are good solutions for this problem. Especially green roofs play an important role in the reduction of heat island effect by adjusting the roof temperature at high outside daily temperature changes. Moreover, they have a role for the biodiversity by cleaning the rain water and refreshing the surrounding air (Cascone et al., 2018).

Despite the fact that it is obvious that green retrofitting has a great impact on the energy policy as it reduces the energy waste, it is also important to develop a strategy for its implementation (Baldwin et al., 2018). As any other concept, green retrofitting creates both opportunities and challenges. One of the main advantages of green retrofitting is the fact that it assures the energy security of a country and besides, it plays a social role by creating a comfortable environment. Among the most important challenges is the financial cost and the long payback-time of green retrofitting projects (Jagarajan et al., 2017). Therefore, besides the necessity of having innovative energy-saving technologies, it is also important to develop economic models and communication tools for motivating the population to use and to invest in the energy-efficient tools (Baldwin et al., 2018). According to Gooding and Gul (2016), the green deal is one of the most efficient methods to determine both the industry and the population to retrofit their homes. The green deal allowed home owners to retrofit their owned buildings, without making and upfront payments (Gooding & Gul, 2016).

Communication Means and Information Sources

Communication is an important trigger of human evolution and is essential for our coexistence. The complexity of human communication and the need to deliver information and messages to large groups of people generated throughout time, the need for mass communication instruments. Communication and information sources are also important for increasing a positive attitude of consumers towards green retrofitting. Knowing more about the advantages of green retrofitting systems, might determine consumers to adopt the more easily. Iwata et al. (2015) also affirm that the pro-environmental behaviour in general is determined by the information provided to the consumer. Moreover, consumers who have access to more education, and hereby to more information, are also more likely to have a more pro-environmental behavior (Pelau & Chinie, 2018).

Communication and information sources can occur on several communication means and channels with the purpose of reaching a high number of target groups, with different backgrounds and demographic characteristics. The traditional communication channels are television, radio, newspapers and print media. In opposition to these traditional means of communication, the modern interaction happens with the help of the internet and mobile devices, through social media networks, search engines and other platform or applications.

With the rising of the Internet, consumers now enjoy the benefits of high-tech mass media, which is not only faster than traditional media, but offers a broader platform portfolio. Mobile phones, computers and the internet are often referred to as "New Age Media." The internet has enabled multiple mass-communication opportunities including e-mail, web pages, podcasts, virtual forums, electronic books, blogging, online TV, and many others instruments that are showing a positive development today. Furthermore, it has facilitated the invention and implementation of social media platforms. These have redefined mass communication. Facebook, Twitter, Instagram and YouTube have simplified and redesigned the communication, access, transmission and dissemination of information (Fill, 2009). This is because the new technologies and applications of social media give the opportunity to publish and share written content, photos, videos and various links. New communities based on common interests can be created. Furthermore, the individual contributions and feedback are promoted and the participation in different campaigns, through comments and information exchange are possible (Tantau et al., 2018). For this reason, for the increased popularity of green retrofitting, it is also important for be present with messages and information in the new-age communication means.

More and more, in recent years, most of the daily social and economic activities will take place online. Traditional communication has already been replaced by online alternatives. Businesses in general and specific for green retrofitting, need to adapt to these changes in order to survive on a constantly changing market, characterized by intense competition. They need to build trust and loyalty. The social media platforms represent a suitable opportunity for a solid positioning on the market and for increasing the customer base, and thereby the revenue. Online marketing provides a direct approach to customers and represents a radical change in how Internet users and businesses discover, read, and share news, information or thoughts (Pickton & Broderick, 2005). Among the most important online communication platforms are Facebook, LinkedIn, YouTube, Twitter and Instagram. Each year a growing number of users are aware of this evolution.

Facebook is the most popular social media platform and at the beginning of 2018 it has registered more than 2.19 billion monthly active users worldwide. In Romania there are about 9.6 million people who are active on Facebook. Gender distribution is in perfect balance, while most users are in the 25-49 age group (Facebrands, 2018). The platform developed in the USA has several applications. On one hand, it enables people living in different parts of the world to communicate and exchange information, written content, graphic and video materials, regardless of geographical or temporal boundaries. It is the best platform to meet old or potential new friends. Furthermore, the registered users have the opportunity to organize themselves in groups of common interests and to adjust their content according to preferences. Companies have quickly recognized the opportunity to connect with customers via Facebook. That's why there are about 16 million Facebook corporate sites worldwide. Many companies are focusing their communication strategies on Facebook, even though, only 49% of marketing specialists think that their Facebook marketing is effective, and about 52% say they have registered a decline in their organic Facebook reach over the last year. Despite these negative results, 62% of the survey participants plan to increase their organic activities and advertising budgets next year (Social Media Marketing Industry Report, 2018).

Twitter, LinkedIn, Instagram and YouTube are other popular social media platforms. Twitter is a micro-blogging site that allows interaction and feedback between different people. At first it gained popularity among celebrities. Today, the number of ordinary users grew a lot and specialists expect a future positive development of the number of registered users. Instagram was acquired by Facebook for $1 billion in 2012 and has been gaining popularity ever since. The platform allows users to share public or private photos and videos on the app or on other social media platforms. At the end of 2017, the application had over 800 monthly active users (Statista, 2018) and 5% of the marketing specialists have considered it the most important social platform (Social Media Marketing Industry Report, 2018). LinkedIn is used for professional networking, so that employers are able to find potential employees and it also enables the active users to expand their professional network. Microsoft bought the platform in 2016 for $26.4 billion. According to the Social Media Marketing Industry Report 2018, LinkedIn is the second most important social media for marketing purposes. 12% of the marketing specialists consider it the most important social media platform for marketing purpose, while this percentage increases for the business-to business field (Social Media Marketing Industry Report 2018).

YouTube is a website where you can upload and watch video content. From music clips to educational materials, users can find their desired videos based on their preferences. Although, all of the above-mentioned platforms meet different needs of active online users, Facebook remains the most popular social media channel. According to the "Social Media Marketing Industry Report" published in 2018, more than two-thirds of marketers have chosen Facebook as their most important social media platform. Facebook is followed by LinkedIn, Instagram and Twitter. Furthermore, online marketing specialists claim that they will increase their future advertising budgets on Facebook, Instagram and Facebook Messenger. A significant 87% of all marketers claim that their efforts on social media platforms have led to an increased online visibility of their companies. A disadvantage of using social media platforms for advertising concerns the uncertainty regarding the measurement of the generated results. According to the mentioned study, only 44% of the specialists agreed that they are able to measure their activity on social media. The ROI problem has always accompanied specialists in recent years. In 2017, 38% said they could measure ROI.

Being present on the social media, allows green retrofitting companies and projects to be more present in the online environment and consequently more present in the results of search engines. Besides their presence on the internet, stimulates different unexperienced and uninformed consumers to find out more information about the advantages of green retrofitting projects. For this reason, it is interesting to observe which has been the interest of consumers towards different key words related to green retrofitting.

Table 1. Key terms used in the research and their German and Romanian translation

Key term in English	Key term in German	Key term in Romanian
Green retrofitting	Grüne Nachrüstung	Modernizare verde/ modernizare ecologica
Energy efficiency	Energieeffizienz	Eficientă energetică
Residential energy saving	Energiesparen Hasushalt	Reducere consum energie
Building retrofitting	Gebäudesanierung	Reabilitatea clădiri
Modernization	Modernisierung	Modernizare
Refurbishment	Sanierung	Retehnologizare
Solar panel	Solarpanel	Panouri solare
Maintenance of equipment	Gerätewartung	Întretinere echipamente
Window retrofitting	Fenster nachrüsten	Modernizarea ferestrer
Green roofs	Dachbegrünung	Acoperis verde
Rain water retention	Regenwasser-rückhaltung	Reciclarea apei de ploaie
Wall vegetation layer	Vegetationsschicht der Wand	Strat vegetal perete
Solar shading	Sonnenschutz Gebäude	Umbrirea solară a clădirilor

METHODOLOGY OF RESEACH

The objective of our research is to make a comparative analysis of the popularity of several concepts related to green retrofitting in the German and Romanian online environment. Therefore, we have aimed to identify the number of searches in the online environment for some green retrofitting concepts. The first step in the analysis has been to identify the key concepts related to green retrofitting and to properly translate these in German and Romanian. The results of this first analysis can be observed in Table 1. For the identified key concepts, we have measured the number of searches with the help of Google Keyword Planner.

In the following the key terms used in the analysis are being presented. Green retrofitting is a method of increasing the energy efficiency of existing housing, by upgrading its characteristics with the purpose of reaching its sustainability (Jagarajan, et al., 2017). With increasing interest for a more efficient and sustainable energy policy, the discussion about the population's possibility to live in comfortable and affordable homes has received increased attention (Gooding & Gul, 2016). Especially for existing and historical buildings, it is important to improve their characteristics in order to correspond to the present expectations of the population (Litti et al., 2018). Energy efficiency is the process by which certain devices, buildings or other products or systems are designed and improved with the purpose of reducing energy consumption. It is a general term that includes several measures of green retrofitting. Residential energy saving describes the process implemented by the final consumer in order to reduce the energy consumption at the final consumer. Several researches have shown that a big part of the current energy consumption is determined at the household level (Filippi, 2015; Gooding & Gul, 2016). Therefore, it is important to find solutions for a more efficient energy use at household level. Building retrofitting refers to the modernization of buildings with the purpose of improving their energy efficiency and reduce the energy losses. New buildings are constructed with innovative materials and modern technologies which preserve energy in a more efficient manner. As it is difficult to demolish and re-build old buildings, it is important to find solutions to improve their energy demand. Building retrofitting allows the modernization of existing buildings, from basic renewal to extensive re-modellation, where several components of the building are replaced (Baldwin et al., 2018). Modernization refers to the replacement of old technologies and materials with newer ones, with improved characteristics and a better energetic performance. Refurbishment is similar to modernization the process by which older technologies are replaced with newer ones. The maintenance of equipment refers to the permanent check of technologies with the purpose of assuring the preservation of their characteristics. Solar panels are green technologies which transform the solar energy into electric energy. Nowadays these are one of the most popular type of renewable energies. Window retrofitting refers to the replacement of old windows with more energy efficient versions. Several studies have shown that a big part of the energy loss of a building is caused by inefficient windows. Therefore, the replacement of old windows with newer versions with double or triple glazed layers can reduce the energy demand in a significant way (Litti et al., 2018). Green roofs are vegetation layers raised on roofs with the purpose of improving the energy performance of a building by an increased water retention. Usually these roofs are formed from several organic and inorganic layers that have a contribution to the isolation of the roof, by reducing the temperature fluctuations of the roof membrane. Besides, green roofs have several benefits for the environment such as the reduction of urban heat island effect, optimization of the bio-diversity and the cleaning of runoff water and surrounding air (Cascone et al., 2018; Feitosa et al., 2018). Rain water retention is the process by of collection and reuse of rain water. It also has an important contribution to climate change as it reduces the demand for clean water and therefore has a positive impact on the environment. Wall vegetation layer describes vegetation grown on walls with the purpose of improving its energetic characteristics. These have similar benefits as the green roofs, with the observation that they are applied on walls (Feitosa et al., 2018). Solar shading is similar to green roofs and wall vegetation layer, a method by which the building temperature is adjusted with the help of the vegetation shadowing.

Taking into consideration the fact that Germany has a larger number of internet users (64.4 million users in October 2018) in comparison to Romania (11.4 million users in October 2018), we have developed a popularity indicator, which allows us to compare the number of searches of the key terms generated by the internet active population. Therefore, the following popularity index has been defined:

$I_{p,ic} = N_{p,ic} / N_{u,c} * 100.000$ (1)
Where $I_{p,ic}$ = Popularity indicator for item i (from Table 1) and country c
$N_{p,ic}$ = Number of searches for item i in country c
$N_{u,c}$ = Number of internet users in country c

In the following chapters, there can be observed the comparative analysis of the popularity of key terms related to green retrofitting based on the values of the popularity indicator $I_{p,ic}$. The results are based on a Google Keyword Planner analysis processed in October 2018, based on data from the year 2017.

Table 2. Descriptive results of the average monthly searches for key terms related to green retrofitting

Key term in English	Key term in German	Average monthly searches Germany 2017*	Average monthly searches Romania 2017*	Key term in Romanian
Green retrofitting	Grüne Nachrüstung	n/a	n/a	Modernizare ecologica
Energy efficiency	Energieeffizienz	5400	260	Eficientă energetică
Residential energy saving	Energiesparen Hasushalt	40	10	Reducere consum energie
Building retrofitting	Gebäudesanierung	320	20	Reabilitatea clădiri
Modernization	Modernisierung	2900	70	Modernizare
Refurbishment	Sanierung	6600	n/a	Retehnologizare
Solar panel	Solarpanel	8100	12100	Panouri solare
Maintenance of equipment	Gerätewartung	1900	n/a	Întretinere echipamente
Window retrofitting	Fenster nachrüsten	170	n/a	Modernizarea ferestrer
Green roofs	Dachbegrünung	5400	n/a	Acoperis verde
Rain water retention	Regenwasser-rückhaltung	140	n/a	Reciclarea apei de ploaie
Wall vegetation layer	Vegetationsschicht der Wand	n/a	n/a	Strat vegetal perete
Solar shading	Sonnenschutz Gebäude	20	n/a	Umbrirea solară a clădirilor

*Data from October 2018; n/a = not available

RESULTS OF THE RESEACH

Descriptive Results of the Search Interest of Consumers for Green Retrofitting and Housing Energy Efficiency in Germany and Romania

A descriptive analysis of the search results in Google Keyword Planner (for October 2018) for data from the year 2017, shows that several key terms related to green retrofitting are more popular in Germany in comparison to Romania, as Table 2 illustrates. It can be easily observed, that for some translations of key terms related to green retrofitting, there are only few or no searches. Especially for key terms referring to vegetation such as green roofs, wall vegetation layer, solar shading or rain water retention, there are no searches for Romania for these terms. Although they appear as search words in Germany their popularity is very low.

Key terms related to innovation such as refurbishment or maintenance of equipment are also not searched in the Romanian online environment. The Google Keyword Planner search has shown no results regarding these concepts. For the German translation of this concept, there have been only a few searches. It is also interesting to observe that although the concept green retrofitting is familiar in the English-speaking world, it has no established translation in German or in Romanian. Therefore, a future research preoccupation could aim to find an adequate concept.

Online Information Sources for Green Retrofitting and Housing Energy Efficiency in Germany and Romania

In order to analyse the importance of the concepts described in Table 1 for the Romanian internet users searching for retrofitting solutions in comparison to the German ones, we have generated and interpreted the search volumes of the above-mentioned terms for the period 1 January – 31 December 2017. In the mentioned period, the average monthly search volume for all analysed terms has been 40.890 in the German online environment and 12.570 times in the Romania one. As Figure 1 shows, the overall search volume for green retrofitting related terms is higher in Germany, but according to the results obtained with the help of the popularity indicator, which allows us to compare the generated search volumes for the two countries, there is a higher interest for green retrofitting solutions among Romanian internet users active on Google.

Surprisingly, the higher overall interest among Romanian users is influenced by an increased search volume for specific green retrofitting related terms like solar panels. In general, the German public has registered a moderate interest for a larger variety of solutions in comparison to the Romanian one, which has an overall higher interest, but for a smaller number of terms. In this context, in 2017, there were no Google searches for the terms refurbishment, maintenance of equipment, window retrofitting, green roofs, rain water retention and solar shading in the Romanian online environment, unlike in the German one, where monthly searches for the mentioned concepts have been registered.

Further on, we have analysed the relative search volumes of terms in both Romanian and German online environment. As can be seen in Figure 2, throughout the year 2017, there has been a higher interest for energy efficiency solutions in Germany in comparison to Romania. The relative calculated average monthly interest among German users regarding energy efficiency is 3.69 higher in comparison to the one registered in the Romanian online environment. It is also interesting to observe that the concept "energy efficiency" has been more popular for both countries at the beginning of the year in comparison to the end of the year. For Germany the popularity index has had its maximum value in March 2017 (I_p=12.58) and it has decreased by the end of the year to a popularity index of 5.59 in December 2017. A similar

Figure 1. Average monthly search results regarding the analysed terms in Germany and Romania

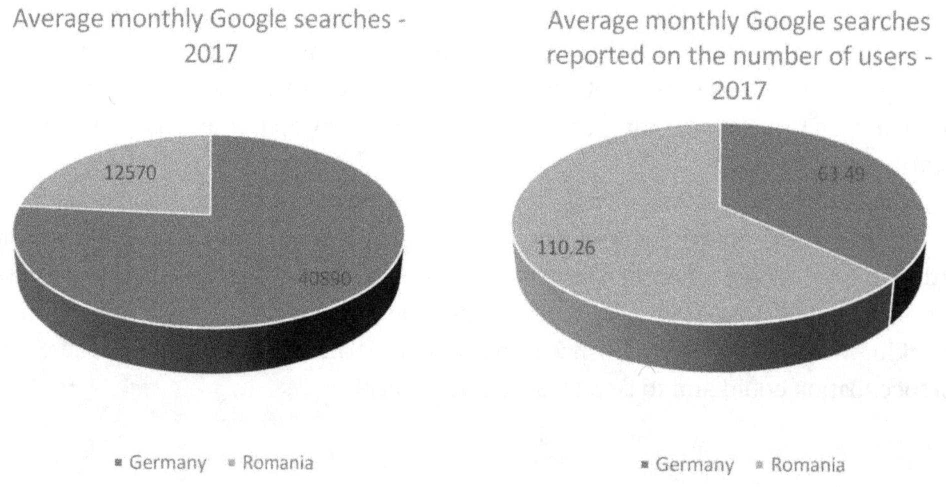

Figure 2. Relative search volume for the term "energy efficiency" in Germany and Romania

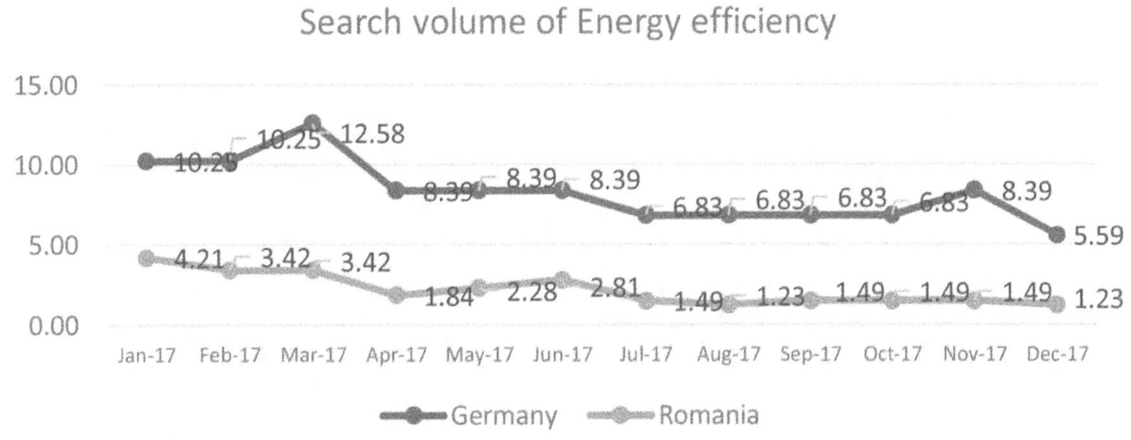

search behaviour has been valid for Romania. The popularity index for searches in Romania has decreased from its maximum value of 4.21 in January 2017 to its minimum value of 1.23 in December 2017.

Table 3 shows the monthly registered number of searches in the analysed period and the relative search volume registered in both countries regarding energy efficiency.

Figure 3 shows the relative search volume associated with the two analysed countries for the key concept "residential energy saving". One can observe, that the interest for energy saving solutions is low for both countries, but it has higher values for Romania in comparison to Germany. For Romania the value of the popularity indicator has an almost constant value of 0.09. The only exceptions have been registered for June, July and December 2017, when no internet user has searched for this concept. One possible explanation could be the fact that these months are often associated with holidays (summer or winter) and probably consumers are less preoccupied by energy saving. For Germany, the highest interest for energy saving has been registered in January 2017 ($I_p = 0.11$), while it reaches its lowest values, similar to Romania in July and December 2017 ($I_p = 0.03$).

In Table 4, there are presented the number of searches associated with energy saving solutions both in relative volume registered in the two countries. In the analysed period, the average relative search volume in Romania was 0,087, while in Germany it only registered 0,062.

Table 3. Google search results regarding energy efficiency registered in Germany and Romania in 2017

Keyword	Jan-17	Feb-17	Mar-17	Apr-17	May-17	Jun-17	Jul-17	Aug-17	Sep-17	Oct-17	Nov-17	Dec-17
Relative search volume in Germany	10.25	10.25	12.58	8.39	8.39	8.39	6.83	6.83	6.83	6.83	8.39	5.59
Relative search volume in Romania	4.21	3.42	3.42	1.84	2.28	2.81	1.49	1.23	1.49	1.49	1.49	1.23
Searches in Germany	6600	6600	8100	5400	5400	5400	4400	4400	4400	4400	5400	3600
Searches in Romania	480	390	390	210	260	320	170	140	170	170	170	140

Figure 3. Relative search volume of the term "residential energy saving" in Germany and Romania

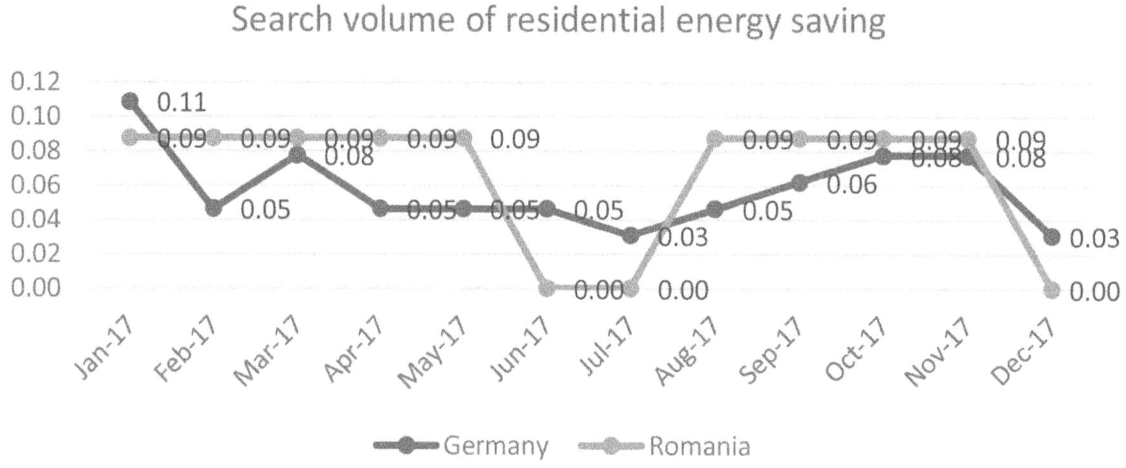

Table 4. Google search results regarding residential energy saving registered in Germany and Romania in the analysed period

Keyword	Jan-17	Feb-17	Mar-17	Apr-17	May-17	Jun-17	Jul-17	Aug-17	Sep-17	Oct-17	Nov-17	Dec-17
Relative search volume in Germany	0.11	0.05	0.08	0.05	0.05	0.05	0.03	0.05	0.06	0.08	0.08	0.03
Relative search volume in Romania	0.09	0.09	0.09	0.09	0.09	0.00	0.00	0.09	0.09	0.09	0.09	0.00
Searches in Germany	70	30	50	30	30	30	20	30	40	50	50	20
Searches in Romania	10	10	10	10	10	0	0	10	10	10	10	0

The relative importance of the building retrofitting is another concept that has been analysed from the perspective of the internet users active in both countries. As Figure 4 illustrates, this concept has registered an average of 2.7 more searches in Germany in comparison to Romania. In 2017, the average relative search volume regarding residential energy saving has registered a popularity index of 0.5 in Germany and 0.18 in Romania. For both countries there can be observed a decreasing trend from January to August 2017 and an increase in the last four months of the year (September-December 2017). In the first part of the year, in the decreasing period, the maximum value of 0.5 is reached in January, March and May, while the minimum value of 0.33 is reached in August and September 2017. The overall maximum value of 0.75 is reached for Germany in November and December 2017. For Romania, there is a more constant trend, having a decrease from the maximum value of 0.35 in February 2017 to the minimum value of 0.09 in July and September 2017. After September the number of searches increases by reaching its maximum value for the last months of the year, in December 2017. The highest difference between the two analysed countries regarding building retrofitting solutions searches has been registered

Figure 4. Relative search volume of the term "building retrofitting" in Germany and Romania

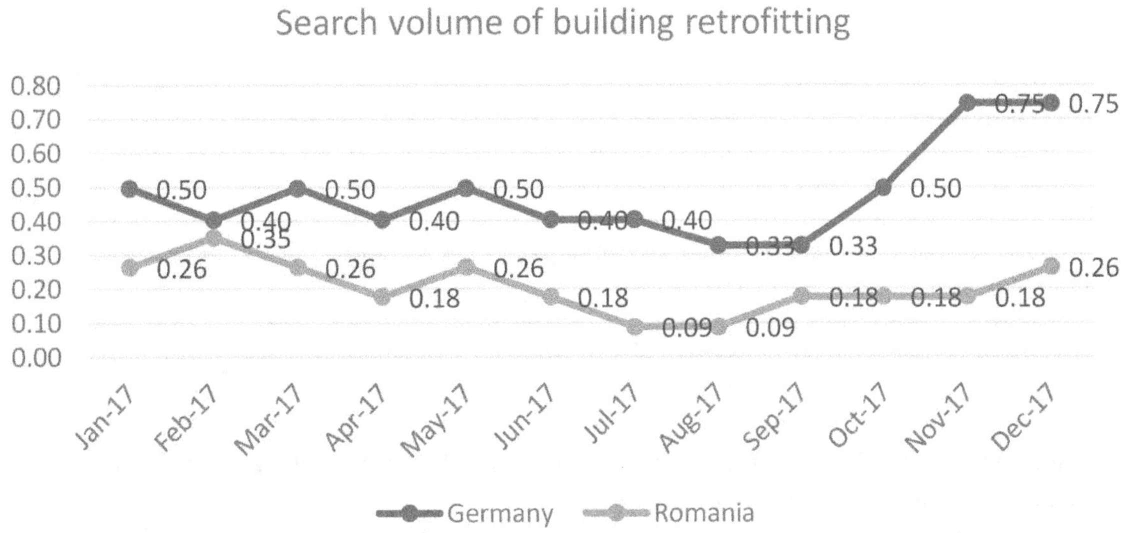

in December 2017, as it can be observed in Figure 4. These results show that the highest interest for building retrofitting has been registered in the cold winter months.

Table 5 presents the number of searches associated with building retrofitting solutions, and also the relative volume registered in the two countries.

For the key term modernization, there can be observed a constant interest through the year. The number of searches for this key concept is higher in comparison to residential energy saving or building retrofitting. For Germany the values of the popularity index range between the minimum value of 3.73 in the summer months June, July and August 2017 and the maximum value of 5.59, reached in March and November 2017. For Romania, there is a higher interest in modernization at the beginning of the year, having a popularity indicator of 1.84 in January and 1.23 in February. The value of the indicator decreases for the summer period, reaching its minimum value of 0.26 in August 2017. For the period September, October and November there is a slight increase to a popularity value of 0.61, followed by a

Table 5. Google search results regarding building retrofitting registered in Germany and Romania in 2017

Keyword	Jan-17	Feb-17	Mar-17	Apr-17	May-17	Jun-17	Jul-17	Aug-17	Sep-17	Oct-17	Nov-17	Dec-17
Relative search volume in Germany	0.50	0.40	0.50	0.40	0.50	0.40	0.40	0.33	0.33	0.50	0.75	0.75
Relative search volume in Romania	0.26	0.35	0.26	0.18	0.26	0.18	0.09	0.09	0.18	0.18	0.18	0.26
Searches in Germany	320	260	320	260	320	260	260	210	210	320	480	480
Searches in Romania	30	40	30	20	30	20	10	10	20	20	20	30

Figure 5. Relative search volume of the term "modernization" in Germany and Romania

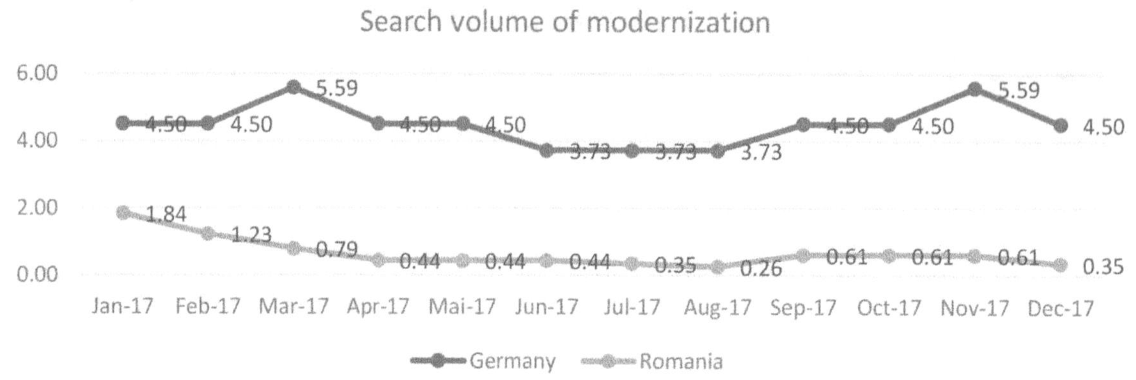

lower value in December 2017. This result also shows that in winter time and especially at the beginning of the year, there is a higher interest for modernization.

In Table 6, there can be observed the number of searches and the relative volume registered associated with modernization for Germany and Romania.

For the concept refurbishment associated with green retrofitting there can be observed a different interest in the two analyzed countries. For Romania, the interest in refurbishment is almost constant along the year, having no big differences between the period of maximum and minimum interest. As it can be observed in Figure 6, the maximum interest of 1.23 has been reached at the beginning of the year in January, February, and March, and in October 2017. For the summer period, the interest of concepts regarding refurbishments has diminished to a popularity index of 0.79 and it has reached the minimum value of 0.61 in December 2017. For Germany, the interest for this concept has been higher in comparison to Romania. Similar to the other analyzed key concepts related to green retrofitting, there is an association to the vacation time. The interest for the key term refurbishment has had a constant value of 15.37 for the period January to Mai 2017 and October and November 2017. The popularity index had a lower value of 12.58 for the vacation time for the period June to September 2017 and in December 2017.

The number of searches and the relative volume associated with refurbishment can be observed in Table 7.

Table 6. Google search results regarding modernization registered in Germany and Romania in 2017

Keyword	Jan-17	Feb-17	Mar-17	Apr-17	May-17	Jun-17	Jul-17	Aug-17	Sep-17	Oct-17	Nov-17	Dec-17
Relative search volume in Germany	4.50	4.50	5.59	4.50	4.50	3.73	3.73	3.73	4.50	4.50	5.59	4.50
Relative search volume in Romania	1.84	1.23	0.79	0.44	0.44	0.44	0.35	0.26	0.61	0.61	0.61	0.35
Searches in Germany	2900	2900	3600	2900	2900	2400	2400	2400	2900	2900	3600	2900
Searches in Romania	210	140	90	50	50	50	40	30	70	70	70	40

Figure 6. Relative search volume of the term "refurbishment" in Germany and Romania

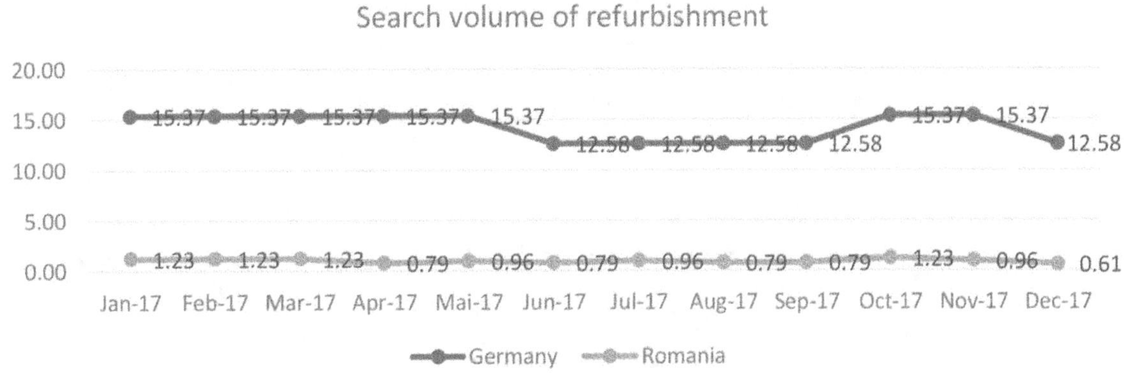

Table 7. Google search results regarding refurbishment registered in Germany and Romania in 2017

Keyword	Jan-17	Feb-17	Mar-17	Apr-17	May-17	Jun-17	Jul-17	Aug-17	Sep-17	Oct-17	Nov-17	Dec-17
Relative search volume in Germany	15.3	15.3	15.3	15.3	15.3	12.5	12.5	12.5	12.5	15.3	15.3	12.5
Relative search volume in Romania	1.23	1.23	1.23	0.79	0.96	0.79	0.96	0.79	0.79	1.23	0.96	0.61
Searches in Germany	9900	9900	9900	9900	9900	8100	8100	8100	8100	9900	9900	8100
Searches in Romania	140	140	140	90	110	90	110	90	90	140	110	70

Solar panel is the only key term related to green retrofitting that has been more popular in Romania in comparison to Germany. Solar panel in Romania has the highest popularity index compared to all terms analyzed in this paper. As it can be observed in Figure 7, the popularity index of the concept solar panel has values between 71.05 in September and December 2017 and the maximum value 129.81, which has been registered in the first half of the year 2017 (January to June). The interest for the term solar panel has been at a high constant level at the beginning of the year 2017. Despite the decrease which occurred during the summer time, the popularity values for this indicator, have been higher compared to other key terms related to green retrofitting. For Germany the popularity of the term solar panel has been lower in comparison to Romania, but at the same time, it has been higher compared to other key terms related to green retrofitting. In opposition to the other analyzed concepts, the interest for solar panels in Germany has been higher during the summer time, achieving its maximum value of 18.97 in June 2017. The minimum value of 8.39 has been reached in December. The high interest for solar panels in the summer time is probably related to the fact that the summer heat determines consumers to think of a more useful ways of using the solar energy. Therefore, they have searched for solar panels in the warm summer period.

In Table 8, there can be observed the number of searches and the relative volume registered for the key term solar panel in the two analysed countries.

Figure 7. Relative search volume of the term "solar panel" in Germany and Romania

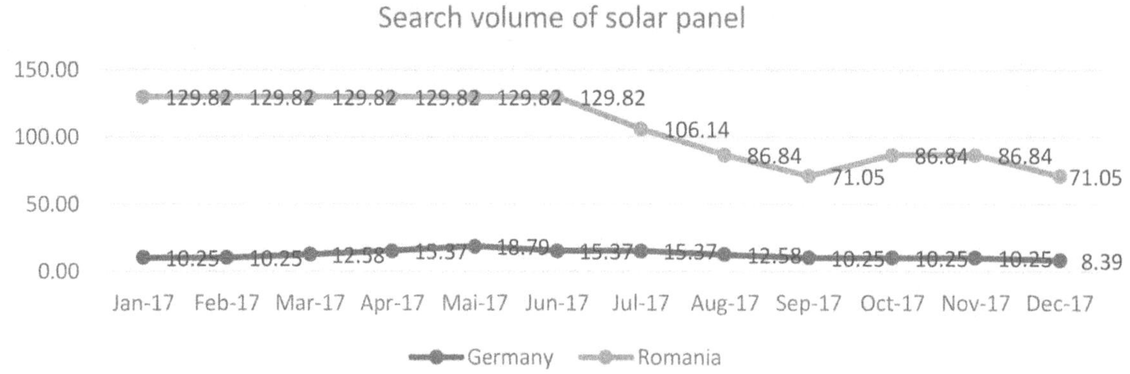

Table 8. Google search results regarding solar panel registered in Germany and Romania in 2017

Keyword	Jan-17	Feb-17	Mar-17	Apr-17	May-17	Jun-17	Jul-17	Aug-17	Sep-17	Oct-17	Nov-17	Dec-17
Relative search volume in Germany	10.25	10.25	12.58	15.37	18.79	15.37	15.37	12.58	10.25	10.25	10.25	8.39
Relative search volume in Romania	129.8	129.8	129.8	129.8	129.8	129.8	106.1	86.84	71.05	86.84	86.84	71.05
Searches in Germany	6600	6600	8100	9900	12100	9900	9900	8100	6600	6600	6600	5400
Searches in Romania	14800	14800	14800	14800	14800	14800	12100	9900	8100	9900	9900	8100

Best Practice Examples for Information Sources for Green Retrofitting and Housing Energy Efficiency in Germany

The 2018 Global Digital reports from *We Are Social* and *Hootsuite* show that the Internet is being used nowadays by more than 4 billion people worldwide. Moreover, the Internet has created a new medium which allows consumers to access information (Ratchford, Talukdar & Lee, 2007). Some researchers have claimed that the ultimate purpose of websites or digital communication channels is the supply of information. In recent decades, there has been tremendous progress in computer technology, electronics and telecommunications. In particular, important steps have been made in the storage, analysis and retrieval of large amounts of data. This, in turn, has led to an increase in database technology that allows companies to gather useful information about customers and their buying behaviour. The technological progress has led to larger databases, which, according to Pickton and Broderick (2005) has led to targeted and personal communication. The new media has not only increased the reach of companies, but also the ability to stay in touch and learn more about their customers.

As we reflect on the evolution of human society in recent decades, we can observe that the energy sector has an increased influence on the development of society as a whole. Emissions of greenhouse gases represent a serious threat with potential catastrophic consequences for humanity. The use of renewable energy sources (RE), combined with improved energy efficiency, can help reduce energy consumption, reduce greenhouse gas emissions and thus prevent an accelerated climate change.

The EU has adopted its own strategy to combat climate change by adopting a European sustainability growth plan, setting ambitious energy targets (the so-called 20-20-20 targets). The road to a low-carbon economy leads to the development of a local public sector that is able to identify and promote economic opportunities. In particular, the local public sector can play a strategic role in public policy. Therefore, in the field of sustainable energy, it is essential to strengthen the knowledge of people working in the local public sector. The 2020 Energy Strategy is a reference for the European strategic thinking regarding competitivity, sustainability and energy security, the first in a series of strategies that aim to place the European economy on a path of decarbonisation, with the ultimate goal of a green and sustainable economy. The strategy proposes a threefold objective: to reduce energy consumption, decrease carbon dioxide emissions and increase energy production from renewable sources. In this context, the consumer is the most important resource of the process (European Commission 2015).

In this regard, a general framework for informing the consumer is necessary. Therefore, in a situation in which the vast majority of the population is still passive to energy related subjects, consumers need to be supported in making informed choices in the energy market. Consumer should be informed about the efficiency of their consumption and about the alternatives they have on the market. Nowadays, the national energy sector in Romania faces challenges at global and local levels: security of the energy supply, increased competition and the reduction of environmental impact. Ensuring the energy supply through clean and competitive energy sources, while taking into account climate change, the rising global energy demand and the uncertain future of traditional energy sources, is one of the EU's biggest challenges.

The implementation of educational programs regarding the energy consumption is one of the most important measures that Romania and Germany, as well as other countries, can take. Not only is this a mean of providing consumers with information on how to improve their consumption, but this can also be used in order to understand the consumption patterns of households or to identify long-term solutions for the community. In the perspective of increasing the visibility of such an idea, the communication should cover an online component. This includes the development of a dedicated website, sustained by a solid presence on the above described social media platforms. The development of website dedicated to the education of energy consumers supports the 2020 Strategy (published by the European Commission in 2015) and its objectives, generating at the same time numerous benefits for the consumer.

In order to increase the consumer's knowledge regarding the reducing of energy consumption and alternatives available on the energy efficiency market, as mentioned above, we recommend the development of a dedicated website, as main communication channel. This development should be sustained by a powerful brand capable of gaining a major visibility in the online environment. In the following there are presented the main aspects which should be taken into consideration when this new communication tool has been developed.

Creating the Website / Gaining Attention

Attracting consumers/ users is a fundamental step which influences the success and performance of the communication tool. Suited keywords are an important part in the developing stage of the website and are implemented in order to attract internet users on the website. To achieve this objective, one tries to create and generate relevant content for the customers in accordance to their interests. Another objective is to retain the users as long as possible on the website and determine them to visit all the website's sections. The domain of the website and of course the campaign's name should be as catchy as possible and transmit the mission of the webpage. For the success of the webpage, the following aspects have to be taken into consideration:

- **Core Concept of the Website:** The development of a website which presents the necessary information in order to encourage consumers to take action and reduce their energy consumption by using renewable energy sources and green retrofitting alternatives. The goal is to help people identify the benefits of energy efficiency and encourage them to participate in an energy transition towards a reduced energy consumption based at national level. For this reason, the presented information must be based on concrete facts, passed on briefly and succinctly.
- **Target Audience:** The website targets household energy consumers by providing solutions to reduce energy consumption and its negative impact on the environment. In addition, the actions will focus on raising awareness and accountability at national level by presenting individual and collective measures that can be implemented in order to protect nature and the environment. Renewable energies sources and green retrofitting solutions will play a major role in the information process.
- **Website Structure:** On the homepage, internet users will discover an interactive map with regions depending on the level of energy produced from green sources. For example, regions coloured in dark green, will represent the highest energy production level from renewable sources. The darker the green colour, the higher the production. This is a good way to represent facts and statistics regarding the energy production in terms of renewable energy in different regions. Furthermore, the homepage aims to increase the interest of the visitors by providing interesting facts and content. The main idea is to encourage visitors to continue their search on the platform and gain more information. By arousing the visitor's interest the website secures a stable traffic which can further generate an increase in visits.
- **Vision:** We aim to reduce the negative impact of energy consumption on the environment by raising awareness regarding renewable energy, energy efficiency solutions and green retrofitting.
- **Mission:** We strive to inform energy consumers regarding green sources of energy and green retrofitting solutions which can determine a reduced energy consumption and energy bill. By responsibly consuming energy and efficiently adjusting the consumption curve, everyone can help to protect nature and implicitly reduce the negative impact on the environment while benefiting of reduced energy bills.

The biggest challenge is to attract users on the website and ensure an optimal information transfer by providing quality content, facts and statistics. In order to achieve this, several steps have to be passed:

- **Conversion Phase/ Arouse Interest:** In this phase the goal is to promote and adapt the key messages on the landing page and include various calls-to-action in the desire of stimulating the interest of the visitors and finally to convert these into "customers". On the website, several topics will be offered with the objective of arousing the visitors' attention and interest. The main sections of the website present news, tips for reducing the energy consumption and the newest trends in the field. By accessing the "Tips" section, the visitors will learn how to reduce their household energy consumption and how they can generate profit by investing in green energy and retrofitting solutions. The tips are divided into several areas adapted to daily activities which can contribute to an energy efficiency increase.

- **Desire Phase/ Stimulate Engagement:** At this stage, customers consider whether the product or service meets their needs and decide if they take action in the next step. This is a critical phase that determines the success of the entire process. In order to ensure a positive outcome, several contextual and relevant information can be sent to users. Consumer knowledge is developed on all levels of the process by presenting useful news, as well as gaining insight regarding the latest technologies available in the field and learning more about the preliminary national laws, but also about energy efficiency and retrofitting solutions. Visitors should be able to find useful information on the website. In the "Trends " section, readers will be able to discover the latest news regarding the energy sector, as well as innovations that will reduce the negative effects of non-renewable energy that prevail throughout the world. A special attention is also given to the national and international legal regulations. The goal is to attract as many users as possible on the website. Furthermore, by presenting the latest innovations associated with energy efficiency, users will understand how using and implementing these solutions can lead to long term benefits.
- **Action Phase:** In this phase, consumrs who had positive experiences by navigating the website will return and build loyalty, but above all, they will become promoters and word of mouth endorsers. In this situation, the social media networks play an important role, because by creating a stable presence on these platforms they can ensure a constant communication through the publishing of relevant information. There are several indicators that can help measure the success of the action phase. The reach delivers information regarding the number of people reached by the post at least once. Moreover, the most important social platforms offer the possibility of managing paid campaigns. The number of users reached by a paid ad can be influenced by budget, audience and competition. To cultivate the relationship with the audience, one can consider alternatives to increase the engagement of the page visitors. Possible interactions are considered: clicks, comments, likes, shares, and so on. The engagement demonstrates the closeness of the relationship with the audience and the relevance of the information to the page visitors. There are also indicators that tell more about the effectiveness of an ad, such as click through rate. This indicator influences the relevance value and it represents the percentage between the number of users who were shown the ad and the number of generated clicks. Another sign of an increased interest for a topic can be analysed through number of website visits or number of active fans on the Facebook page. This indicator gives an overview regarding the community evolution and its activity as well as the popularity of the brand. By analysing the above mentioned indicators you can find out if people are actually interested in the content of the pages and if they are eager to learn more about the topic.

DISCUSSIONS AND CONCLUSION

In spite of the fact that there is a certain interest for several key terms related to green retrofitting and energy efficiency, their popularity could be increased even more. The results of our research have shown that none of the researched key terms has reached 10.000 searches within one month. This shows that, in spite of the importance of these terms for the energy strategy for the two analyzed countries, the products and services which increase the energy efficiency are not familiar to the large public. Therefore, several communication and information instruments should be implemented in order to increase their popularity and the consumer's interest for them. In these communication tools, it is important to develop topics that are appealing to the consumers so that they are attracted by the webpage and that they are determined to act.

The results of our research have shown that the consumers have a higher interest for several key concepts which are related to increasing the temperature for housing and less for the cooling process of housing, which can occur in hot summer periods. One of the most familiar and most searched key terms is solar panel. Although the concept has been very popular for both analyzed countries, it is the only term which is more popular in Romania compared to Germany. Refurbishment and energy efficiency are also more popular in comparison to other key terms related to green retrofitting. As mentioned before, several terms referring to different cooling procedures of houses are rather unfamiliar. Especially terms related to roof or wall vegetation, as well as solar shading are rather unknown, as the results have indicated few searches for Germany and no searches for Romania in the analyzed period. Even our main term "green retrofitting" should be more promoted among consumers, as there is no familiar translation in German or Romanian for this term. In this sense, future research should focus on defining and establishing a translation for this important term related to increasing the energy efficiency of housing.

The limitation of our research refers to the fact that the analysis has been done for only one year, therefore it is difficult to establish the dynamic of these terms in time. Despite this aspect, it was easy to observe that, with the exception of solar panels, most of the key terms related to green retrofitting are more searched in the winter period or at the beginning of the year and less during summer or Christmas vacation. Solar panel has been the only term which was more searched during the sunny months. Therefore, it is important to know the periods of the year, when to start different information campaigns for increasing the awareness and usage of energy efficient methods by the consumers. Green retrofitting and energy efficiency are key concepts for assuring a sustainable energy strategy for a country. Therefore, it is important to increase the promotion of these concepts and their importance for the energy security as well as for the consumers' housing thermic comfort. Only by doing this, consumers can be convinced to implement such instruments in their homes and by doing this to assure the energetic stability of their country.

Despite some initial barriers related to high initial cost, reduced policy support, lack of experience for the implementation and low consumer's awareness, green retrofitting will have an important role in the future developments. The need for a higher energy performance of buildings will determine both decision makers, municipalities, businesses and private households to invest more in several retrofitting measures. In spite of the increased initial cost, green retrofitting measures will show their financial efficiency throughout their lifecycle and therefore the awareness about its importance will increase. Moreover, the increasing expectations of the population will also determine a higher demand for retrofitted buildings.

REFERENCES

Baldwin, A. N., Loveday, D. L., Li, B., Murray, C., & Yu, W. (2018). A research agenda for the retrofitting of residential buildings in China – A case study. *Energy Policy*, *113*, 41–51. doi:10.1016/j.enpol.2017.10.056

Beunder, A., & Groot, L. (2015). Energy consumption, cultural background and payment structure. *Journal of Cleaner Production*, *94*, 137–143. doi:10.1016/j.jclepro.2015.01.083

Cascone, S., Catania, F., Gagliano, A., & Sciuto, G. (2018). A comprehensive study on green roof performance for retrofitting existing buildings. *Building and Environment*, *136*, 227–239. doi:10.1016/j.buildenv.2018.03.052

Cayla, J. M., Maizi, N., & Marchand, C. (2011). The role of income in energy consumption behaviour: Evidence from French households' data. *Energy Policy*, *39*, 7874–7883. doi:10.1016/j.enpol.2011.09.036

Claudy, M. C., Peterson, M., & O'Driscoll, A. (2013). Understanding the Attitude-Behavior Gap for Renewable Energy Systems Using Behavioral Reasoning Theory. *Journal of Macromarketing*, *33*(4), 273–287. doi:10.1177/0276146713481605

Dixon, T., Lannon, S., & Eames, M. (2018). Reflections on disruptive energy innovation in urban retrofitting: Methodology, practice and policy. *Energy Research & Social Science*, *37*, 255–259. doi:10.1016/j.erss.2017.10.009

El-Darwish, I., & Gomaa, M. (2017). Retrofitting strategy for building envelopes to achieve energy efficiency. *Alexandria Engineering Journal*, *56*, 579–589. doi:10.1016/j.aej.2017.05.011

European Comission. (2015). Europe 2020 Strategy. Retrieved from: https://ec.europa.eu/info/business-economy-euro/economic-and-fiscal-policy-coordination/eu-economic-governance-monitoring-prevention-correction/european-semester/framework/europe-2020-strategy_en

Facebrands. (2018). Retrieved from www.facebrands.ro

Feitosa, R. C., & Wilkinson, S. J. (2018). Attenuating heat stress through green roof and green wall retrofit. *Building and Environment*, *140*, 11–22. doi:10.1016/j.buildenv.2018.05.034

Filippi, M. (2015). Remarks on the green retrofitting of historic buildings in Italy. *Energy and Building*, *95*, 15–22. doi:10.1016/j.enbuild.2014.11.001

Fill, C. (2009). *Marketing Communications: Interactivity, Communities and Content*. Financial Times/Prentice Hall.

Gooding, L., & Gul, M. S. (2016). Energy efficiency retrofitting services supply chains: A review of evolving demands from housing policy. *Energy Strategy Reviews*, *11-12*, 29–40. doi:10.1016/j.esr.2016.06.003

Hast, A. Alimohammadisagvand, & B, Syri, S. (2015). Consumer attitudes towards renewable energy in China - The case of Shanghai. *Sustainable Cities and Society*, *17*, 69–79. doi:10.1016/j.scs.2015.04.003

Iwata, K., Katayama, H., & Arimura, T. H. (2015). Do households misperceive the benefits of energy-saving actions? Evidence from a Japanese household survey. *Energy for Sustainable Development*, *25*, 27–33. doi:10.1016/j.esd.2014.12.005

Jagarajan, R., Asmoni, M. N. A. M., Mohammed, A. H., Jaafar, M. N., Mei, J. L. Y., & Baba, M. (2017). Green retrofitting – A review of current status, implementations and challenges. *Renewable & Sustainable Energy Reviews*, *67*, 1360–1368. doi:10.1016/j.rser.2016.09.091

Lauren, N., Fielding, K. S., Smith, L., & Louis, W. R. (2016). You did, so you can and you will: Self-efficacy as a mediator of spillover from easy to more difficult pro-environmental behavior. *Journal of Environmental Psychology*, *48*, 191–199. doi:10.1016/j.jenvp.2016.10.004

Litti, G., Audenaert, A., & Lavagna, M. (2018). Life cycle operating energy saving from windows retrofitting in heritage buildings accounting for technical performance decay. *Journal of Building Engineering*, *17*, 135–153. doi:10.1016/j.jobe.2018.02.006

Mukherjee, S., & Chakraborty, D. (2013). Is environmental sustainability influenced by socioeconomic and sociopolitical factors? Cross-country empirical evidence. *Sustainable Development*, *21*(6), 353–371. doi:10.1002d.502

Pelau, C., & Al Pop, N. (2018). Implications for the energy policy derived from the relation between the cultural dimensions of Hofstede's model and the consumption of renewable energies. *Energy Policy*, *118*, 160–168. doi:10.1016/j.enpol.2018.03.042

Pelau, C., & Chinie, A. C. (2018). Econometric Model for Measuring the Impact of the Education Level of the Population on the Recycling Rate in a Circular Economy. *Amfiteatru Economic*, *20*(48), 340–355.

Pickton, D., & Broderick, A. (2005). *Integrated Marketing Communications*. Prentice Hall.

Sardianou, E., & Genoudi, P. (2013). Which factors affect the willingness of consumers to adopt renewable energies? *Renewable Energy*, *57*, 1–4. doi:10.1016/j.renene.2013.01.031

Social Media Marketing Industry Report. (2018). Socialmediaexaminer. Retrieved from https://www.socialmediaexaminer.com/social-media-marketing-industry-report-2018/

Soon, J. J., & Ahmad, S. A. (2015). Willingly or grudgingly? A meta-analysis on the willingness-to-pay for renewable energy use. *Renewable & Sustainable Energy Reviews*, *44*, 877–887. doi:10.1016/j.rser.2015.01.041

Statista. (2018). Number of monthly active Instagram users from January 2013 to September 2017. Retrieved from www.statista.com/statistics/253577/number-of-monthly-active-instagram-users/

Stoppok, M., Jess, A., Freitag, R., & Alber, E. (2018). Of culture, consumption and cost: A comparative analysis of household energy consumption in Kenya, Germany and Spain. *Energy Research & Social Science*, *40*, 127–139. doi:10.1016/j.erss.2017.12.004

Tantau, A., Pelau, C., & Pop, M. (2018). Fake news in the energy industry. In *Fourth BASIQ International Conference on New Trends in Sustainable Business and Consumption (BASIQ)*, Editura ASE (pp. 204-210).

Tantau, A., & Staiger, R. (2018). Business Models for Renewable Energy Initiatives: Emerging Research and Opportunities. Hershey, PA: IGI Global. doi:10.4018/978-1-5225-2688-9

Volland, B. (2016). The role of risk and trust attitudes in explaining residential energy demand: Evidence from the United Kingdom. *Ecological Economics*, *132*, 14–30. doi:10.1016/j.ecolecon.2016.10.002

Yoo, S. H., & Kwak, S. Y. (2009). Willingness to pay for green electricity in Korea: A contingent valuation study. *Energy Policy*, *37*, 5408–5416. doi:10.1016/j.enpol.2009.07.062

KEY TERMS AND DEFINITIONS

Building Retrofitting: The modernization of buildings with the purpose of improving their energy efficiency and reduce the energy losses.

Energy Efficiency: The process by which certain devices, buildings or other products or systems are designed and improved with the purpose of reducing their energy consumption.

Green Retrofitting: Method of increasing the energy efficiency of existing housing, by upgrading its characteristics with the purpose of reaching sustainability.

Green Roofs: Vegetation layers raised on roofs with the purpose of improving the energy performance of a building by an increased water retention.

Modernization: The replacement of old technologies and materials with newer ones, with improved characteristics and a better energetic performance.

Section 2

Social Aspects of Green Retrofitting

Chapter 5
Smart Homes as a Solution for Sustainable and More Inclusive Retrofitting of Existing Buildings

Doru Alexandru Plesea
https://orcid.org/0000-0002-0678-0170
The Bucharest University of Economic Studies, Romania

Bogdan Cristian Onete
https://orcid.org/0000-0002-7271-7663
The Bucharest University of Economic Studies, Romania

Ion Daniel Zgură
The Bucharest University of Economic Studies, Romania

ABSTRACT

All around the world, taking stock of buildings older than 1970, when the first energy efficiency arose, is an important one. Today, all urban agglomerations confront themselves with environmental problems generated by air pollution, cars, and buildings, vie for the first place as the highest polluters. In the future, cities will become increasingly populated, and as a result these problems will more and more affect the quality of life. A way to prevent this scenario is a gradual transition to smart cities. In this approach, smart houses will become indispensable. In order to maintain unaltered cities' specificity, the only acceptable solution is to retrofit old buildings, especially of those classified as historical monuments or as iconic buildings. By retrofitting these buildings, they could be prepared for integration in future cities.

DOI: 10.4018/978-1-5225-9104-7.ch005

INTRODUCTION

At the Paris Climate Conference (Conference Of Parties - COP 21), 195 countries adopted the first-ever universal, legally binding global climate deal. This agreement sets out a global action plan to avoid dangerous climate change by limiting global warming to well below 2°C above pre-industrial levels. Also, governments agreed to aim to limit the increase to 1,5°C since this would significantly reduce risks and impacts of climate change.

The main cause of climate change is the human activities, especially those that imply burning fossil fuels, cutting down rain forests and farming live-stock. These activities add enormous amounts of greenhouse gases to those naturally occurring in the atmosphere, increasing the greenhouse effect and the global warming. Many of these gases are naturally produced but some of human activities contribute to the increase of their concentrations in the atmosphere.

The most important greenhouse gases directly emitted by humans include carbon dioxide (CO_2), methane (CH_4), nitrous oxide N_2O) and several others. Carbon dioxide is the primary greenhouse gas that is contributing to recent climate change. Atmospheric CO_2 concentrations have increased by more than 40% since pre-industrial times being responsible for 64% of man-made warming (European Commission, 2018).

Residential and commercial sector has a contribution of 9.83% to the global carbon dioxide emissions (Table 1). But this value does not include emissions due to electricity consumption for household purposes (heating of dwellings, domestic hot water production etc.) (Ritchie & Roser, 2018)

In the world household electricity use varies within wide limits. For example, the average American or Canadian household in 2010 used about two to three times more than a typical European home and about twenty times more than the typical Nigerian household (Wilson, 2018)

Also, the share of using electricity for different household purposes varies within wide limits between different parts of the world. For example, in Europe in 2012, 19% of the total electricity consumption for household purposes were used for heating systems and electric boilers, 9% for water heating and 5% for ventilation, while in the USA, in 2015, 15% of the total residential electricity consumption were used for space heating, 14% for water heating and 17% for air conditioning (Bertoldi et al., 2012; US Energy Information Administration [EIA], 2016).

Table 1. Global carbon dioxide emissions by sector in 2010

Sector	Global carbon emissions	
	Millions gigagrams CO_2 per year	Percent of total (%)
Energy	20.330	59,06
Transport	5.530	16,06
Residential & commercial	3.380	9,83
Agriculture, land use & forestry	2.620	7,61
Industry	2.480	7,20
Waste	0.032	0,09
Other sources	0.047	0,14
Total	34.420	100,00

Source: (FAO, 2017)

As a conclusion, around 33-46% of the total electricity consumption for household purposes is used for ensuring proper living conditions in residential buildings. Finally, on average, over one-third of all final energy and almost half of electricity consumption for household purposes are consumed in buildings and this consumption generates over one-third of global carbon emissions (International Energy Agency [IEA], 2013).

In these conditions improving the energy efficiency in buildings is therefore of critical importance. The authors consider that smart homes, harmoniously integrated in smart cities are solutions relying on the new technologies that would be able to achieve the ambitious objectives stated at Paris climate conference.

Of course, that new buildings are realized with new more energy efficient materials. The problem we face to is what should we do with older buildings, built before the advent of first energy standards. For these buildings, authors consider that retrofitting could be a proper solution because in most cases it is much cheaper to retrofit an older building than demolish it and built a new one. So, by retrofitting those buildings they could be ready to enter in a new era, the smart city era.

SMART HOME: A HOLISTIC APPROACH RELATED TO FUTURE CITIES

Smart house, home automation, intelligent building, integrated home system, smart living, domotics are all synonymous terms related with the smart home concept.

A smart home is not smart because it is well-built, or it uses its surface in a "clever" way. Also, it isn't smart because it uses energy or water in an efficient way, but what makes it smart is the interactive technologies and home appliances that are working together being able to take decisions by user's request (Harper, 2003).

Ricquebourg et al., (2007) considers that:

A smart home can be described by a house which is equipped with smart objects, a home network make it possible to transport information between objects and a residential gateway to connect the smart home to the outside Internet world. (p.1)

According to Bendel (2018), the term "Smart Home" aims at the information and sensory upgraded home, networked in itself and externally.

Aldrich (2003) defines the smart home:

...as a residence equipped with computing and information technology which anticipates and responds to the needs of the occupants, working to promote their comfort, convenience, security and entertainment through the management of technology within the home and connections to the world beyond (p. 17).

While the first two definitions are limited to a technical approach, the third focuses on the advantages offered by the smart home to their tenants.

The authors consider that a more complete definition would be: a smart home uses automation, communication technologies and artificial intelligence in order to offer their tenants an increased safety and quality of life.

Today most of household appliances and electronic equipment have to a higher or lower extent a degree of automation and in many cases the capacity to override human commands in order to avoid dangerous situations. Some examples are: the flame control circuit at a boiler or gas oven, the protection circuit of an electric iron which disconnects it from mains after a period of inactivity exceeding 1-2 minutes, various protection systems in wash machines. Also, there are household appliances which use sensors or timers in order to control their state. Examples are automatic light switches or programmable timers found in microwave ovens.

To be considered as a smart home must have all appliances connected together through an internal network, an integrated management system that supervises all appliances and systems as whole and outside connections via Internet. A Smart home offers to its tenants the possibility to command and supervise from everywhere all house appliances, equipment and systems.

The foundations of the Smart Home were laid in 1960s with the widespread use of household and electronic appliances, but only in 1970s were laid the foundations of home automation with the advent of X10 technology. X10, the first general purpose home automation technology, emerged in 1975 and by 1978 first X10 products arose on the market. It was developed in order to allow remote control of home devices and applications and for a long period of time remained the most widely available. Relaying on the power line communication, this technology has limitations, such as: poor performance, distance limitations, compatibility problems with some devices, signal collision proneness, slowness, limited functionality, lack support for encryption, sporadic reliability. These major drawbacks are the ones that pushed this technology out of sight. However, because many X-10 devices still exist and few home automation enthusiasts are ready to throw them off yet, hybrid systems were developed (Fritz, 2018).

The term "smart home" was first used by American Association of House Builders in 1984 (Harper, 2003). Further the smart home concept evolved together with IT and communication technologies. The 1980s and 1990 marked the development of the computing techniques and the digital technologies that gradually replaced analog technologies. These technologies played a important role in developing digital systems and components suitable for use in domestic buildings. Also, from power line transmissions the domestic communication technologies evolved to the ones using coaxial and twisted pair cables and nowadays to high-speed technologies, based on fiber optics and 4G/5G communications.

So far, despite the advantages offered, smart homes haven't achieved a great commercial success. According to Aldrich (2003) and Gann et al. (1999) the reasons of the reduced widespread of smart homes are related to barriers like:

- **Low Cost-Benefit Ratio:** People perceive smart homes as luxury things they can easily dispense with. Smart homes advantages like: higher safety, security and convenience, a better energy and environmental management, improved internal and external communications, better healthcare support services, new forms of working and learning, house controlling and entertaining are considered by most tenants quite important but the price paid for it makes the smart houses to be considered unaffordable for most of them.
- **Multiplicity of Standards:** In the field of home automation is also a barrier because there are several technologies mutually incompatible. To complicate matters even more the new digital technologies are incompatible with the older ones.
- **The Existing Stock of Old Buildings Which are Less Suitable for Turning Them in Smart Homes:** Barlow and Gann (1998) consider that in Europe, where the old housing stock is high, the only solution to turn these dwellings in smart homes is retrofitting, which is a more expensive so-

lution than building new ones. Authors consider that WiFi technologies evolved very much from 1998 till now and today wireless solutions for retrofitting are viable at a low cost.

- **Smart Home Devices and Equipment are not Linked With User's Needs:** And that's why consumers aren't motivated to buy them. For a long period of time the technologies and devices related with the smart homes were perceived to be too sophisticated for the ordinary user. Authors consider the evolution of technologies related to IT and communications obliged the consumers to learn to use increasingly more sophisticated appliances but at the same time, the emergence of the touch screens and the development of visual technologies, especially those related with the graphic interface have created prerequisites for the acceptance of smart home technologies. Nowadays young and middle-aged people are using on a daily basis and on a large-scale devices with touch screens, being familiar with their graphical interfaces. That's why authors consider that current consumers are ready to accept smart home devices controlled via a mobile phone, a tablet or similar devices.

New concepts like IoT and smart city that evolved in the last decade will create, on a mid-term period, the prerequisites to move to the smart home as a standard solution for living. The authors point of view is that smart home, as parts of smart cities will be a solution for future human settlements. This solution will be required in order to obtain higher energy efficiencies and so, to achieve at a greater extent, de-carbonised cities.

Of course, it will be standard smart homes with minimum home automation, for those people who didn't afford higher costs and luxury smart homes with a larger scale home automation. For example, a standard smart home will have only systems that will add more control points to home's systems at very little additional expenses, such as: HVAC, lighting and security systems. The reason why these systems would be considered as necessarily for a standard smart home is that it can be integrated in smart city's energy management system in order to increase city's energy efficiency. Luxury smart homes will have additional automation systems that will provide value-added services, such as: automated monitoring and control of domestic appliances and domestic robots, interactive and multimedia information systems and a greater number of user interfaces that will offer more possibilities for remote controlling of home's systems.

A smart home is not a collection of disparate smart devices and appliances, but ones that work together to create a remotely controllable network. From this point of view, a major drawback that hindered the development of smart homes was the incompatibility between the communication protocols used for connections between devices, appliances and equipment. X10, Universal Powerline Bus (UPB), INSTEON, Z-Wave, ZigBee, Thread and also the well-known Wi-Fi and Bluetooth are some of home automation protocols that are available now on the market.

Being born sometime between 2008 and 2009 after CISCO Internet Business Solution Group's estimations (Evans, 2011), Internet of Things (IoT) creates the conditions for all devices, appliances and equipment surrounding us to be connected together in one network, the Internet.

The Internet Engineering Task Force (IETF) a large, open, international community of network designers, operators, vendors and researchers concerned with the evolution of the Internet architecture provides its own description of IoT: "IoT will connect objects around us (electronic, electrical, non-electrical) to provide seamless communication and contextual services provided by them. Development of RFID tags, sensors, actuators, mobile phones make it e to materialize IoT which interact and co-operate to make the service better and accessible anytime, from everywhere" (Internet Engineering Task Force [IETF], 2010)

The Institute of Electrical and Electronics Engineers (IEEE), a global professional engineering organization in its report on Internet of Things described it as: "A network of items – each embedded with sensors – which are connected to the Internet" (Institute of Electrical and Electronics Engineers [IEEE], 2014).

The International Telecommunication Union (ITU), the United Nations specialized agency for information and communication technologies, in its 2005 IoT Report, describes the IoT as "a ubiquitous network" in which the concept of ubiquitous networks is founded upon the all-inclusive use of networks and networked devices (International Telecommunication Union [ITU], 2005). With other words, ITU defines the IoT as a network that is available anytime, by anything and for anyone.

As pointed out above, there are several definitions of IoT, all of them converging to the following idea: IoT is a concept and a paradigm that consider all surroundings electronic devices as things/objects connected each other through wireless and wired connections and communicating and interacting together over the Internet. A "thing" in the IoT could be every electronic object that can be assigned an IP address and is able to transfer data over a network. Some examples of such things are: sensors, actuators, switches, RFID tags, devices, home appliances and industrial equipment, but also people with a heart monitor implant or animals with a biochip transponder.

Through the Internet these things make themselves recognizable in the network, been able to send information about themselves and receive information about other things; so they obtain intelligence by making or enable context related decisions.

IoT creates the prerequisites for developing smart environments. These smart environments could be residential and industrial buildings, cars, urban infrastructures and, finally, the whole cities.

Internet of Everything (IoE), a new concept used for the first time by Cisco, would be a possible further development for IoT. While the concept of IoT focuses predominantly on physical objects communicating with each other, the new concept of IoE lies in the intelligent connection, where network intelligence is used to bring all of the concepts together to a more united system.

As Luke Simmons, online marketing manager at Cloud Rails explains (as cited in Morelo, 2019):

In some ways, you can see the Internet of Things as the equivalent of a railroad line, including the tracks and the connections, whereas the Internet of Everything is all of that, and the trains, ticket machines, staff, customers, weather conditions, etc.

The authors consider IoT as a disruptive innovation because it will generate knock-on effects that will influence our way of living, working, riding and perceiving the reality and thus the societal organization.

For ordinary people IoT will create new opportunities in a long list of domains such as: smart homes, e-health, automotive, transport and logistics, retail, green energy, manufacturing, environmental monitoring, smart cities, personalized end-user applications (van Kranenburg et al., 2011).

Today, to achieve an objective like realizing a smart city could be considered a "mission impossible". The only things that cities building share together are some of its networks: electricity, gas, water and, in some cases, district heating. Smart homes are very rare and, where they exist, use different incompatible technologies. But almost every city around the world has a fixed telephone network which, in some cases, is superseded by a fiber-optic network.

As you can see in Table 2, the rate of Internet penetration around the globe on 31 December 2017 was 55%.

Table 2. World internet usage

World regions	Population[1]	Internet users[2]	Penetration rate (% of population)
Africa	1,287,914,329	453,329,534	35.2
Asia	4,207,588,157	2,023,630,194	48.1
Europe	827,650,849	704,833,750	85.2
Latin America/ Caribbean	652,047,996	438,248,446	67.2
Middle East	254,438,981	164,037,259	64.5
North America	363,844,662	345,660,847	95.0
Oceania/Australia	41,273,454	28,439,277	68.9
World Total	7.634.758.428	4.199.194.131	55.0

Source: adapted from Internet World Stats

In order to develop an IoT based network, these facts are very important because now most cities around the world have already the necessary IT infrastructure. And where this infrastructure does not exist, a mobile network solution may be taken into account.

It is important to develop in cities IoT ready networks because this is the first step toward smart cities implementation.

According to Musa (2016) "a smart city is defined as a city that engages its citizens and connects its infrastructure electronically" (p. 2).

Smart Cities Council (as cited in Eremia et al., 2017, p.14) is defining the Smart City as a city that:

...uses information and communications technology (ICT) to enhance its livability, workability and sustainability.

From authors' point of view, a smart city uses the power of IoT, as a ubiquitous communication network, to interconnect all urban infrastructures and public services in order to implement an intelligent management system that will help managing the city in a unitary way, thus ensuring optimal living conditions and the well-being of community members.

Smart cities and smart homes are concepts that complement one another. We can hardly speak about smart cities without taking in account the smart homes, because the buildings in a city are ones of the most important parts and also these buildings are largely responsible for their tenants' life quality.

In future, new technologies related to artificial intelligence (AI) will contribute to a further development of smart homes and also smart cities. Being first introduced in 1950's AI is being revitalized with the availability and power of Big Data. The use of new AI-enabled systems is expanding rapidly (Miller, 2018) and, also, they transform the way of doing business (Daugherty & Wilson, 2018).

Older definitions of AI are more closely related to robots, seeing it more as humanoid entity. For example, according to Wiener (1966) AI constructs are able to "learn from their own mistakes" and von Neumann (1974) considers that AI can "imitate the behavior of any other machine".

Modern definition of AI refers also to computer-based systems able of interpreting data collections and taking in real-time the optimal decisions. According to Encyclopedia Britannica, AI means "the ability of a digital computer or computer-controlled robot to perform tasks commonly associated with intelligent beings" (Copeland, 2019).

According to Duan et al. (2019) there is no commonly accepted definition of AI. A common definition refers to the ability of machine to learn from experience, adjust to new inputs and perform human-like tasks.

In Gartner's 2018 technology trend survey, AI is considered as the most important strategic technology (Panetta, 2018). The ability of AI systems to work with Big Data environment will be a real help for cities management systems, like the Intelligent Urban Management System, presented later in this chapter. Only AI in conjunction with Big Data Warehouses will be able to use the huge volumes of data that the IoT devices will generate. According to Roven (2018):

AI gives computers the ability to see, hear, move, and reason. AI-powered computer vision systems, for example, could allow computers to identify millions of elements of urban life simultaneously, such as people, cars, public workers, trash, accidents, fires, disasters, etc. The system could allow not only for autonomous monitoring but to make decisions based on how each of these elements behave, how behaviors change over the course of each day or over time, and how elements are responding to city systems. In short, AI will understand how cities are being used and how they are functioning and could assist city planners in understanding how the city is responding to various changes and initiatives.

The combination of AI, machine learning and contextually rich, real-time data streams delivered by IoT sensors and networks will be the main pillar of the further smart cities. The ability of AI systems to work with Big Data environment will be a real help for cities management systems, like the Intelligent Urban Management System, presented later in this chapter. Only AI in conjunction with Big Data Warehouses will be able to use the huge volumes of data that the IoT devices will generate.

POSSIBILITIES FOR A SUSTAINABLE DEVELOPMENT OF FUTURE CITIES BASED ON THE SMART CITY CONCEPTION

As the United Nations mention in its study "World Urbanization Prospects: the 2018 Revision – key facts", globally, more people live in urban areas than in rural ones. In 2018, 56% of the world's population resides in urban areas, the most urbanized regions being North America (with 82% of its population living in urban areas), Latin America and the Caribbean (81%), Europe (74%) and Oceania (68%). By 2050, 68% of the world's population is estimated to be urban and by 2030 world is projected to have 43 megacities, most of them in developing regions.

As the world continues to urbanize, sustainable development depends increasingly on the successful management of urban growth.

A major problem for future's cities will be the greenhouse gas (GHG) emissions because, according to the World Bank (2010), they are the major contributors to this type of pollution. Many sources including World Bank (2010), Bloomberg (2007), O'Meara (1999), Stern (2007), Munich Reinsurance (2004) estimates that cities are responsible for 75-80% of global GHG emissions and also consumes as much as 80% of energy production worldwide. In future, GHG emissions will be driven lesser by industrial activities and more by energy used in residential sector for household activities, especially for heating, cooling and lighting.

Since the time of Industrial Revolution, cities have often been blamed for causing environmental problems. Today, cities' concentrations of population, economic activities and infrastructure creates heightened risks from heat waves, foods, sea level rise and other impacts that climate change will aggravate. Many urban centers around the world experience and increased heat stress in summers, respiratory distress generated by air pollution, power failures especially in summer or winter, when overloading of electricity mains are more and more frequent. Less visible effects for urban agglomeration residents, but with a worldwide impact on the global biosphere are: direct loss in vegetation biomass from areas with high probability of urban expansion, direct threats to ecosystems by reducing biodiversity etc.

Another major challenge for the future cities will be urban infrastructure including here those for: energy, gas, water, communications, transportation, public lighting, district heating and cooling (where available) and also public services like: education, healthcare, emergency services (police, firefighters), social services, waste management etc. that will be increasingly difficult to manage on the current urban governance models.

For a sustainable development of future cities, the following three dimensions must be taken in account: the economic, the social and the environmental one.

The Economic Dimension

From ABI analyst Dominique Bonte's point of view, "economic development should be seen at the top of a hierarchy of benefits" (as cited in Hamblen, 2018), because smart cities will act as a public investment multiplier (will have a public investments multiplier effect): $ 10 trillion in incremental Gross Domestic Product (GDP) for all cities globally in the next decade and recurring, sustainable growth of 2,8% by 2026 (Hamblin, 2018).

Sustained economic growth is always accompanied by urbanization. In its book "Prosperity of Cities: State of the World Cities 2012/2013", UN-Habitat states that:

While share of urban populations worldwide increased from 33% to 51% between 1960 and 2010, per capita income increased by 152 per cent from US $ 2382 to US $ 6006 over the same period (p.43). The only exception from this rule is Africa, where urban population grew without economic growth (Fay & Opal, 1999).

According to the World Bank (2015) "a competitive city successfully facilitates its firms and industries to grow jobs, raise productivity and increase incomes of citizens" (p.2). Analyzing data on 750 cities for 2005 to 2012, World Bank found that 72% of it outperformed their countries in terms of economic growth. Also, by 2025, the world's top 600 cities are expected to account for 60% of global GDP (World Bank, 2015).

While cities are related with economic growth, the advent of smart cities will boost it, through the widespread use of information technology and Internet, which will link together employees, businesses and the global economy. In other words, while traditional cities act locally, smart cities will act regionally and will compete globally.

The evolution toward the smart cities is not easy to achieve because it requires financial resources that are not available to the most urban communities. But city governance needs to understand that investment in technologies related to the smart cities is strictly necessary in order to keep their cities competitive and attractive for the business community. An accessible solution for almost every city is the development of public-private partnerships.

Cities that will not invest in smart city technologies will thus enter in a vicious circle that ultimately will lead to economic stagnation and their decline. An example can be considered Phoenix, Arizona but also most American cities are endangered to decline because they use more resources than necessary (Walters, 2018). In future, cities with low income levels will be mainly production centers, while those with higher income levels will become centers for financial, high-tech and creative industries, having the opportunity to evolve themselves by shifting economic activities into higher value-added sectors.

Smart cities will attract more businesses because their smart infrastructure will offer advantages, such as: reduced operating costs, improved profit margins and creates new opportunities for growth.

Another important advantage is the ubiquity of the businesses. In a globalized world a business that can operate all around the world will have more economic opportunities than a traditional one, acting locally or regionally.

Being focused mainly on those economic activities related to information and other high value-added sectors the smart cites will offer a higher flexibility of labor market. From employees working from home for a local company to those who work for companies located all around the world, all will take benefit from smart cities' high-speed informatics highways. This will bring advantages both to citizens, by reducing the unemployment rate, and to the community, by reductions in traffic and congestion. A study by Juniper Research (2018) estimates that: "heavy congestion in cities can result in drivers spending more than 70 hours per year in gridlock" (p.2). By implementing smart mobility projects, it could be possible to "give back" to each citizen roughly 60 hours per year from the time spent in traffic (Juniper Research, 2018).

The Social Dimension

To be accepted by its inhabitants, smart city strategies must start with people, not with the technology. To be smart, it is not enough that a city has an intelligent monitoring system, supervising city's operations or that all household appliances and devices are connected to the Internet. For its citizens a smart city is about using IoT and communication technologies for making better decisions and for delivering a better quality of life. With other words, the final goal of a smart city is to respond more effectively and dynamically to the needs and desires of its inhabitants.

Quality of life has many aspects beginning with dwellers comfort, continuing with the quality of municipal services and finishing with the quality of the environment.

Another aspect of social impact of smart cities is its governance. Traditional governance lacks in terms of transparency, accountability, efficient use of resources and equal access to information and decisions related to city administration. Smart cities aim to foster more aware, educated and informed citizens, offering them a new type of governance, the so-called e-governance or smart governance. This new type of governance will offer to its citizens the opportunity to involve comprehensively in decision making process, in public policies making and in public governance. The base to all these will be the Internet and the mobile devices that will keep citizens informed and also giving them the possibility to act to public aspects that will affect their personal lives. From other point of view, the new channels for communication like social networks will push city governments to be more responsive to the grievances of the general public.

The main challenges that smart cities will face with, will be to obtain and maintain social equity and social cohesion, to fight against poverty and unemployment and to offer a fair access to city's resources for all citizens. It will be a challenge because of vulnerable groups of population like those from poorer socio-economic backgrounds, people with poor IT skills, etc.

The Environmental Dimension

From cities' environment point of view, air pollution is a major hazard. According to the World Bank (2016) "air pollution is not just a health risk; it is also an economic burden" (p. 2). It causes a loss of productive labor and also raises the expenses with healthcare. The health risk poses by air pollution is the fourth greatest risk factor in upper-middle income countries and the third leading risk factor in low-middle income countries (World Bank, 2016).

There are different measures that city administration could take in order to reduce air pollution in cities by:

- A widespread use of less polluting vehicles for public transport by gradual phasing out Diesel and gasoline buses and replacing them with electric vehicles.
- Encouraging citizens to give up their old cars and replacing it with electric vehicles. City administration offer financial incentive to its citizens to do so.
- Retrofitting older buildings in order to achieve a highly energy efficient and de-carbonised building stock. Also, a solution in this connection is to implement an intelligent management system in every building.
- A better management of municipal wastes by organizing an integrated waste system involving separate collection, a reduction in waste produced, the construction of modern facilities and restoration of former landfill sites.
- Using new low carbon transport solutions like car sharing, bike sharing and, in future, autonomous vehicles.
- IoT will offer solutions to reduce traffic congestions by using intelligent traffic signals, intelligent re-routing solutions and sending real time traffic information and real time public transport information.

All around the globe there are currently efforts to realize smart cities. According to IoT Analytics (as cited in Scully, 2018) in 2018 most of the IoT projects were in Smart City (367 projects), followed by Connected Building IoT projects (193 projects). Most of the smart cities projects are located in Europe (45%), followed by America (34%) and Asia (18%).

The Smart City initiatives started by governments and municipalities all around the world underlie the multitude of IoT projects. Examples of such initiatives include cities, such as: Singapore, Barcelona, Los Angeles, Dubai, Hamburg, Birmingham, Berlin, Rio de Janeiro etc. The most popular Smart City applications are related to Smart Traffic, including projects, such as: traffic monitoring, parking systems, car and bike sharing. Another direction for the Smart Cities initiatives evolves around the utilities, lighting, environmental monitoring and public safety (Scully, 2018). According to IoT Analytics (as cited in Scully, 2018) the distribution of the most IoT Smart City projects are related to traffic solutions (39%), followed by utilities applications (22%), lighting solutions (21%), environmental monitoring (18%) and public safety (17%). All projects offer solutions for some issues affecting specific parts of city's infrastructure and public services: smart city lighting, smart traffic, intelligent parking systems, environmental and infrastructure sensing, remote video kiosk solutions for citizens etc.

None of these projects started from a global, integrated vision about smart cities. The solution presented further in the case study part of this chapter offers a holistic point of view for building a smart city preserving old buildings and offering them through retrofitting a new life in future's cities.

CASE STUDY: A POSSIBLE SOLUTION FOR HARMONIOUSLY RETROFITTING OF EXISTING BUILDINGS IN THE FUTURE'S SMART CITIES

All around the world there are residential building stocks built before 1970 as shown in Tables 3 and 4.

The first building efficiency standard appeared in the US in the seventh decade of the 20th century, but there was concern to reduce the energy losses of buildings (Table 5).

U-value or thermal transmittance shows the heat loss in a building element such as walls, floors or roofs. Also known as heat transfer coefficient it measures how well parts of a building transfer heat. In other words, the U-value show how effective is a building element at preventing heat from transmitting between the inside and the outside of a building. As energy prices increase there is a greater awareness of efficient use of energy and a de-carbonised economy, lower and lower U-values are required.

As a conclusion, most of residential buildings from USA and Europe have a lower energy efficiency; most of them fall under the current standards. From the building stock, which do not meet the current standards for energy efficiency, are part of both houses built in the years after World War II (WW II) and older dwellings, some of which are historical and architectural monuments.

Table 3. Stock of US housing units by year of construction

Period	Before 1950	1950 to 1959	1960 to 1969	1970 to 1979	1980 to 1989	1990 to 1999	2000 to 2009	2010 to 2015
Share of dwellings (%)	17.59	10.65	10.82	15.48	13.53	14.21	14.38	3.21

Source: calculated on the basis of the data from 2015 Residential energy consumption survey; US Energy Information Administration

Table 4. EU-28 stock of residential buildings by age

Period	Before 1945	1945 to 1969	1970 to 1979	1980 to 1989	1990 to 1999	2000 to 2009	After 2010
Share of dwellings (%)	22.69	26.19	15.98	12.54	9.34	9.56	3.69

Source: European Commission (2014). EU Buildings Database

Table 5. U-Value of building envelope for residential buildings by age

Period	Before 1945	1945 to 1969	1970 to 1979	1980 to 1989	1990 to 1999	2000 to2009	After 2010	Average
U-Value $W/m^2 \cdot °C$	2.42	2.26	1.85	1.46	1.20	0.94	0.28	1.69

Source: European Commission (2014). EU Buildings Database

In Europe most of the buildings built after WW II are in most cases multi-story blocks of flats. They were built to meet the need to recover the building stock after it was affected by damages caused by war. The motto under which they were built was; "to be built quickly and to be cheap". For these buildings there are two solutions: to be demolished and to rebuild new ones or to retrofit them. The chosen solution depends on the state in which they are; many of these dwellings are in advanced degradation stages.

Most of the buildings before WW II are located in the city centre and demolition would affect city's appearance and its iconic parts. That's why demolition is not a solution, the only acceptable being retrofitting them.

That's why in order to meet environmental objectives, such as reduced greenhouse gas emissions, many governments introduced new policies and programmes with the scope to encourage households in order to improve their energy efficiency and to reduce their environmental "footprint" (Judson & Maller, 2014). Initiatives like 'The Green Deal' in the UK and the German "CO_2 Building Rehabilitation Programme" provide loans for homeowners to overcome the upfront investment barrier in renovating or retrofitting their homes (*Science for Environment Policy*, 2015).

The terms renovation, refurbishment and retrofitting are often taken to mean the same thing and used as if they are interchangeable. However, they have different specific meanings. According to the Cambridge Dictionary (2019):

- **Renovation:** "The act or process of repairing and improving something, especially a building".
- **Refurbishment:** "Work such as painting, repairing, and cleaning that is done to make a building look new again".
- **Retrofitting:** "To provide a machine with a part, or a place with equipment, that it did not originally have when it was built".

Being more complex, retrofitting brings a more substantial change at the building level than the other two. In some cases, retrofitting process implies also some renovation and refurbishment tasks.

The main problems that must be solved through retrofitting buildings older than 1970, in order to meet the requirements of a smart home, are:

- Reducing heat loss by using high efficiency thermal insulators for walls, roofs, doors, windows, skylights. These materials will act also as sound insulator between the inside and the outside of a building.
- Improving the sound insulation between rooms in order to increase the acoustic comfort of the dwellers.
- Reducing building's energy demand by using on-site renewable energy, produced by solar PV, solar water heaters or, where is possible, by using wind mills. All these could be mounted on building's roof or on sun-exposed parts. Also, modern technologies like the one used in Luminescent Solar Concentrator are the most promising technologies to continue harvesting electricity and assuring transparency needed by a window (Wigman & van der Kolk, 2012).
- Upgrading the existing installations (plumbing/water, gas, heating, electrical). In many cases these are quite old installations and replacing it with new ones could be a good solution, especially when the upgrade is made to a smart building.

- Installing the wires and Wi-Fi access point in order to implement the network needed by the smart building, and also by the IoT for household appliances. This network will also connect the building with smart city's environment.
- Installing all the sensors, actuators and home automation devices in order to achieve a smart building.
- Installing the computer and the software required by the building monitoring system.

According to Earnes et al., (2014): "at an urban or city scale retrofit means something much larger and more comprehensive, more integrated". That's why, for a sustainable retrofitting of old buildings, the process must be seen as a whole at the city level. The retrofitting process must take in account any foreseeable changes that could be done. It is a more expensive, but it is better to intervene once, making all foreseeable changes, than making from time to time small changes, but which affect the whole building, and intervene as few as possible in order to keep expenses as low as possible.

That's why authors consider that, by retrofitting process, old buildings will be prepared for the future's smart cities. By seeing retrofitting process as a whole at the city level, the retrofitted buildings will be harmoniously integrated with the new ones and together they will be prepared for smart homes and smart cities solution. Equipped with IoT devices and modern communication equipment, connected through wires or wireless to urban information networks, the retrofitted buildings will be a part of smart cities environment. As a result, their tenants will enjoy better living conditions at higher energy efficiency and their lives will be safer due to buildings monitoring systems. Also, they will be more informed due to the information provided by the intelligent urban management system.

The Buildings' Management System

The buildings' management system acts like a general supervisor of all buildings in the city, setting the optimum operating parameters for all buildings in the city. In contrast to this system, the building monitoring system acts at a local level, setting the best operating parameters for all building systems. These two systems are interlinked, sending and receiving data and commands via city's IT infrastructure.

The building monitoring system periodically receives, from the traffic management system, data about the air conditions in the vicinity of each building and so the HVAC systems work accordingly. For example, if the air pollution near the building is high, the HVAC system is disposed automatically to use a higher number of filters for a better air cleaning and filtration. This aspect is very important for buildings like hospitals and schools, because they house vulnerable people like patients and children.

The building monitoring system receives data from sensors and actuators distributed in houses and offices. This data can be used to make life more comfortable in several aspects: rooms heating can be adapted to tenant's preferences and to the weather; the room lighting can be changed according to the time of the day; domestic incidents can be avoided with appropriate monitoring and alarm systems; and energy can be saved by automatically switching off the electrical equipment when not needed. The building monitoring system is also responsible for the building's security. Gas, smoke and presence sensors, gives information to this system and if it is necessary the building monitoring system automatically alerts the proper public service (ambulance, police, firefighters).

To get better out of the energy consumption peaks buildings may have on roofs or sun-exposed parts solar cell panels. The provided energy it is stored in rechargeable cells and used when the buildings management system finds out that there is an energetic peak in the city. So are avoided situations of power failures especially in summer or winter when overloading of electricity mains are more and more frequent.

As shown in Figure 1, the buildings' management system acts like a decision factor at the city level: it receives all the data from building monitoring systems and, in conjunction with the historical data stored in its cloud, decides the best energetic solution for each situation. Consequently, the building monitoring system acts like a decision factor at building's level, by data received from own sensors and orders received from the buildings' management system.

The buildings' management system is well suited if the energy providers adopt the policy to dynamically change energy prices to influence the overall energy consumption in a way that smoothes load peaks. The buildings' management system optimizes the power consumption costs throughout the day by observing prices and switches the power sources (electricity mains, solar or building's batteries) accordingly.

The Intelligent Urban Management System

The retrofitted buildings are now ready to be a part of smart city's environment. Further, the authors will present their solution for sustainable retrofitting old buildings in the future's smart cities. This solution bases on the knowledge and achievements in IoT and smart cities field.

The city's "brain" is the intelligent urban management system. As presented in Figure 2, the intelligent urban management system is a concept that relies mainly on artificial intelligence, telecommunications and information. All these main components work as a whole and are inextricably linked.

The artificial intelligence component may be considered as the "brain" of the whole system, because it takes the proper decisions for the city, starting from the existing conditions. It can be considered as a "virtual" mayor, with an executive role only. Compared to its human equivalent the "virtual mayor" has a better analysis and synthesis capacity and also a higher reaction speed.

The telecommunication component is like a spinal cord and nerve fibers in the human body. Its role is to transmit information and commands through the fixed and wireless networks. It includes all kind of

Figure 1. The buildings' management system
Source: authors

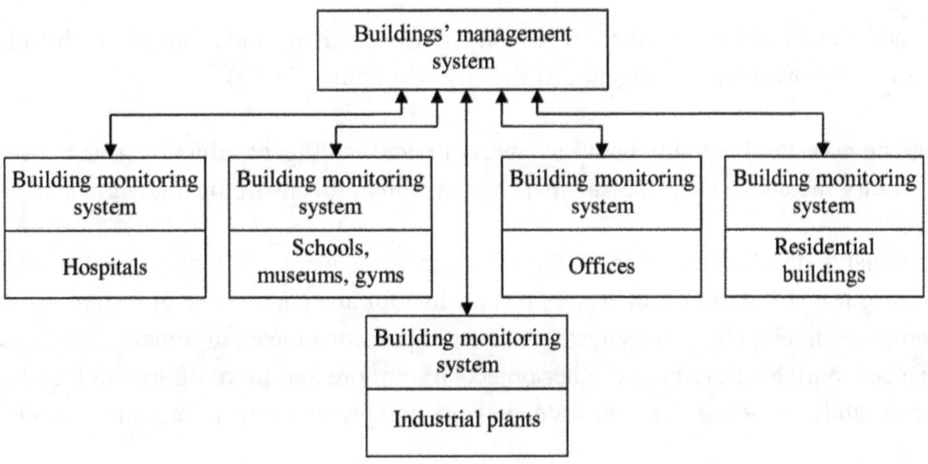

Figure 2. Main components of an intelligent urban management system
Source: authors

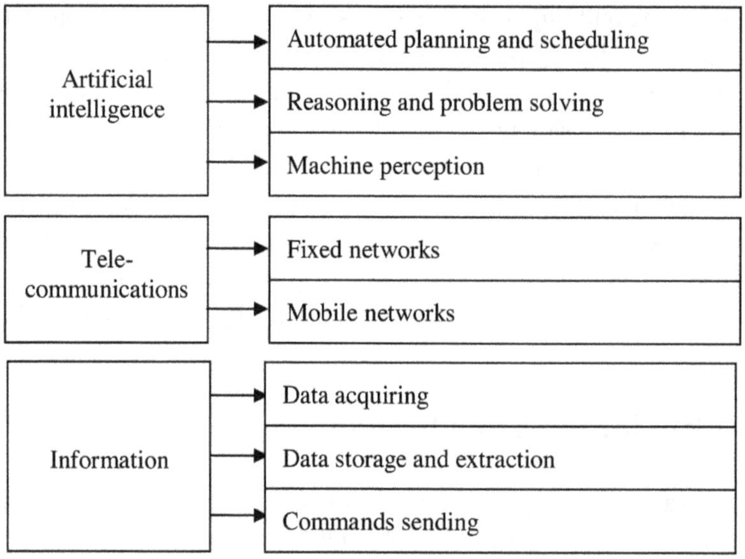

networks located in the city area. Because parts of these networks are public one a very important issue is assuring networks security. Strong encryption and password solutions are required for this purpose.

The information component is responsible for acquiring input data from various sources like: sensors, surveillance cameras, RFID receivers, public services and utility operators and citizens, for putting it in the proper format, to store it in clouds, t6o extract through data mining techniques and finally to send data as commands.

The intelligent management system may be considered as a main part of a smart city. As it is generally accepted, that a smart city has the role to:

- Make a more efficient use of physical infrastructure through artificial intelligence to support a strong and healthy economic, social, cultural development (Hollands, 2008).
- Engage effectively with local people in local governance and decision (Komninos, 2013; Gooch et al., 2015).
- Learn, adapt and innovate and thereby respond more effectively and promptly to changing circumstances by improving the intelligence of the city (Komninos, 2013).

There can be no sign of equality between smart city and intelligent urban management system, because a smart city includes also a "social" part by engaging citizens in local governance and decision (e-governance).

For a sustainable development of the future cities, the intelligent urban management system plays a key role because it is that part of a smart city responsible for an efficient use of resources.

Development of intelligent management solutions implies that the environment, buildings, vehicles, home appliances, portable devices and other objects have more and more information associated with them and/or the ability to sense, communicate, network and produce new information. In addition, non-sensing things (i.e. things that may have functionality, but do not provide information or data) can be added but only as execution elements.

As shown in Figure 3, the proposed intelligent urban management solution has 2 main components: the traffic management system and the buildings management system. Both systems are controlled, commanded and coordinated as a whole by the intelligent urban management system which acts like a "brain" of the smart city.

These systems are working in conjunction with systems with a limited autonomy like the street lighting network or smart homes appliances. For example, the street lighting system has the capacity to self-adjust the amount of emitted light depending on ambient light. However, in the event of a failure, the street lighting system transmits further the information to the traffic management system, which announces the intervention team.

The Traffic Management System

For a good road traffic monitoring the license plates of the motor vehicles contain an RF-ID tag and an antenna. This type of license plate allows a remote electronic identification of vehicles. Road signs and traffic lights on every corner of the street have a triple role:

- Contains a RF-ID reader which allows vehicle identification and therefore traffic monitoring.
- Contains a LED dot matrix panel which acts a road sign and/or traffic light.
- Contains also gas sensors for nitrogen oxides (NO or NO_2), carbon oxides (CO and CO_2), oxygen and temperature. These sensors are needed for monitoring the local air pollution and, also, temperature. Through the information obtained from these sensors the traffic monitoring system realizes periodically maps of city air quality.

Surveillance cameras, mounted at road intersections, are very useful to know if problems such as: car accidents or traffic congestions occur.

The traffic monitoring system receives information from RF-ID readers, gas sensors and surveillance cameras. This information allows an estimation of the traffic volume and pollution level in every point

Figure 3. The main components of the intelligent urban management system
Source: authors

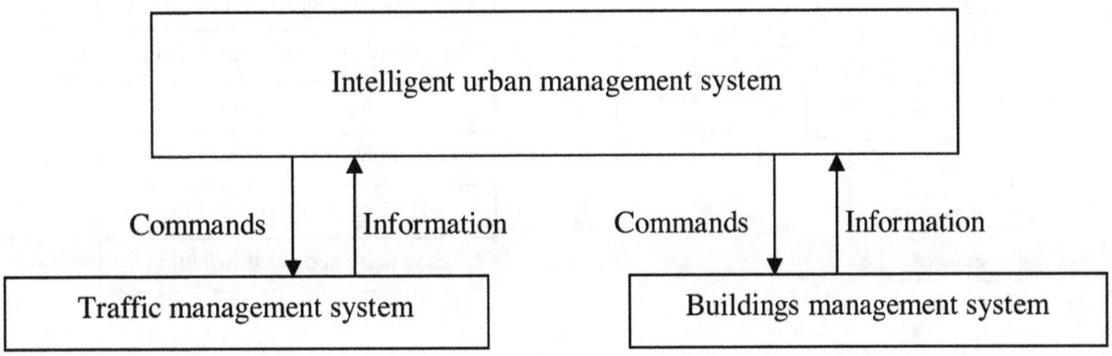

of the city. Also, through object detection techniques and with the video data received from surveillance cameras it is possible to know when and where traffic congestion occurs.

Because all these devices are mounted on road signs a wide area network based on optical fiber can be used for linking them with the server used as traffic monitoring system.

As shown in Figure 4, the traffic management system is the decision system for traffic component of the intelligent urban management system. It has the role to collect all relevant data from the traffic monitoring system and from local and national public services and relying on this information to estimate the optimum traffic solution for the whole city. This system receives notifications from local public services, such as: emergency services (ambulance, police, fire), public transportation, water delivery and wastewater collection services, waste collection services, traffic and road operations services, electricity distribution services etc. and also from national services like weather forecast service.

Based on this information the traffic management system decides the optimum traffic model for the given conditions and sends instructions for the traffic monitoring system. This system, which is a data collection and execution system, sets up the traffic lights and the road signs. For example, if street of a road intersection is congested the traffic monitoring system, using the data received from the traffic management system, changes the traffic light timing. If some unforeseen events occur (accidents, firefighter's interventions, road operations etc.) the traffic management system assures a re-routing of traffic to avoid the area with problems and to ensure the optimum traffic flow. It is a hard problem for this system because it must take in account all possible routes based on historical data stored in its cloud. Once a solution is given the traffic monitoring systems modifies accordingly the road signs.

Figure 4. The traffic management system
Source: authors

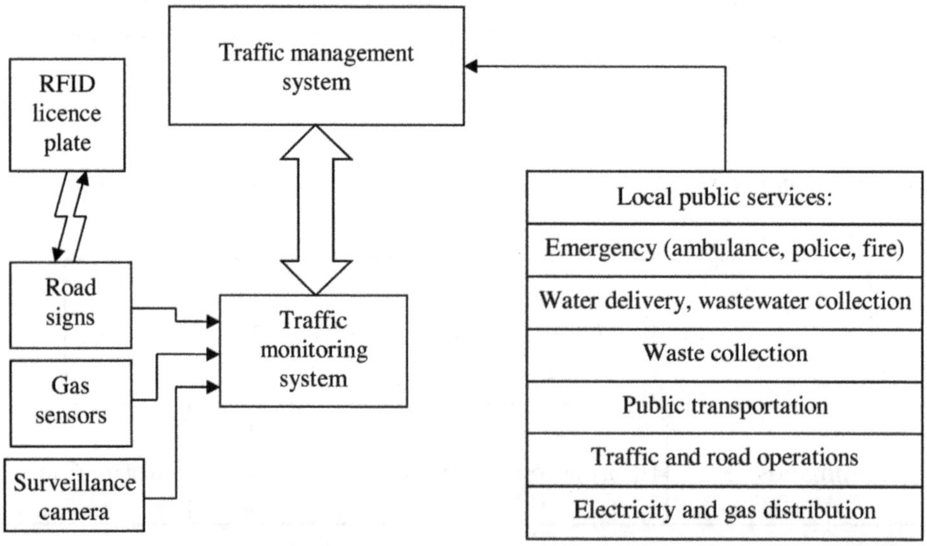

The citizens are informed in real time about these changes through two channels, depending of the car's features:

- Through a dedicated data channel in conjunction with the automotive navigation system.
- Through the RDS channel of local FM broadcasting station (for older cars and for ones without navigation system).

If local transportation network uses tickets, cards or passes with a RFID tag and all vehicles are wireless connected to the traffic monitoring system, the traffic management system estimates in real time the vehicle load on each route. Finally, the traffic management system will give solutions for a better load of all public transportation vehicles.

Through local administration policies citizens can be encouraged to travel from home to their jobs and back by using a car sharing system. If someone wants to use this system, he can ask the traffic management system and it gives the proper solution, if the other part is enlisted in the group of car- sharing users.

CONCLUSION

Retrofitting old buildings offers the opportunity to increase comfort and dwellers' quality of life and giving also the possibility to implement the changes required by the migration to smart homes, a necessary step toward the future smart cities. This could be the right solution especially for European cities, which, at greater extend, consist of old buildings, most of them iconic for city's architecture.

REFERENCES

Aldrich, F. K. (2003). Smart Homes: Past, Present and Future. In R. Harper (Ed.), *Inside the Smart Home* (p. 17). London: Springer. doi:10.1007/1-85233-854-7_2

Bendel, O. (2018). *Smart Home Definition*. Gabler Wirtschaftslexikon. Retrieved from https://wirtschaftslexikon.gabler.de/definition/smart-home-54137/version-277191

Bertoldi, P., Hirl, B., & Labanca, N. (2012). *Energy Efficiency Status Report 2012. Electricity Consumption and Efficiency Trends in the EU-27*. Retrieved from https://setis.ec.europa.eu/sites/default/files/reports/energy-efficiency-status-report-2012.pdf

Cambridge Dictionary. (n.d.). Retrofitting. Retrieved from https://dictionary.cambridge.org/dictionary/english/retrofitting

Cambridge Dictionary. (n.d.). Renovation. Retrieved from https://dictionary.cambridge.org/dictionary/english/renovation

Cambridge Dictionary. (n.d.). Refurbishment. Retrieved from https://dictionary.cambridge.org/dictionary/english/refurbishment?q=+refurbishment

Copleand, B. J. (2019). *Artificial Intelligence*. Encyclopedia Britannica. Retrieved from https://www.britannica.com/technology/artificial-intelligence

Daugherty, P. R., & Wilson, H. J. (2018). *Human + machine: Reimaging work in the age of AI*. Harvard Business Press.

Dougall, S. (2018). *Smart city projects to watch in 2018*. Cbronline. Retrieved from https://www.cbronline.com/internet-of-things/smart-cities/smart-city-projects/

Duan, Y., Edwards, J. S., & Dwivedi, Y. K. (2019). Artificial intelligence for decision making in the era of Big Data – evolution, challenges and research data. *International Journal of Information Management*, *48*, 63–71. doi:10.1016/j.ijinfomgt.2019.01.021

Eames, M., Dixon, T., Lannon, S., Hunt, M., De Laurentis, C., Marvin, S., . . . Georgiadou, M. C. (2014). Retrofit 2050: Critical Challenges for Urban Transitions. Retrieved from http://centaur.reading.ac.uk/36187/1/critical%20challenges%20briefing-March%202014.pdf

Eremia, M., Toma, L., & Sanduleac, M. (2016). The Smart City Concept in the 21st Century. *Procedia Engineering, 181*, 14. Retrieved from https://www.sciencedirect.com/science/article/pii/S1877705817309402

European Commission. (2015). *EU Buildings Database*. Retrieved from https://ec.europa.eu/energy/en/eu-buildings-database#eu_buildings_countries_tab

European Commission DG Environment News Alert Service. (2015). Energy efficiency policies for home renovations and retrofitting should consider the social factors. *Science for Environment Policy*. Retrieved from http://ec.europa.eu/environment/integration/research/newsalert/pdf/energy_efficiency_policies_home%20renovations%20_retrofitting_should_consider_social_factors_49si5_en.pdf

Evans, D. (2011). *The Internet of Things*. How the Next Evolution of the Internet Is Changing Everything. CISCO.

Fay, M., & Opal, C. (1999). Urbanization without Growth: A not so uncommon Phenomenon. World Bank. doi:10.1596/1813-9450-2412

Fritz, R. (2018). Is X-10 an obsolete technology? *Lifewire*. Retrieved from https://www.lifewire.com/is-x-10-obsolete-818408

Fritz, R. (2018). What Is the Best Home Automation Technology? *Lifewire*. Retrieved from https://www.lifewire.com/best-home-automation-technology-817601

Gann, D., Barlow, J., & Venables, T. (1999). *Digital futures – making homes smarter*. Coventry: Chartered Institute of Housing.

Gooch, D., Wolff, A., Kortuem, G., & Brown, R. 2015. Reimagining the Role of Citizens in Smart City Projects. In *Proceedings of the 2015 ACM International Joint Conference on Pervasive and Ubiquitous Computing and Proceedings of the 2015 ACM International Symposium on Wearable Computers UbiComp '15*. New York, NY: ACM, pp.1587–1594 10.1145/2800835.2801622

Hamblen, M. (2018). *Economic Benefits Of A Smart City*, Retrieved from https://www.smartresilient.com/economic-benefits-smart-city

Harper, R. (2003). Inside the Smart Home: Ideas, Possibilities and Methods. In R. Harper (Ed.), *Inside the Smart Home*. London: Springer. doi:10.1007/1-85233-854-7_1

Hollands, R. G. (2008). Will the real smart city please stand up? *City*, *12*(3), 303–320. doi:10.1080/13604810802479126

IEA. (2013). *Transition to Sustainable Buildings. Strategies and Opportunities to 2050.* Retrieved from https://www.iea.org/publications/freepublications/publication/Building2013_free.pdf

IEEE. (2014). *The Institute Special Report: The Internet of Things.* Retrieved from http://theinstitute.ieee.org/static/special-report-the-internet-of-things

IETF. (2010). *The Internet of Things - Concept and Problem Statement.* Retrieved from http://tools.ietf.org/id/draft-lee-iot-problem-statement-00.txt

Internet World Stats. (2018). *Internet Usage Statistics.* Retrieved from https://www.internetworldstats.com/stats.htm

ITU. (2005). *The Internet of Things Executive Summary, ITU Internet Reports 2005.* Retrieved from http://www.itu.int/osg/spu/publications/internetofthings/InternetofThings_summary.pdf

Judson, E. P., & Maller, C. (2014). Housing renovations and energy efficiency: Insights from homeowners' practices. *Building Research and Information*, *42*(4), 501–511. doi:10.1080/09613218.2014.894808

Juniper Research. (2018). Smart cities – what's in it for citizens? Retrieved from https://newsroom.intel.com/wp-content/uploads/sites/11/2018/03/smart-cities-whats-in-it-for-citizens.pdf

Komninos, N. (2013). What makes cities intelligent? In *M. Deakin (Ed.), Smart Cities: Governing, Modelling and Analysing the Transition* (p. 77). Taylor and Francis.

Miller, S. (2018). AI: Augmentation, more so then automation. Asian Management Insights, 5(1), 1-20.

Morelo, D. (2019). *What is Internet of Everything.* Retrieved from https://www.bporiver.com/how-to/%20technology/what-is-internet-of-everything

Musa, S. (2016). *Smart City Roadmap.* Retrieved from https://www.academia.edu/21181336/Smart_City_Roadmap

Panetta, K. (2017). Gartner top 10 strategic technology trends for 2018. *Gartner.* Retrieved from https://www.gartner.com/smarterwithgartner/gartner-top-10-strategic-technology-trends-for-2018/

Ricquebourg, V., Menga, D., Durand, D., Marhic, B., Delahoche, L., & Loge, C. (2007). The Smart Home Concept: our immediate future. In *2006 1ST IEEE International Conference on E-Learning in Industrial Electronics.* doi:10.1109/ICELIE.2006.347206

Ritchie, H., & Roser, M. (2018). CO_2 and other Greenhouse Gas Emissions. *OurWorldInData.* Retrieved from https://ourworldindata.org/co2-and-other-greenhouse-gas-emissions

Roven, A. (2019). *AI in Smart Cities.* Retrieved from https://medium.com/neuromation-blog/ai-in-smart-cities-dfe2fa7d2829

Scully, P. (2018). The Top 10 IoT Segments in 2018 – based on 1600 real IoT projects. *IoT Analytics.* Retrieved from https://iot-analytics.com/top-10-iot-segments-2018-real-iot-projects/

UN-Habitat. (2013). State of the World's Cities 2012/2013: Prosperity of Cities. Routledge, New York, and UN-Habitat.

United Nations. (2018). *World Urbanization Prospects: The 2018 Revision – key facts*. Retrieved from https://population.un.org/wup/Publications/Files/WUP2018-KeyFacts.pdf

US Energy Information Administration. (2018). *Residential energy consumption survey (RECS). 2015 RECS survey data*. Retrieved from https://www.eia.gov/consumption/residential/data/2015/index.php?view=characteristics#house

Van Kranenburg, R., Anzelmo, E., Bassi, A., Caprio, D., Dodson, S., & Ratto, M. (2011). The Internet of Things. In *1st Berlin Symposium on Internet and Society: Exploring the Digital Future.* Retrieved from https://www.researchgate.net/profile/Matt_Ratto/publication/228360933_The_Internet_of_Things/links/0912f513755ebd1e87000000.pdf

Vinod Kumar, T. M., & Dahiya, B. (2017). Smart Economy in Smart Cities. In T. Vinod Kumar (Ed.), *Smart Economy in Smart Cities. Advances in 21st Century Human Settlements*. Singapore: Springer; doi:10.1007/978-981-10-1610-3_1

Von Neumann, J. (1974). *The Computer and the Brain*. New Haven: Yale University Press.

Walters, J. (2018, *March 20*). Plight of Phoenix: how long can the world's 'least sustainable' city survive? *The Guardian*. Retrieved from https://www.theguardian.com/cities/2018/mar/20/phoenix-least-sustainable-city-survive-water

Wiegman, J. W. E., & van der Kolk, E. (2012). Building integrated thin film luminescent solar concentrators: Detailed efficiency characterization and light transport modelling. *Solar Energy Materials and Solar Cells*, *103*, 41–47. doi:10.1016/j.solmat.2012.04.016

Wiener, N. (1966). *God and Golem, Inc* (pp. 20–22). Cambridge, MA, USA: MIT Press.

Wilson, L. (2011). Average household electricity use around the world. *Shrink That Footprint*. Retrieved from http://shrinkthatfootprint.com/average-household-electricity-consumption

World Bank. (2010). Cities and Climate Change: An Urgent Agenda. Retrieved from https://openknowledge.worldbank.org/handle/10986/17381

World Bank. (2015). Competitive Cities for Jobs and Growth: What, Who, and How. Retrieved from https://openknowledge.worldbank.org/handle/10986/23227

World Bank. (2016). The Cost of Air Pollution. Strengthening the Economic Case for Action. Retrieved from http://documents.worldbank.org/curated/en/781521473177013155/pdf/108141-REVISED-Cost-of-PollutionWebCORRECTEDfile.pdf

ADDITIONAL READING

Ma, Z., Cooper, P., Daly, D., & Ledo, L. (2012). Existing building retrofits: Methodology and state-of-the-art. *Energy and Building*, *55*, 889–902. doi:10.1016/j.enbuild.2012.08.018

McKinsey Global Institute. (2018). *Smart Cities: Digital Solutions for a More Livable Future*. Retrieved from https://www.mckinsey.com/~/media/mckinsey/industries/capital%20projects%20and%20infrastructure/our%20insights/smart%20cities%20digital%20solutions%20for%20a%20more%20livable%20future/mgi-smart-cities-full-report.ashx

Vermesan, O., & Friess, P. (Eds.). (2013). *Internet of Things – Converging Technologies for Smart Environments and Integrated Ecosystems*. Aalborg: River Publishers.

KEY TERMS AND DEFINITIONS

Air Pollution: is a type of environmental pollution usually caused by harmful substances in the air like carbon, sulphur and nitrogen oxides. These substances with a poisonous effect are waste mainly from human activities.

Buildings' Management System: is a system that acts like a general supervisor of all buildings in the city, setting the optimum operating parameters for it. The Buildings' Management System sees the city buildings as a whole and acts accordingly.

Building's Monitoring System: acts at a local level, setting the best operating parameters for all building systems. The scope of the building monitoring system is to improve the quality of life of its tenants and also to offer them a safer place for living and working.

Intelligent Urban Management System: is a concept that relies mainly on artificial intelligence, telecommunications and information. All these main components work as a whole and are inextricably linked. Intelligent Urban Management System's role it to supervise all city's systems and associated infrastructure, optimizing and improving its operation. It is the "brain" of a smart city.

Internet of Things (IoT): is a concept and a paradigm that consider all surroundings electronic devices as things/objects connected each other through wireless and wired connections and communicating and interacting together over the Internet. A "thing" in the IoT could be every electronic object that can be assigned an IP address and is able to transfer data over a network. Some examples of such things are: sensors, actuators, switches, RFID tags, devices, home appliances and industrial equipment, but also people with a heart monitor implant or animals with a biochip transponder.

Smart City: is an urban area that integrates Internet of Things (IoT) devices, Artificial Intelligence (AI) and information and communication technologies in order to optimize the efficiency of city operations, services and infrastructure use, keeping citizens permanently informed and increasing contact between citizens and government.

Smart House: a house that uses automation, communication technologies and artificial intelligence in order to offer their tenants a increased safety and quality of life. A smart house has all appliances connected together through an internal network, an integrated management system that supervises all appliances and systems as whole and outside connections via Internet. A Smart home offers to its tenants the possibility to command and supervise from everywhere all house appliances, equipment and systems.

Smart Government: is about using technology to facilitate and support better planning and decision making. It is about improving democratic processes and transforming the ways that public services are delivered. It is a new way of governance relying on information and communication technologies and it is citizen centric, data driven and performance focused.

ENDNOTES

1 Demographic values are based on data from United Nations Population Division (http://www.un.org/en/development/desa/population/)

2 Internet usage information comes from data published by Nielsen Online (http://nielsen.com/us/en.html), by International Telecommunications Unit (http://www.itu.int), by GfK, by local ICT regulators

Chapter 6
Phenomena Implied by Sustainable and Green Retrofitting: A Quantitative Approach

Giani Ionel Gradinaru
The Bucharest University of Economic Studies, Romania

Alina Paula Moise
Ion Mincu University of Architecture and Urbanism, Romania

Raluca Dana Caplescu
The Institute of National Economy, Romania

ABSTRACT

With all the new technology and scientific progress, the human population is significantly increasing, and the planet will hardly support the overpopulation. The increase of the carbon footprint and natural resources decrease with each human. Society must find solutions to use resources in a smart way, reducing the ecological impact and stopping the cities from sprawling. Instead of extending the manmade environment and using more resources, we can act on what we have, upgrading our buildings by using sustainable and green retrofitting and making them energy efficient and environmentally friendly, while covering our cultural, social and psychological needs. Overpopulation, as well as depopulation/migration, greatly affects architecture. This study aims at finding retrofit solutions and strategies in an attempt to improve the existing urban environment quality and buildings' energy efficiency, provide interior comfort and contribute to aesthetic urban image, as well as doing so in a cheap and efficient way.

DOI: 10.4018/978-1-5225-9104-7.ch006

INTRODUCTION

Sustainable and green retrofitting has been proposed to upgrade the building's stock in many countries with different climatic conditions. A state-of-the-art review of sustainable and green retrofitting emphasizing current implementations, technologies, and benefits is presented in this chapter. By reviewing the benefits related to the reduction of building energy consumption, mitigation of urban heat island effect and improvement of air pollution, this chapter shows how sustainable and green retrofitting may contribute to more sustainable buildings and cities. Economic considerations related to the life-cycle cost of the retrofitted buildings are presented together with policies promoting retrofitting worldwide.

The aim of the chapter is to define the concept of sustainability and to identify solutions and strategy for retrofitting an existing building, analysing the current situation, the investment and the benefits of retrofitting. The aim is to understand the necessity of retrofitting and the methods that can be used to improve the quality of the existing built environment.

Sustainability is a term that defines the ability of a system to function for a long time, often related to the concept of sustainable development. The most widespread definition is the one of Brundtland report in 1987, which defines "[s]ustainable development [as] a development that meets the needs of the present without compromising the ability of future generations to meet their own needs." So it suggests that we should carefully manage what we have, taking into consideration the future effects. The term refers not only to the environmental development - which includes the natural resources - but also to the economic, social, cultural, safety aspects (Dabija, 2018).

The contemporary principles of sustainability and resilience imply the use of what it already has, focusing on how to adapt existing built sites to the society's needs with less impact on the natural environment and without destroying the balance of the network that connects the material to the immaterial spheres of the specific place (culture, religion, economy, tradition, etc.) (Dabija, 2018). In an epoch of consumerism, fast fashion and massive migration, people are easily abandoning places and move to other new places. So, it should permanently adapt the places that are no more suitable or used through conversion, update, integration, extend, restoration and retrofit.

BACKGROUND

Looking at the vernacular houses it can see that those buildings are the most sustainable. They were made with biodegradable local materials -material that can easily return to their soil and which don't affect the nature. They were perfectly adapted to their climate and geographical context, with the suitable roof slope, oriented according to the cardinal point which made the interior cool in the summer and warm in the winter. Those strategies that adapt the house to its context and users' needs without consuming energy are called passive strategies. More than that, it was a biophilic house – a house which was creating a perfect symbiosis between human-made environment and nature without affecting the nature.

Unfortunately, that type of vernacular house is not perfectly suitable for the current society. The new society doesn't have time to bring firewood from the forest, it doesn't have time to think about the protection of nature, it doesn't have enough local materials; there are millions of companies that produce construction systems and products for buildings; the people are moving from house to house and they want to build cheap and fast; they are dangerous and irresponsible consumers; they are absorbed by technology. It is hard to turn back into a vernacular house, but it can get inspired by its sincerity and its relationship with the natural environment.

Throughout the history of urbanization, there has been tension between the economic, environmental, social and security benefits of dense cities and the human desire to be connected to nature. Starting with the industrial revolution in the 19th century lots of people moved from the rural areas into the cities and the cheap and dense buildings, the industry and the huge infrastructure took the place of nature. The people were fascinated about the new technologies and they believed that they have the whole control over the world, over nature. The factories have been surrounded by crowded and unhygienic workers housing. The aim was to bring as many people was possible in to the city to feel the industry. In the 20th century, the modernist architects tried to improve the quality of life of the industrial cities. Le Corbusier, the father of the modernism, proposed the replacement of older cities with "towers in a park", an ultimately heartless, placeless strategy that, after World War II, gave rise to urban renewal. The strategy wiped out whole neighbourhoods, eliminating their historic fabric, and replaced them with a failed monoculture of isolated low-income public housing projects. The traditional houses have been replaced by impersonal and cheap white boxes spread in the city. The modernism was a response to the problems of the crowded and unhygenic industrial cities, but it ends as a failed social experiment because the society has changed. The energy-efficiency wasn't a subject to have in mind at that time. Some theoreticians like Ebenezer Howard and later the architects of the Art and Crafts Movement were trying to bring back nature and the traditional techniques into the design, but the breath wasn't that big as modernisms.

The design of the modern urban built environment has encouraged the massive transformation and degradation of natural systems and increased human separation from nature. Unfortunately, this design paradigm has resulted in unsustainable energy and resource consumption, major biodiversity loss, widespread chemical pollution, atmospheric degradation and climate change. It must assume that and fixing the existing cities. Sustainable development, with its increasing economic efficiency, protection and restoration of ecological systems and improvement of human well-being, is a tool in the action against climate change and its disastrous consequences. (Sinha et al., 2012).

The balance between resources and needs it should be found. With the excessive energy consumption and the excessive production of greenhouse gas emission worldwide, the demand for saving strategies increases. The buildings are one of the main sources of those negative consequences. Energy consumption in buildings increased drastically over the last decade. Different policy actions towards the promotion of energy efficiency in the building sector have been developed in different zones. Retrofitting is an important tool in those strategies. The objective of this chapter is to study and find sustainable and green retrofitting strategies for existing buildings.

According to a study made in 2010 by United Nations Environment Programme (UNEP), the biggest generators of greenhouse gas emission were China (23%), USA (14%), Europe (10%), India (5%) and Russia (5%). The sectors responsible for greenhouse gas emission are: Energy (35%), Industry (18%), Transport (13%), Agriculture (11%), Forestry (11%), Buildings (8%) and Waste (4%).

By intervening in the buildings sector, it can automatically intervene in almost all the polluting sectors. The greenhouse gas emission from the energy sector can be reduced by building efficiently from an energy perspective, from transport sector by using local materials and integrating mixed function on the district level, in the case of the building, from industry sector using environmentally friendly materials, and from waste by using the recycled and/or recyclable product and using the existing buildings or product. Carbon dioxide from cement production and fossil fuel combustion is the largest contributor to total greenhouse gas emissions (UNEP, 2014). Buildings are pollutants in the construction phase, and especially in the use and maintenance phase, as well as in the demolition/ transformation phase. A green building is a building which is environmentally responsible, throughout its life cycle: design, construction, use, maintenance, renovation and demolition. Usually, people don't take into consideration what happens after the house will no longer be suitable for its initial use. Life cycle analysis of a product or building is a tool of environmental management that identifies the flows of materials, waste and energy in the life of the product and its impact on the environment. Many studies have shown that the operation phase accounts for the majority of the energy consumption in the building lifetime.

The aesthetical quality is also an important part of a sustainable building. A building which is aesthetical pleasant will be more protected by the people and will improve the general image of the city. As many studies and historical theories proved about the concept of beauty trough the time, "beauty" is not about the taste of the observer, is about the well-made. A well-made building in terms of physical and psychological comfort will be considered beautiful by the occupant. A beautiful building will be more durable in time because of its qualities that make it been appreciated and protected by the people.

Besides overpopulation, there is another contemporary problem: migration. People are no longer attached to a place. The meaning of the word "home" does not have the intensity and the power that it had in the past. People are moving from place to place and gaps appear in the cities– abandoned places, abandoned buildings. It cannot simply ignore them and extend the city in the natural environment with new buildings. It should start with a macro view and first retrofit the city and stop urban sprawling and the consumption of more resources. First of all, it needs to look at the urban area as a WHOLE. The city is a system with interconnected elements. If one element changes, it will have an impact on another element and furthermore on the whole system. It can be considered the building as covalent part of the city.

In order to control the consumption of the cities and to improve environmental quality, a conceptual model arose somewhere in the 1990s promoting compact energy-efficient urban development. Green Urbanism is a concept which develops many patterns in time, but some principles stand up in every context: minimize the energy consumption and concentrate the city creating sustainable communities. Green Urbanism principles are based on the triple-zero framework: zero fossil-fuel energy use, zero waste and zero carbon emissions.

The German architect and urban planner Steffen Lehmann developed a strategic case study of green urbanism for an Australian city, presented in his book "Green Urbanism: Formulating a Series of Holistic Principles." One of the 15 principles of Green Urbanism, which is referring specifically to retrofitting, is Principle 8: Density and Retrofitting of Existing Districts. In order to bring people back to the city centre and minimize the distances between home and work and amenities, he proposes the densification of the city centre through mixed-use urban infill, centre regeneration, buildings vertically rather than horizontally and retrofitting or adaptive-reuse of the inefficient building stock. The aim is to create compact and sustainable communities and gradually reduce the carbon footprint. The city should also be self-sufficient and use Energy Services Company principles for self-financing energy efficiency as part of the retrofitting programmes (Steffen Lehmann, 2010).

SUSTAINABLE AND GREEN RETROFITTING SOLUTIONS FOR EXISTING BUILDINGS

The development of the green building movement is led by the global sustainability goals. Sustainable development and green buildings are concepts that are interconnected. The Green Program is having an unprecedented success as is providing a measurement unit for people's efforts towards sustainable development. (Sinha, Gupta, & Kutnar, 2012). A sustainable building is more than an energy-efficient building and more than a building with a green roof, green facades and integrated photovoltaic systems (Dabija, 2018). A sustainable building should be energy-efficient, self-sufficient, made with as much as possible green and local materials, integrated into its context.

By making a smart investment at the beginning or retrofitting the existing building, it can reduce the costs and the energy consumption from use and maintenance. In time, the resident will recover the investment and will save money and at the same time will benefit from the comfort, as well as protecting the environment. Addressing the need for human satisfaction is very important. According to the Commission Agency, people spend up to 90% of the day indoors — at home, work or school and if the thermal comfort parameters are not proper that can affect human health and productivity. That makes us consume energy to ensure the interior thermal comfort. At building level, research on the human benefits of sustainable design has centred on three primary topics: health, comfort and satisfaction. Thermal comfort is a subjective notion correlated with the well-being and psychological state according to the environment (ASHRAE). To be more precise, it is the psychological-physiological sensation that corresponds to the balance between domestic heat production and the ambient heat of the human-environment heat when this equilibrium is achieved without adaptation efforts that generate disagreeable sensations. So, the thermal comfort could be measured in terms of the human adaptation efforts. The main factors that influence thermal comfort are those that determine the metabolic rate, which is reflected in heat gain and loss, clothing insulation, air temperature, mean radiant temperature, air speed and relative humidity. Also, the psychological parameters, such as individual expectations can influence thermal comfort (de Dear, Richard; Brager, Gail, 1998). The indoor thermal comfort parameters values are indicated by regulations and by Green Building Certification Systems as LEED, BREEAM, DGNB, HQE.

Improving the energy efficiency of a building it can also contribute to the improvement of the thermal comfort parameters. Energy-efficiency is the quality of a building to consume energy to the necessities of normal utilisation of a building (heating, cooling, ventilation, lighting, etc.).

The European Union (EU) is a very active organisation in the field of sustainability and developed an entire program in order to reduce the greenhouse gas emission and address climate pollution by setting a binding legislation to ensure the EU meets its climate and energy targets for the year 2020. The package sets three key targets:

- 20% cut in greenhouse gas emissions
- 20% of EU energy from renewable sources
- 20% improvement in energy efficiency

Based on the 2020 climate and energy package, EU adopted in October 2014, the 2030 climate and energy framework, that goes further and sets three key targets for the year 2030:

- At least 40% cuts in greenhouse gas emissions (from 1990 levels)
- At least 27% share for renewable energy
- At least 27% improvement in energy efficiency

It is also in line with the longer-term perspective set out in the Roadmap for moving to a competitive low carbon economy in 2050, the Energy Roadmap 2050 and the Transport White Paper. The main tool in achieving those targets it's that the developed countries have to act as a group, because humans live in a system and they cannot change the world act as individual.

The EU Legislation provides a calculation methodology of an energy performance, minimum requires for energy performance, the obtaining of an Energy Performance Certificate, from which the heritage, religious buildings and holiday houses are excluded. The Energy Performance of Buildings Directive (EPBD) is the main legislative instrument at EU level to achieve energy performance in buildings. As they say, "the objective of the Directive is to promote the improvement of the energy performance of buildings within the Union, taking into account outdoor climatic and local conditions, as well as indoor climate requirements and cost-effectiveness." EPBD, 2010/31/EU, recast in 2010, mentions that starting from 2018 the public buildings will not receive the Construction Permit if they are not at least nearly zero-energy buildings (nZEB) and from 2020 that will be valid also for the residential buildings. The EPBD define the minimum requirements of energy performance for buildings. The strategies are adapted to each EU country depending on their cultural, social, climatic and economic context so each EU country has their National Energy Efficiency Action Plan (NEEAP) based on the EPBD, 2010/31/EU. These plans may include targets differentiated according to building category.

A **nearly zero-energy building (nZEB)**, is a building that has to lead, in time, to reduced consumption of energy from fossil sources and use renewable (non-fossil) sources of energy. This is a concept approved and implemented by the European Union with the goal of having all buildings in the region under nZEB standards by 2020. In the directive 'nearly zero-energy building' means a building that has a very high energy performance. The nearly zero or very low amount of energy required should be covered to a very significant extent by energy from renewable sources, including energy from renewable sources produced on-site or nearby. Similar concepts are developed by U.S. Department of Energy Office of Energy Efficiency & Renewable Energy. These buildings consequently contribute less to overall release of greenhouse gas into the atmosphere than similar non-ZNE buildings (European Commission). The energy modernisation programmes are applicable for both new and old buildings. An existing non energy-efficient building can be transformed in nZEB by changing the design in order to add passive elements, by changing the old installation and by adding new technology. The aim is to meet the minimum cost-related requirements set up and to ensure the technical solutions adopted in accordance with the provisions of Delegated Regulation (EU) No 244/2012. All EU state-members are required to establish a comparative methodological framework for the calculation of cost optimality levels for the energy performance requirements of buildings. Cost optimal performance level means the energy performance in terms of primary energy leading to minimum life cycle cost (Kurnitskia, 2011). Energy simulation results for various construction concepts and building technical systems are used as input data to conduct economic calculations. These calculations resulted in cost optimal and nZEB energy performance levels (Marszal, 2010). Energy performance is defined in the Article 2(4) of the Commission Delegated Regulation (EU) No 244/2012 as "...the amount of energy needed to meet the energy demand associated with a typical use of the building which includes, inter alia, energy used for heating, cooling, ventilation, hot water and lighting."

Under Annex I(3) to the Regulation, the calculation of energy performance starts with the calculation of final energy needs for heating and cooling, and ends with the calculation of the net primary energy.

Because each country has their climatic, geographical, social and cultural context, the Member States of the EU are advised to set national definitions of NZEB and their guideline in order to achieve the

EU's targets. The benchmarks for the energy performance of nZEB are different on each climatic zone. A building from a Mediterranean zone will not follow the same principles as a building from a Nordic zone. As an example, according to the Commission Recommendation (EU) from 2016, a new Mediterranean single family house has this requirement: 0-15 kWh/(m^2.y) of net primary energy with, typically, 50-65 kWh/(m^2.y) of primary energy use covered by 50 kWh/(m^2.y) of on-site renewable sources while a new Nordic family house has this requirement: 40-65 kWh/(m^2.y) of net primary energy with, typically, 65-90 kWh/(m^2.y) of primary energy use covered by 25 kWh/(m^2.y) of on-site renewable sources.

Some general advices of the commission to its Member States are:

- To use renewable energy sources in an integrated design concept to cover the low energy requirements of nearly-zero energy buildings.
- To ensure proper indoor environment and to avoid deterioration of indoor air quality, comfort and health conditions in the building.
- To set their mechanism that will be used to monitor the fulfilment of the nZEB targets and to consider the possibility to set up differentiated sanctions for new buildings after the nZEB deadlines have passed.
- To set a consistent mixtures of policy instruments (policy packages) to provide the required long-term stability to investors in efficient buildings, to stimulate cost-effective deep and nZEB renovations.

An energy efficiency building follows some basic principles:

- Right dimensions – don't make your house too big if you'll not use it
- Performant envelope – a bigger investment will help you save money long term
- Active strategies, energy efficiency equipment – try to find equipment that is optimally designed, is not big energy consumer and it is suitable for your building
- Renewable energy sources such as sunlight, wind, water, biomass
- Passive strategies – which don't consume energy - as a well-designed building in accordance with the compass points, windows shutters, ensuring of natural ventilation, use of sustainable materials, etc.
- Efficient lighting (natural+ LED) – try to use the natural light as much as possible because is much better for health
- Building Management System (BMS), otherwise known as a Building Automation System (BAS)
- Users education – people should be taught about and helped to make their homes more energy-efficient
- Interior comfort – ensure the optimal thermal comfort parameters with a minimal effort
- Minimum consumption of energy (Ghiță, 2017)

For an existing building it could be hard to apply some of the passive strategies because it is already designed. Concerning that, there is a big stock of existing buildings which do not meet the energy-performance requirement and it can focus on the rehabilitation and modernisation of existing buildings. EPBD also sets minimum energy performance standards for buildings undergoing renovation. Their energy performance should be upgraded in order to meet minimum energy performance requirements. Just building new buildings, even if they are nZEB, and leaving the existing buildings, which are big consumers, doesn't solve the problems at the level of the city, and even less at global level in terms of building sustainability community and reducing the greenhouse gas emission. The existing tissues should adapt as well. For existing buildings, there is already a structure and an envelope so they must be upgraded by adding new technology to make them be energy-efficient. Buildings Performance Institute Europe (BPIE), has prepared a guide in order to assist the Member States in the process of developing their renovation strategies entitled "A Guide to Developing Strategies for Building Energy Renovation" and coordinated by Dan Staniaszek. This guide provides a "long-term strategy for mobilising investment in the renovation of the national stock of residential and commercial buildings, both public and private" (Staniaszek, 2013).

Retrofitting is the term which defines the action of adapting something older by adding new technologies or features in order to make the building energy efficient. An exhaustive study is required in the process of designing an effective building retrofit. The solutions could involve planimetric and volumetric changes and exclusion of the obsolete building elements. Housing retrofitting should reduce the environmental impact (e.g., energy and resource consumption, emission of air and water pollutants, waste generation, and noise), increase the indoor comfort, and improve the architectural appearance of the building facades (Ardente, Beccali, Cellura, Mistretta, 2010).

In order to reduce the energy consumption of the buildings there are some basic retrofitting measures:

- Envelope insulation (walls, floors, roofs).
- Upgrading of windows and doors quality.
- Solar shading - notably to reduce the requirement for air conditioning.
- Reducing air infiltration – the thermal bridges.
- Upgrade of heating ventilation air conditioning (HVAC) system.
- Installation of combined heat and power systems.
- Connection to district heating system.
- Installation of mechanical ventilation heat recovery.
- Upgrade of lighting systems - Switching to compact fluorescent light bulbs and/or LED light bulbs.
- Improved controls (Building Management System- BMS).
- Installation of renewable energy measures (solar hot water, solar photovoltaic panels, heat pumps, biomass boilers, mini wind turbines, solar hot water, biogas, etc.) (Staniaszek, 2014).

The most significant benefits of energy consumption assessment are achieved if the envelope thermal insulations, lighting systems and glazing quality are improved. The thermal performance of a building relies on its ability to minimize heat exchange through the envelope which implies to prevent air penetration through the envelope by ensuring suitable layers for it. (Ardente, Beccali, Cellura, & Mistretta, 2010). The main aim is to prevent heat transfer from the exterior of a house to its interior in summer and vice versa in winter. In general, building envelopes should include the resistance to air, water, heat, light, and noise transfer. As for thermal envelopes, they include outer walls, roof, foundation, windows and doors. Designing packages of measures for buildings of the same category would be more efficient. For example, grouping more blocks of flats to the same installation of a renewable source of energy and make a connection to district heating system.

The Building Management System (BMS) is an extra upgrade to retrofit an existing building. This is an Automation System which controls and monitors the building's systems and equipment responsible for lighting, heating, ventilation, humidity as well as households, multimedia systems, shutter systems and security systems (Ghiță, 2017). It requires a big investment, but it can significantly reduce the energy consumption. It could be difficult to introduce this aspect in retrofitting policies, which need the action of the local authority, for the residential sector because at the moment it is seen rather as something fancy and not even indispensable. On the other hand, it could be very useful, especially for public buildings, most of which are relatively new, already having this control system.

The issue is that a retrofit is not binding to be sustainable and green. It's true that achieving energy efficiency and indoor thermal comfort are goals of a sustainable design but sustainability is more than that. As well as achieving energy efficiency, sustainability implies the increase of economic efficiency, protecting and restoring ecological systems, and improving human well-being.

The materials that are chosen for retrofitting are very important. Life cycle performance of the material should be analysed before deciding to use it. They consume resources and energy, therefore having an environmental impact.

The materials there are used for sustainable and green retrofit should at least:

- Have a minimum environmental impact and embodied energy, taking all its life cycle into consideration, from extraction to decomposition.
- Be local: to reduce the pollution resulting from transport and to improve the local economy.
- Be performant: resistant in time.
- Be reusable and recyclable or biodegradable: if you can't use again the product or its material, insure that its "death" it's not polluting.

Resource management, pollution or indoor environmental quality, and performance are the most general criteria for evaluating building materials are (Milani, 2005; Spiegel & Meadows, 2006). The resources used by a material should be considered in all its life-cycle, including all the components and energy used to extract, process, transport, use, dispose of/recycle and decomposition/incineration. The buildings are very big consumers of materials. Every product used in construction is resource and energy intensive. The energy used to produce or process the construction products to a usable form is known as embodied (or embedded) energy. Pollution includes all the emissions used to extract or produce the material, as well as the emissions from use and maintenance – emissions from products used to clean and maintain the material– along with the pollution resulting from its final incineration or land-filling (Milani, 2005). Performance refers to how well the material does its intended job. Even if the material doesn't pollute too much on its production phase, a material with low durability can hardly be qualified as green. Durability is defined as the ability of a building or any of its components to perform the required functions in a service environment over a period of time without too much cost and effort for maintenance or repair. An example of the green material is wood. Wood is a durable material, should be used extensively in construction. Its processing is easy and doesn't pollute much, can be easily procured if it's a local material, it is recyclable and biodegradable, it has very good thermal performance and it is also a warm natural material good for psychological comfort. The only minor issue is that it needs special treatment against insects and other exterior factors to ensure its durability. On the other hand, the most ungreen material is, as probably everyone knows, the plastic and all the "poly" family (polypropylene, polycarbonate, polyester, polystyrene, polyurethanes, etc.).

The product is the result of processing the materials. For a product, the life cycle starts with procuring the raw material, primary processing, secondary processing or manufacturing, packaging, shipping and handling, installation, in-use energy consumption, maintenance, and end-of-life strategies. For buildings, the life cycle generally starts with extraction of raw resources from the natural environment or recovery of materials from a previous use. The raw resources are then manufactured into usable products, such as steel, concrete etc. (Sinha, Gupta, Kutnar, 2012).

Another issue is that the existing buildings are built in different periods and in different contexts. It is essential that buildings meet energy performance requirements in line with their characteristics (buildings elements and function), the local climate and cultural, social and economic context when major retrofit works are planned. For example, in the case of a historical building, retrofitting for improving the energy-efficiency could cause façade damage and would be more important to preserve the building as heritage than make it energy-efficient. The heritage buildings are not obliged to comply with these requirements if their architectural and historic value might be compromised. Also, a building which is used from time to time (e.g. a theatre or a wedding hall) maybe doesn't need so strict retrofitting as a residential building which is permanently used. According to the European Directive 2010/31/EU (EPBD), "[w]ith low demolition rates (0.1% per year) and low refurbishment rates (1.2% per year), the EU needs to boost the number of deep renovations (60% savings or more) and to speed up the renovation rate of the existing stock to above 2% annually."

For undertaking major renovation actions that may have an impact on the building structure, it is necessary to have a permit attesting that the renovation project meets all the legal requirements. If the building is listed as or is located in a historical area, an additional permit confirming that the renovation of the façade does not change the main characteristics of the building and/or area is required.

The Energy Efficiency Directive mentions in Article 4 – Building Renovation that all the Member States shall establish a long-term strategy for mobilising investment in the renovation of the national stock, both residential and commercial buildings, both public and private. This strategy includes a statistical study of the national building stock, identification of cost-efficient approaches to renovations and policies and measures to stimulate this cost-efficient deep renovation of buildings, including staged deep renovations, a long-term perspective to guide investment decisions for individuals, construction industry and financial institutions and provide an evidence-based estimate of expected energy savings and wider benefits. According to the dictionary, "renovation" means "to make something old to look new again by repairing and improving it, especially a building", while "retrofit" means "to change or improve something such as a machine or a building by adding parts, equipment, or features to it" (MacMillan Dictionary). So, the terms are quite similar, but renovation is rather repairing, while retrofit is an upgrade.

Not in the least, education of the occupants and investors is very important because the building doesn't use energy: people do. Without information, just by force, the action plan of retrofitting will not be a success. The aim is to create a mentality that makes people see that their actions influence the whole system and the negative effects turn back to them – the world is a complex system. One of the important pillars of sustainability is education – teaching people how to act in order to protect their environment; the ability to educate and change the behaviours to act for sustainable development. Some means that are emphasized are critical thinking, imagining future scenarios and making decisions in a collaborative way (Dabija, 2018).

BENEFITS OF SUSTAINABLE AND GREEN RETROFITTING

Based on the triple bottom line structure of sustainable development or green development, economic, environmental and social aspects are the three parts of sustainability claimed to have similar level of criticality (Wilkinson, 2012).

There are benefits of building stock retrofitting for both investors and for citizens. Usually, the only benefit that is seen by the potential investors is the energy cost-saving, yet doing so undervalues the full impact. However, many benefits accrue to society at large and hence are not valued by individual investors. The aims of retrofitting are to improve the energy efficiency of buildings in the use and maintenance phases, but also to create more pleasant places and improve the indoor thermal comfort parameters for the wellbeing and health of the occupants. All of these while also minimizing the environmental impact through reducing the greenhouse gas emission and the consumption of resources. So, there are economic, social and environmental benefits.

The impacts of undertaking sustainable energy renovation of buildings can be summarised as follows:

- **Economic Benefits:** Energy Cost Savings; Economic Stimulus; Impact on GDP; Property Values; Industrial Competitiveness; Impact on Public Finances; Energy Import Bill. It also has a major impact on the affordability of housing and in the construction industry, which generates about 9% of Europe's GDP and directly accounts for 18 million direct jobs. Usually to retrofit an existing building can be cheaper and faster than demolished it and build a new one; the adaptation of a building can increased its values without a really effort; the energy consumption will be decreased as well as the costs.
- **Societal Benefits:** Reduction in Fuel Poverty; Health Benefits; Preserves the societal and cultural assets of the society and increase the physical and psychological comfort without the effort of building a house from 0. Increased Comfort/Productivity: The studies show that proper lighting, especially natural light, would considerably improve the occupant's capacity of concentration and productivity. Natural ventilation, proper values of air humidity and a recommended temperature depending on the activity, are also important factors for health, comfort and productivity.
- **Environmental Benefits:** Minimize the greenhouse gas emission; Minimize de carbon footprint; Air quality improvement, Minimize the land utilization by using renewable sources of energy and sustainable and green materials taking into consideration the life-cycle of the product/ building.
- **Energy System Benefits:** Energy Security; Avoided New Generation Capacity; Reduced Peak Loads. Achieve energy independence will also make the local economy stronger.

This list represents some general benefits that can be obtained from renovating buildings to higher energy performance standards. Some are tangible and readily quantifiable, while others, like the social cohesion and sustainable communities, are less so and may be difficult to be quantified, but they are visible.

If retrofitting is considered at district or city level, then it needs to evaluate the renovation strategies. All the urban actors should be part of the plan for achieving sustainable and green retrofitting. Providing funding for renovation may be seen as being of strategic national importance in terms of the savings it provides for public health expenditure and reduced social security payments, in addition to being a cost-effective mean of cutting energy import bills and meeting climate commitments. Put simply, renovating the nation's buildings improves the health and the wealth of its citizens.

Long-term strategies for mobilising investment in the renovation of national building stocks provide multiple benefits but the strategy needs to be seen as a living document that is regularly reviewed and updated. To impose limits for 2020 (the package 20/20/20), and further for 2050, will help national authorities to develop personalised strategies and create a policy framework in accordance with the directive. In Europe, the period until 2020 is a period in which policy packages will need to be tried and tested, and learning and experience shared across the EU. Building owners will need to be sensitised to the importance of building renovation.

THE BIGGEST CONSUMERS OF RESOURCES

The European Union Building Stock Observatory monitors the energy performance of buildings across Europe. The information is contained in a database, a data mapper and factsheets. It assesses improvements in the energy efficiency of buildings and the impact of this on the actual energy consumption of the buildings sector overall.

The Observatory tracks many different aspects including:

- Energy efficiency levels in buildings in individual European Union countries and the European Union as a whole.
- Different certification schemes and how they are implemented.
- Financing available for renovating buildings.
- Energy poverty levels across the European Union. (Building stock observatory).

On a global scale, the buildings are responsible for the consumption of 40% of all energy, 55% of materials and resources, and are also responsible for 33% carbon dioxide emissions. Around 70% of buildings are residential and 30% represent the tertiary sector. Globally, infrastructure and building construction consume 60% of the raw materials extracted from the Earth (Bribian et al., 2011; MMSD, 2002).

Case Study: Retrofitting in Eastern Europe, Romania

A big problem of building's energy performance and an emergency in retrofitting them is represented by Eastern Europe. Large housing estates made in the Soviet era are an important, almost iconic, common element of the building stock in Eastern Europe. These large, monotonous blocks of flats, mostly using prefabricated sandwich panels, apply uniform solutions and similar standards. Unfortunately, similarities extend to other periods as well. They have a lot of problems and are the very opposite of a sustainable and green building. The buildings have very low energy performance, no thermal and phonic insulation, no energy from renewable sources, sometimes insufficient light and a desolate aspect. It needs to talk about Eastern Europe because here is where the retrofitting action should be of the utmost large-scale and should represent an emergency. Due to poor economy and lack of education about sustainability the retrofitting actions are almost absent here.

The share of such dwellings in the countries described in this paper is 16.7% on average, representing a moderate figure in the region. Just to provide a frame of reference: Slovakia has 778,000 such buildings, representing 40% of the residential building stock; the same figure in Poland is 6,171,000, (49%).

Estonia, 45% and in Romania, 29.4% of the dwellings are in apartment buildings, 68.3% of which consist of large panel buildings (Csoknyai, 2015).

According to ODYSEE, the household sector and the tertiary sector (i.e. offices, retail premises and other non-residential buildings) is responsible for 46% of total national energy consumption.

Based on Law 372/2005 (recast 2013) provisions, all new buildings should be NZE buildings starting 1 January 2019 for public buildings and 1 January 2021 for all others (the deadlines represent the starting date of the construction works). But it shouldn't ignore the enormous amount of existing building which cannot simply be demolished and rebuild new fancy sustainable buildings. The strategy for the energy renovation of Romania's building stock is very well documented and provides all the necessary information but, unfortunately, is more theory than practice.

About 25% of the building stock in Europe was built in the middle of the 20th century – the modernist era (Tadeu, Rodrigues, & Freire, 2015). After the wars and urban demographic boom, the cities had to be reconstructed fast and cheap. The buildings from that period are among the largest contributors to the poor energy performance of the building sector, especially the socialist-modernist buildings from Eastern Europe.

According to the analyses made in 2014 by the Ministry of Regional Development and Public Administration, the biggest consumers are the dwellings and the educational buildings. These are usually buildings made in the communist period with cheap prefabricated panels, made by typical projects that are not adapted to the climate zones. These buildings represent the typical blocks of flats and schools of the socialist-modernist era, present in all the Eastern Europe, in the former Soviet countries. There were no aesthetical purpose and functional conditions were minimal. The buildings built mainly between 1950 and 1990 and have poor energy performance due to the lack of technical regulations before 1973 (when the energy crisis started) regarding the building envelope and its thermal protection design. Thus, they do not comply with current energy performance requirements. The final energy consumption of these buildings varies between 150 and 350 kWh/m^2.year. It may also be noted that buildings constructed in the early years after 1990 have a low energy performance (150 - 250 kWh/m^2.year), while buildings constructed after 2000 have a better energy performance (120 - 230 kWh/m2 year). The relatively new offices and commercial buildings have new technologies. Their façades are made of glass or ventilated facades systems, so they don't need envelope retrofitting. They can be upgraded by implementing active systems, such as photovoltaic, led illuminating, water collector, etc. The problem is that they have private owners and it is their decision if they want to invest in retrofitting.

So, the programs which are the most present on the retrofitting lists are the block of flats and educational buildings. Law 372/2005 (recast 2013) and the national regulation of certification of energy auditors for buildings make no certification distinction for public buildings. Energy auditors must comply with the general requirements and the cost of energy performance C is similar to those of residential buildings. The control procedure for public buildings is also similar. But the control, generally, has focused on the display of the EPC in public buildings because those are viewed as examples for the public. Also, to encourage owners and tenants to improve the energy performance of their building or building unit, a certificate of energy efficiency is mandatory to be made when you want to sell a property.

Data and Methods

Given the novelty of the domain and the slow pace at which retrofitting may be applied to existing building stock, available data is rather limited. However, the most recent census recorded some valuable information, from which the following variables were extracted, at county level, for the analysis:

- Total number of buildings and number of residential buildings.
- Total number of dwellings and number of occupied dwellings, including information on the structure of dwellings (placement in residential buildings, whether the buildings are blocks of flats or individual houses).
- Information on all retrofitted buildings by type (wall insulation, insulating balcony doors and windows, closed balcony, roof thermal insulation) and number (one of the aforementioned or more) of improvements, as well as number of new dwellings with all of the above and number of dwellings without any of those improvements.
- Same information as above, but only for dwellings that have insulated double (or even triple) glazing units (IGU) for energy efficiency.

The source of the data is the 2011 census results, published by the Romanian National Institute of Statistics (NIS) at:

http://www.recensamantromania.ro/noutati/volumul-iv-cladiri-locuinte-gospodarii/.

Additionally, the following macroeconomic variables (all for 2011) were collected from the NIS website: GDP/capital (computed from the county GDP, divided by the permanent population at July 1st), average monthly net salary, employment rate and unemployment rate.

Starting from these data, the variables were analyzed for underlying factors, using Principal Component Analysis (PCA) in order to obtain components for a multiple linear regression and the counties were grouped into clusters using hierarchical clustering.

The first method was applied on the combination of the macroeconomic variables and percentages in total existing dwellings of the following types of dwellings:

- Dwellings with insulated double-glazed windows that have at least one type of improvement.
- New dwellings with insulated double-glazed windows with all improvements.
- Dwellings with insulated double-glazed windows and no improvements.
- Dwellings in blocks of flats.
- Dwellings in individual houses.

For facilitating the interpretation of results, the Varimax rotation was used, together with eliminating from the output all coefficients with absolute values below 0.5. Also, the Anderson-Rubin scores were saved. These scores were then used in multiple linear regressions in order to determine the factors that influence the percentage of dwellings that underwent thermic rehabilitation.

The second analysis aimed at creating clusters of counties based on the same variables used in the PCA. The cluster method used was Ward's and the measure was Squared Euclidian Distance. The labelling variable was county name.

RESULTS

Principal Component Analysis

The PCA KMO, measuring the adequacy of the sample, has a value of 0.71, which is above the recommended value of 0.5, thus indicating that the sample is adequate. Bartlet's measure tests the hypothesis that the correlation matrix of the variables is an identity matrix (namely, that all variables are independent), which in this case is rejected (Sig.<0.001). The result leads to the conclusion that the variables analyzed have common variance, which indicates the presence of underlying factors (Table 1).

After extraction, all variables have communalities above 0.5, indicating a significant amount of common variance in the data. This is also confirmed by the factors extracted: four factors account for a total explained variation of about 78% (Table 2).

Rotation was used for easier interpretation and the four components resulted could be described as existent stock structure (component 1), general improvements (component 2), economic development of the county (component 3) and specific improvements (component 4), as shown in Table 3.

Table 1. KMO and Bartlett's test

Kaiser-Meyer-Olkin Measure of Sampling Adequacy.		,710
Bartlett's Test of Sphericity	Approx. Chi-Square	561,495
	Df	91
	Sig.	,000

Table 2. Total variance explained

Component	Initial Eigenvalues			Rotation Sums of Squared Loadings		
	Total	% of Variance	Cumulative %	Total	% of Variance	Cumulative %
1	5,712	40,802	40,802	3,843	27,447	27,447
2	2,483	17,734	58,537	2,815	20,106	47,553
3	1,401	10,010	68,546	2,807	20,048	67,602
4	1,355	9,679	78,225	1,487	10,624	78,225
5	,912	6,515	84,740			
6	,731	5,222	89,962			
7	,424	3,025	92,987			
8	,286	2,042	95,029			
9	,260	1,859	96,889			
10	,171	1,219	98,108			
11	,148	1,058	99,166			
12	,084	,603	99,769			
13	,028	,203	99,972			
14	,004	,028	100,000			

Table 3. Rotated component matrix

	Component			
	1	2	3	4
Percentage of blocks of flats	,890			
Percentage of individual houses	-,873			
Percentage of dwellings with insulated double-glazed windows and closed balcony	,801			
Percentage of dwellings with insulated double-glazed windows and no improvements	,746			
Percentage of conventional dwellings occupied	,692			
Percentage of dwellings with insulated double-glazed windows and insulated walls		,944		
Percentage of dwellings with insulated double-glazed windows and multiple improvements		,894		
Percentage of new dwellings with insulated double-glazed windows and all improvements		,770		
Employment rate			,923	
GDP/capita (thousands of lei)	,509		,739	
Average monthly net salary (lei)			,606	
Unemployment rate			-,584	
Percentage of dwellings with insulated double-glazed doors and windows for the balcony				,850
Percentage of dwellings with insulated double-glazed windows and insulated roof				,732
Extraction Method: Principal Component Analysis. Rotation Method: Varimax with Kaiser Normalization.				

a. Rotation converged in 6 iterations.

All variables load highly onto a factor, most of them being strongly correlated to it. The only inverse correlation in component 1 is for percentage of individual houses. This indicates that when the percentage of individual houses in total dwelling stock is high, values for the other four variables are low. This is especially true for percentage of blocks of flats and dwellings with closed balcony. A possible explanation for the latter is that houses tend to be rather traditional, in the sense that they are either only one level in height or have no balconies in the upper level. Another interesting finding related to component 1 is that a high percentage of occupied individual houses are paired with a medium-low percentage of dwellings with insulated double-glazed windows and no improvements. Paired with the trend of migrating from urban to rural areas, especially after retirement or not long before it, this finding indicates that in doing so, people also renovate and retrofit the existing building stock in rural areas (mostly old family homes). There is also the possibility that they invest in their parent's house or that they build new ones in the place of the old houses.

Given the implementation of the national strategy, through which buildings associations can insulate their building with subsidies from the city hall, it comes as no surprise that the most common retrofitting solution applied is insulated walls, which correlates very highly with component 2. Also, since building permits are not issued for projects that do not comply with the minimum standards, as well as because people began to be increasingly interested in the energy certificate of the building, new dwellings are built with modern technologies embedded and are more energy-efficient than the existing stock.

Improvements correlated with the last component are rather rare for the time being, but one must keep in mind that the data is from the 2011 census, when the initiative was still relatively new. For this reason, newer information might show notable improvements in this regard as well.

The most prominent variable for the third component is employment rate. This indicates higher living standard, as well as higher expectations, but also possibility to have the financial power to invest in retrofitting. The GDP/capita, which was used as proxy for the county's economy, and the average monthly salary (proxy for living standard) have upper-middle correlations with the component. This might suggest that the aspect that is most important for people is the stability that a job offers. This conclusion might also be supported with the argument that if people have stable jobs, they may take credits with which to pay for their investments in retrofitting, which otherwise would be inaccessible to them. GDP/capital also has a middle correlation with the first component, thus linking the economy with the size and structure of the dwelling stock.

Multiple Regressions

The scores for the four components were introduced in a multiple linear regression having as dependent variable the percentage of dwellings that underwent a retrofitting process. Several models were tested, starting from introducing only the four factors (macroeconomic factor was not significant), to introducing only the three factors related to retrofitting, plus the macroeconomic variables and the percentage of occupied dwellings separately and removing the non-significant variables (the change in R^2 is not significant upon removal of those variables). The final model comprises factors 1, 2 and 4 (retrofitting), percentage of occupied dwellings and employment rate (Table 4). The OLS hypotheses hold trues for the sample, the model is statistically significant (F=122,85; Sig. < 0.0001) and the total variance explained is 94.5%.

The coefficients obtained suggest a direct relationship between the percentage of retrofit dwellings in total existing stock and the independent variables, with the exception of percentage of occupied dwellings. Predictable though as the results might be, they have the merit of quantifying to a certain extent the relationship between the dependent variable and the regression. Given the loading factors for each component, only the increase in the percentage of individual houses would result in a negative impact on the dependent variable, while all other variables have positive impacts. A notable contribution is made

Table 4. Coefficients of the final regression model

Model		Unstandardized Coefficients		Standardized Coefficients	T	Sig.	95,0% C.I. for B	
		B	Std. Error	Beta			Lower	Upper
3	(Constant)	,520	,160		3,247	,003	,195	,846
	Existent stock structure	,027	,005	,293	5,278	,000	,016	,037
	General improvements	,087	,004	,949	23,165	,000	,079	,094
	Individual house specific improvements	,017	,004	,187	4,736	,000	,010	,024
	Employment rate	,001	,000	,094	2,391	,022	,000	,002
	Percentage of occupied dwellings	-,441	,192	-,131	-2,294	,028	-,830	-,051

a. Dependent Variable: Percentage of retrofit dwellings in total existing stock

by the percentage of dwellings with insulated double-glazed windows and insulated walls or multiple improvements (Table 3), which have the biggest impact.

On the other hand, the more occupied dwellings, the lower the percentage of improved dwellings. This suggests that unoccupied dwellings have very low energy efficiency, in contrast with the occupied ones, where the owners invest in making them more efficient, most likely to reduce their costs with heating/cooling.

A small, but statistically significant effect is also that of the employment rate, the increase in which indicates an increase in percentage of retrofit dwellings in total existing stock. Related to the point made above, employed people have not only higher living standards, but also more disposable income to invest in retrofitting, although the relatively small impact suggests that education to this aim is stringently necessary in order to improve the energy efficiency of buildings in Romania.

Hierarchical Cluster

Based on the same variables as for the PCA, the 42 Romanian counties were grouped into four clusters. Before clustering, all variables were standardized. The cut value used was 10, since above it lays the first large difference between clusters.

The last two clusters obtained are the city of Bucharest and the county Ilfov, which are very different both from each other and from the rest of the country. Their main characteristics are presented in Table 5, together with a synthetic description of the other two clusters, based on average values taking all counties in the cluster into consideration. All counties have dwelling occupancy rates above 80%, with 81% in Ilfov and 89% in Bucharest. Also, the dwellings with insulated double-glazed windows and closed balcony or insulated with energy-efficient doors and windows or with insulated roofs are very rare as unique improvements for a Romanian dwelling. That is to say, there are few such dwellings that have exactly one of the three.

The four clusters basically synthesize the situation in Romania at county level. On the one hand, there is the capital city, Bucharest, which is a highly urban area with mostly blocks of flats. A significant number of dwellings in Bucharest are built before 1989, thus with low energy-efficiency. Among these buildings, the ones from the second half of the 20th century stand out as particularly intensive resource consumers. About half of the dwellings that have insulated double glazed windows have no other type of improvement. Even though the macroeconomic indicators suggest there are sufficient resources to deal with this problem (both GDP/capita and average monthly net salary are highest in the country) and that there is sufficient workforce (80% employment rate and 2% unemployment rate), one must take into account the large gap between the classes. Bucharest is home mostly to people living at extremes, with either very high or very low living standards. Also, it is the city where cooperation seems least probable, possible also because people work long and hectic hours and they don't form communities in their buildings.

On the other hand, there is the Ilfov county, which is mostly a rural area surrounding Bucharest. This area has mostly individual houses, but since its development began especially after 2000, it benefited from developing infrastructure and new buildings, as multinational companies established their Romanian headquarters in and around the capital. Thus, the houses built are mostly suburban houses, relatively new and with several energy-efficient technologies. This is explained by the great number of residential projects developed in the past two decades, when the building market boomed. Since within the borders of Bucharest building opportunities were rare and expensive, developers targeted the margins of the city,

Table 5. Hierarchical cluster results

cluster 1	cluster 2	cluster 3	cluster 4
28 counties	12 counties	Bucharest	Ilfov
Most dwellings are in individual homes	About half dwellings in individual home and half in blocks of flats	Most dwellings are in blocks of flats	Most dwellings are in individual homes
Low percentage of the dwellings with insulated double-glazed windows and insulated walls		Low percentage of the dwellings with insulated double-glazed windows and insulated walls	High percentage of the dwellings with insulated double-glazed windows and insulated walls
Lowest percentage of dwellings with insulated double glazed with more than one improvement			Highest percentage of dwellings with insulated double glazed with more than one improvement
			High percentage of new dwellings with insulated double glazed and all thermic facilities
		Almost half of the dwellings with insulated double-glazed windows do not have any kind of improvement	A fifth of the dwellings with insulated double-glazed windows don not have any kind of improvement
Smallest GDP/capita		Highest GDP/capita	High GDP/capita
Smallest average monthly net salary		Highest average monthly net salary	High average monthly net salary
Low employment rate	Low employment rate	Highest employment rate	High employment rate
High unemployment rate	High unemployment rate	Low unemployment rate	Low unemployment rate

which are part of Ilfov, but are very close to Bucharest. Dwellings with no improvements are most likely those further from the capital, which are older.

With regard to the rest of the country, there is, on the one hand, the mostly rural, less developed part of Romania (28 counties), which did not attract big investors, thus having lower living standards and less efficient economy, as well as lower employment rate and higher unemployment. Being rural to a high extent, many people work in agriculture, be it for subsistence or large scale. On the other hand, a less performing economy is associated with higher unemployment rates. Also, this cluster contains several counties that experienced mass emigration. For these reasons, retrofitting of the existent dwelling stocks is slower in these counties, which, on average, have registered the least progress.

The remaining 12 counties, that form the second cluster, are more developed, more urbanized, with half the dwelling stock in blocks of flats. The economies of these counties is sustained by investors, thus the GDP per capita and the average monthly net salary are higher than those of the first cluster. As the rest of the province, these counties also have low employment rates, high unemployment rates and high emigration, though not as much as the previous cluster. For these counties, the retrofitting process is quite advanced, compared to the rest of the country (except for Bucharest and Ilfov).

The key message to take is that things are heading in the right direction and the speed with which the new technologies are adopted greatly depends on the current situation (existing stock of dwellings, structure of the building stock, accessibility and affordability of retrofitting solutions and macroeconomic conditions).

CONCLUSION

Despite EU framework and national strategy in force, the analysis indicates a rather reduced level of retrofitting in Romania. On the one hand, these results may be due to the fact that the data analyzed was from 2011, when the legislation had only just been developed, thus the census might have caught only some early changes. It will be possible to assess this situation after the next census, by comparing the results.

On the other hand, the few results in the area of adapting and reusing the existing buildings stock might also be explained by an underlying cultural factor: it is common in Romania that people do things on their own, instead of getting together and collaborating. In this sense, the results seen so far might be given more by individual households finding solutions on their own in order to reduce expenses, rather than a collective (county or national-wise) approach.

Having this in mind, the need to educate people about the benefits of retrofitting, as well as those of collaboration becomes increasingly more stringent. Education is an important pillar of sustainability and teaching people how to act in order to protect the environment is crucial, as is the ability to educate and change behaviours to act for sustainable development. Some means that are emphasized are critical thinking, imagining future scenarios and making decisions in a collaborative way (Dabija, 2018).

REFERENCES

Ardente, F., Beccali, M., Cellura, M., & Mistretta, M. (2011). Energy and environmental benefits in public buildings as a result of retrofit actions. *Renewable & Sustainable Energy Reviews*, *15*(1), 460–470. doi:10.1016/j.rser.2010.09.022

Atanasie, B. (n.d.). Net-Zero Energy Building (nZEB) Implementation In Romania. *Buildings Performance Institute Europe (BPIE)*. Retrieved from www.bpie.eu

Csoknyai, T., Hrabovszky-Horvátha, S., Georgiev, Z., Jovanovic-Popovic, M., Stankovic, B., Villatoro, O., & Szendro, G. (2015). Building stock characteristics and energy performance of residential buildings in Eastern-European countries. *Energy and Buildings,* 132(15), *39-52*. Retrieved from www.sciencedirect.com

Dabija, A.-M. (2018). Sustainability - A Principle of Education in Architecture (and not only). *Paper presented at Conference on Sustainable Energy*. Retrieved from www.researchgate.net

de Dear, R., & Brager, G. S. (1998). Developing an adaptive model of thermal comfort and preference. UC Berkeley. Retrieved from www.escholarship.org

El-Darwish, I., & Gomaa, M. (2017). Retrofitting strategy for building envelopes to achieve energy efficiency. *Alexandria Engineering Journal*, *56*(4), 579-589. Retrieved from www.sciencedirect.com

European Union. (n.d.). European Commission Directives. Retrieved from www.ec.europa.eu

Ghită A.-M. (2017). Training platform "Sustainability Consultant". *Solar Decathlon Bucharest*. Retrieved from www.efden.org

Kurnitskia, J., Saarib, A., Kalameesc, T., Vuolled, M., Niemeläd, J., & Tark, T. (2011). Optimal and nearly zero (nZEB) energy performance calculations for residential buildings with REHVA definition for nZEB national implementation. *Energy and Buildings, 43*(11). Retrieved from www.sciencedirect.com

Marszal, A. J., Heiselberga, P., Bourrelle, J. S., Musall, E., Voss, K., Sartori, I., & Napolitanoe, A. (2010). Zero Energy Building – A review of definitions and calculation methodologies. *Energy and Buildings, 43*(4). Retrieved from www.sciencedirect.com

Milani, B. (2005). *Building materials in a green economy: Community-based strategies for dematerialization* [PhD Dissertation]. University of Toronto, Canada

Sinha, A., Gupta, R., & Kutnar, A. (2012). *Sustainable Development and Green Buildings.* Retrieved from www.researchgate.net

Staniaszek, D. (2013). *A Guide to Developing Strategies for Building Energy Renovation.* Buildings Performance Institute Europe (BPIE). Retrieved from www.bpie.eu

United Nations Environment Program -UNEP. (2007). *Buildings and Climate Change: Status, Challenges and Opportunities.*

Wilkinson S., (2012). Analysing sustainable retrofit potential in premium office buildings, *Struct Survey/ International Journal of Building Pathology and Adaptation, 30*(5).

ADDITIONAL READING

Adalberth, K. (2000). Energy use in four multi-family buildings during their life cycle. *International Journal of Low Energy and Sustainable Buildings, 1*, 1–22.

Dall'O, G., Galante, A., & Pasetti, G. (2012). A methodology for evaluating the potential energy savings of retrofitting residential building stocks.

Furuto, A. (2013). *'New Acadia, Retrofitting Urban Decay' Winning Proposal / Garrett Rock.* Retrieved from www.archdaily.com

Gustavsson, L., & Joelsson, A. (2010). Life cycle energy analysis of residential buildings. *Energy and Building, 42*, 210–220.

Reffat, R. (2004). Sustainable construction in developing countries. *Paper presented in the Proceedings of First Architectural International Conference*, Cairo University, Egypt.

Sartori, I., & Hestnes, A. G. (2007). Energy use in the life cycle of conventional and low-energy buildings. *Energy and Building, 39*(3), 249–257.

Thormark, C. (2007). A low energy building in a life cycle. Its embodied energy, energy need for operation and recycling potential. *Building and Environment, 37*(4), 429–435. doi:10.1016/S0360-1323(01)00033-6

Walsh, N. P. (2017). Harvard HouseZero - A Retrofit Response to Climate Change. Retrieved from www.archdaily.com

KEY TERMS AND DEFINITIONS

Green Building: A building which is environmentally responsible, throughout its life cycle: design, construction, use, maintenance, renovation and demolition.

Carbon Footprint: The amount of carbon dioxide released into the atmosphere trough the life-cycle of a product or as a result of an activity.

Energy Efficiency: The characteristic of a product, process or activity to consume the energy in useful work and not wasted as useless heat. In the building sector represent the goal to reduce the amount of energy required to provide products and services.

Life-Cycle: The series of changes that a product, process or activity goes through during its existence.

Nearly Zero-Energy Building (nZEB): A building that has to lead, in time, to reduced consumption of energy from fossil sources and use renewable (non-fossil) sources of energy.

Renewable Energy: Energy that is collected from renewable resources, which are naturally replenished on a human timescale, such as sunlight, wind, rain, tides, waves, and geothermal heat.

Sustainability: The process of maintaining change in a balanced environment, in which the exploitation of resources, the direction of investments, the orientation of technological development and institutional change focuses on meeting the needs of the present without compromising the ability of future generations to meet their needs.

Chapter 7
Considerations Regarding the Green Retrofitting of Residential Buildings From Human Wellbeing Perspectives

Raluca Andreea Felseghi
https://orcid.org/0000-0001-6310-5932
Technical University of Cluj-Napoca, Romania

Teodora Melania Şoimoşan
https://orcid.org/0000-0002-5929-1071
Technical University of Cluj-Napoca, Romania

Constatin Filote
https://orcid.org/0000-0001-6640-5308
Stefan cel Mare University of Suceava, Romania

Maria Simona Răboaca
National R&D Institute for Cryogenic and Isotopic Technologies, Romania

ABSTRACT

Currently, buildings are considered to be a continuously evolving organism that over time has to be treated, rehabilitated, and upgraded to meet the requirements set by the user at a certain stage. Buildings are a central element of the EU member states' energy efficiency policies, accounting for about 40% of final energy consumption, and 36% of greenhouse gas emissions, and about 75% of buildings are not energy efficient. Recent applications and studies establish that green retrofitting has maintained older existing buildings to increase energy efficiency, optimize building performance, increase occupants' satisfaction, and boost economic return while decreasing greenhouse gas emissions. In this regard, this chapter aims to address the main factors that negatively affect the performance of residential buildings and presents the common green retrofitting measures that can be taken to ensure the state of human well-being in residential buildings.

DOI: 10.4018/978-1-5225-9104-7.ch007

INTRODUCTION

At a time when energy, environmental, economic and social concerns are becoming more important, being represented by climate change or those that jeopardize energy security, resource exhaustion or human health, reducing energy consumption along with human comfort and well-being in the buildings sector is of strategic importance, both at national and international level. Besides the efforts to development of new high-quality buildings which assures a high level of comfort to the building's occupants while complying with the conditions of energy and environmental efficiency, it is essential to address an attitude towards human well-being and healthy environments, along with the decrease in high energy consumption levels in the context of rehabilitation of existing buildings.

Green retrofitting of the existing building stock is essential, not only to meet national energy efficiency targets, but also to meet the medium and long-term climate change objectives and move towards a sustainable, low-carbon economy by 2050.

By making a significant contribution to the EU's energy consumption, conventional energy use and carbon dioxide emissions, as well as a number of factors that can adversely affect occupants' health, the buildings sector is the subject of many policies and objectives in the medium and long term to reduce the negative impact on the environment. Aims formulated by the 20-20-20 target by 2020 are the set of three key objectives for:

- Reduction by 20% from EU's greenhouse gas emissions compared to 1999 levels.
- Increasing the share of energy produced from renewable sources in the EU by 20%.
- Improving the energy efficiency in the EU by 20%. (MDRAP, 2017b)

As the European energy system faces an increasingly pressing need for sustainable, affordable and competitive energy supply for all citizens, the European Commission adopted on 30 November 2016 the legislative package "Clean Energy for All Europeans", through which is aimed the implementation of strategies and measures to achieve the objectives of the Energy Union for the first ten-year period (2021-2030), notably for the EU's 2030 energy and climate objectives, and refers to: energy security, energy market, energy efficiency, de-carbonisation, research, innovation and competitiveness.

In a wider perspective, the EU set a set of long-term objectives in roadmaps until 2050 (COM, 2011). Regarding the building sector, the main three roadmaps are:

- The EU's objective for transition to a competitive low carbon economy by 2050 (COM, 2011a), which identified the need to reduce by 88% - 91% the carbon dioxide emissions from the residential and service sectors (collectively called the real estate sector) by 2050 compared to 1990 levels.
- The Energy Perspective 2050 (COM, 2011b), through which "increasing the energy efficiency potential of new and existing buildings is essential" for a sustainable future.
- An Energy Efficient Europe Plan (COM, 2011c) identifying the real estate sector as one of the top three sectors responsible for 70% -80% of the overall negative impact on the environment. Achieving better constructions and optimizing their use would reduce by over 50% the amount of raw materials extracted from the underground and could reduce water consumption by 30%.

These roadmaps are a long-term aspiration that is not only desirable from a social and economic point of view, but also ecologically essential and human health, safety and well-being in buildings.

Existing EU Buildings Fund: It is estimated that there are 24 billion m^2 of useful floor space in the EU, Switzerland and Norway. According with the Buildings Performance Institute Europe (BPIE, 2011), for the analytical purposes, the European countries have been divided based on climatic, building typology and market similarities into three regions, illustrate in Table 1. In comparison to China and the US, Europe has the highest 'buildings density' (building floor space over land area) followed by China and then US.

Half of the total estimated floor space is located in the North & West region of Europe while the remaining 36% and 14% are contained in the South and Central & East regions, respectively.

Non-residential buildings account for 25% of the total stock in Europe.

The BPIE survey has classified buildings in different specific chronological periods for regions. The age profiles of the residential building stock of different regions for the floor area data has been consolidated into three representative age periods:

- Old: typically representing buildings up to 1960;
- Modern: typically representing buildings from 1961 to 1990;
- Recent: typically representing buildings from 1991 to 2010.

Table 1. UE regions (BPIE, 2011)

			Land area (km^2)	Building Floor Space	Population
UE	North & Vest	AT, BE, CH, DE, DK, FI, FR, IE, LU, NL, NO, SE, UK	4 324 782	24 billion m^2	281 million
	Central & East	BG, CZ, EE, HU, LT, LV, PL, RO, SI, SK			102 million
	South	CY, GR, ES, IT, MT, PT			129 million
US			9 826 675	25 billion m^2	309 million
China			9 598 080	35 billion m^2	1338 million

Figure 1. Residential building stock (BPIE, 2011)

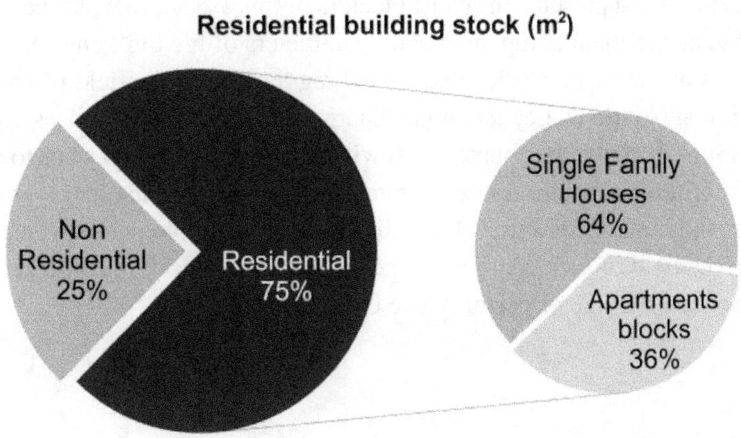

Figure 2. Classification of buildings at different time periods specific to the regions (BPIE, 2011)

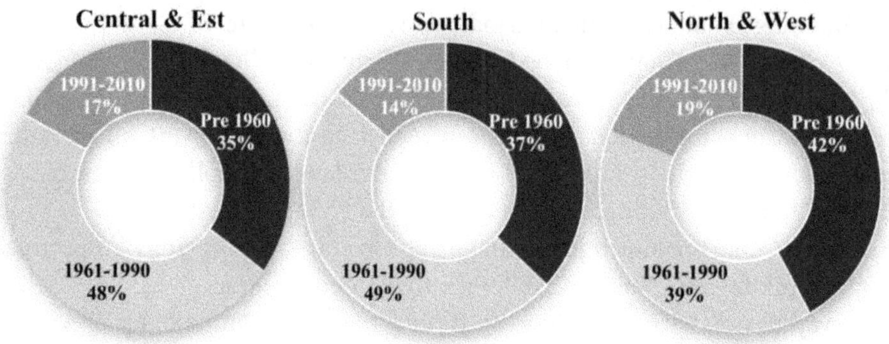

Romania has an important heritage of buildings, mainly built between 1960 and 1990, with a low degree of thermal insulation, the consequence of which, prior to the energy crisis of 1973, there were no regulations regarding the thermal protection of the walls and perimeter closure elements and are no longer suitable for the purpose for which they were built (MDRAP, 2017b).

26 years after the Revolution, housing in Romania is among the most crowded, degraded, expensive and inadequate to the quality of living between the countries of Europe and exposes a large number of people to various risks, the real estate situation having significant negative effects on housing conditions (MDRAP, 2017a).

Currently, development of green and smart buildings through innovative technologies applications has registered increasing interest. Future buildings, including intelligent, smart, green and sustainable characteristics, have become a common object of various academic debates, research projects and practical implementations related to the design, standardization, engineering in construction field (Ghaffarianhoseini et al., 2018).

Human health, safety and well-being in the building are a particularly important state of building design centered on the occupant and on the optimum comfort parameters of the built environment. In the general context presented, also taking into account the size of the housing stock inadequate to the dwelling, authors propose that through this chapter to provide useful synthesis of the factors with impact on occupant health, well-being and productivity. In this respect, the chapter defines the concept of Sick Building Syndrome (SBS) and presents the causal factors of this aspect, then defines Well-Being Building concept and presents the main components and parameters of the built environment in accordance with the specifications of the main entities that issued regulations in the field of the well-being building: U.S. Green Building Council's Leadership in Energy and Environmental Design and International WELL Building Institute. Finally, the chapter ends with a series of common ways to renovate buildings and general green retrofitting solutions for existing residential buildings in order to improve their energy efficiency and improve housing conditions (in light of recent technological advances), that will be of interest to experts in the field of architecture, engineering, design, building construction, sustainability, public health, outdoor and the indoor performance of buildings.

SICK BUILDING SYNDROME CONCEPT

Human in the developed countries spend around 90% of their time inside their workplace or in their own homes (Yang, 2007), so the design, construction or retrofitting of healthy and comfortable buildings represents a critical fundament (Evans, 1998; WHO, 1999).

Edification of green energy-efficient buildings with high air tightness of the building envelope and inadequate maintained ventilation systems are often fallows in unhealthy and uncomfortable indoor quality conditions as well as in Sick Building Syndrome (SBS).

Sick Building Syndrome (SBS) has attracted worldwide attention from the public, engineers, architects, physicians and research communities, because of the rise in cases of overheating, deficient ventilation and poor indoor air quality (IAQ) and has increased the global awareness of the built environment's role in human health, safety and well-being.

Definitions of Sick Building Syndrome

SBS is the common denomination for buildings with defects in their construction or functionality that can cause people to face health issues. The SBS concept was initially developed by the World Health Organization (WHO) in 1983, afterwards then various scientific, research, public health communities have provided a series of definitions for SBS, which are found in the literature, and have been synthetically presented in Table 2.

The common aspect that is frequently repeated by the presented definitions confirmed the association existed between the perceptions of weak physical indoor environmental conditions and appearance of common medical symptoms. SBS is recognized by the medical community, but it can be very difficult to diagnostic because persons can expose such a wide array of symptoms.

The increase of SBS during the last 40 years is assigned to the oil embargo in 1973, which established a high influence for construction of energy efficient buildings. This has led to design and realization of the tight, sealed buildings, which were appreciable for controlling temperature, but critically affected indoor air quality. Building occupants lost the possibility to open windows, control temperature and lighting and the ability to access clean air. Simultaneously, building owners have tried to economies money by cutting back on building services and installations, especially on ventilation systems (Hamilton, 2010).

Recent approaches treat SBS from the perspective of a collection of factors that contribute negatively to human physical health and psychological well-being.

Factors That Can Cause Sick Building Syndrome (Contributors)

The World Health Organization (WHO, 1983) appraised that up to 30% of new and renovated buildings worldwide may be associated to SBS. Comprehensive study (Burge, 1987) achieved in the UK on 4373 office workers in 42 public buildings illustrated that 29% of those analyzed experienced five or more of the specific features SBS symptoms. An examination carried out by Woods et al., (1987) on 600 office workers in the USA established that 20% of the employees get through SBS symptoms and most of them were confirmed that this decreases their working efficiency. Additionally, a study on 1390 employees in 5 office buildings in Quebec, Canada (Bourbeau, 1997) revealed that 50% of workers experienced SBS symptoms. SBS may also take place in other environments such as schools, kindergartens and residential buildings (Araki, 2010; Engvall, 2001; Sahlberg, 2013; Scheel, 2001; Takigawa, 2012). In analyses on

Table 2. Defining the SBS concept

	Reference	Definition
2011 2018	Abdul-Wahab Clements-Croome	SBS is referred to scenarios where over 20% of buildings users are concerned about a similar medical issue while being in the building as a result of an unidentified reason during a minimum cycle of 2 weeks.
2010	Crook and Burton	SBS is a complex spectrum of ill health symptoms associated with the indoor environment. In broad terms, these symptoms can be divided into mucous membrane symptoms related to eyes, nose and throat; dry skin; general symptoms of headache and lethargy. These symptoms should improve within hours of leaving the problem building.
1991	EPA	SBS is used to describe situations in which building occupants experience acute health and comfort effects that appear to be linked to time spent in a building, but no specific illness or cause can be identified.
2018	Ghaffarianhoseini et al.	Sick building syndrome (SBS) is a collection of factors that can negatively affect physical and mental health of people in several ways.
1996	Goldman	SBS is where the occupants of a building are affected over an indefinite period, and is directly connected with the building itself.
2009	Gomzi and Bobić	An environmentally related condition with increased prevalence of nonspecific symptoms among the population of certain buildings, often without clinical signs and objective measures of symptoms.
2007	Greer	A group of nonspecific symptoms with a temporal connection to a particular building, but with no specific or obvious cause.
2011	Imai & Imai (2011)	SBS symptoms are related to the environments, with a focus on chemical exposure and indoor moulds, where mucosa symptoms and general symptoms such as headache, dyspnoea, loss of consciousness, and visual disturbance start appearing shortly after people move into a new house or into a renovated house.
2011	Jansz	SBS is the density of "workers complaints," in which the WHO defined this density to be 20% of the building occupants presenting with the symptoms of SBS. SBS is a clinical diagnosis without any cause, or causes, being specifically identified.
1990	Molina et al.	SBS is the name given to a set of varied symptoms experienced predominantly by people working in air-conditioned buildings, although it has also been observed in naturally ventilated buildings.
2013	Norhidayah et al. (2013)	SBS is the exposure of common symptoms amongst certain individuals working or living in an environment, where the appearance of these signs depends on their presence in that place.
2009	Passarelli	SBS is a range of nonspecific illnesses that are experienced by an occupant while inside a particular building or within a specific area of the inside environment. The symptoms experienced usually disappear in hours, or in some cases days, after the occupant is away from the enclosed environment.
2008	Rostron	SBS is a syndrome of complaints covering nonspecific feelings of malaise, the onset of which is associated with occupancy of certain modern buildings.
2016	Shan et al. (2016)	SBS is a group of symptoms related but not limited to the irritation of the eyes, nose, throat, skin, breath, and other general symptoms such as headache and lethargy that temporally occur among occupants of a certain building.
2012	Takigawa et al. (2012)	SBS is characterized by various nonspecific subjective symptoms, including irritation of the eyes, nose, and throat, headache and general fatigue, in rooms with deteriorated IAQ.
1983	WHO	A collection of nonspecific symptoms including eye, nose and throat irritation, mental fatigue, headaches, nausea, dizziness and skin irritations, which seem to be linked with occupancy of certain workplaces.

residential buildings (Araki, 2010; Sahlberg, 2013; Takigawa, 2012), 12% to 30.8% of inhabitants were identified as having SBS.

Identification of contributory risk factors for SBS and their relevant characteristics represents an important stage towards efficacious prevention and control of SBS symptoms. Based on the comprehensive literature review the contributory risk factors for SBS were classified into six main groups: physical, biological, chemical, psychosocial, individual and others, summarized in Table 3 and synthetically illustrated in the following subchapter.

Physical Contributors

These factors are related with the physical parameters of buildings that can stimulate the appearance of SBS symptoms. These contributors can appreciably affect the human health status, well-being and comfort of residential building occupants. Comfort in residential buildings can be defined as being the whole of conditions and factors that ensure a pleasant, agreeable, comfortable and hygienic existence. It can also represent the set of optimal conditions necessary for the physiological life of human to reside in his living spaces.

Comfort in buildings has two components: indoor air quality and thermal comfort. The most relevant main parameters in the category of physical risk factors are environmental parameters of thermal comfort, parameters of building ventilation, indoor air quality: noise, vibrations, illuminance level, degree of ionization, electromagnetic radiation fields. Parameters of thermal comfort refer to: indoor air temperature; relative humidity of indoor air; indoor air speed; the average radiation temperature of the room delimitation perimeter elements; clothing and the metabolism of the human body.

Air Quality

Air quality can be defined as the total of secondary factors that, in addition to thermal comfort factors, provide a pleasant microclimate in an enclosure. The subjective comfort of people in a closed space depends on many factors, which can be grouped as follows (according to Blanchere's principle):

Table 3. Main Contributors and Risk Factors of Sick Building Syndrome

Factors & contributors that can cause SBS					
Physical	**Biological**	**Chemical**	**Psychosocial**	**Individual**	**Others**
* indoor parameters * parameters of ventilation * air quality * noise and vibrations * illuminance level * degree of ionization *electromagnetic fields	* moulds, mites and fungi * viruses * cockroaches * bacteria * house dust * pollen * microbes volatile organic compounds (MVOC)	* volatile organic compounds *formaldehyde * phthalates * pesticides * constructional products * household products * mineral fibers * tobacco smoke * odors * other pollutants	* social status * stress * loneliness and helplessness * communication * monotonous environment	* genetic tendency * own atopy * gender * health status * individual characteristics	* ergonomic * universal design * location * geographic placement * geopatogenic zones building characteristics * ownership * building characteristics

- Temperature, humidity and air circulation - thermal comfort.
- Smell and breathing - olfactory comfort.
- Acoustic factors - acoustic comfort.
- Sight and color effects - visual comfort.
- Special factors (solar inputs, ionization, vibrations and movements of the building, etc).

Indoor Air Temperature

Air temperature inside a room (t_i) has a strong influence on well-being and health (Amina, et.al., 2015). That's why, when it comes to where we live, the comfortable indoor environment plays an important role for the occupant. The notion of thermal comfort has a subjective character, depending on the behavior of each individual in relation to the environment (Hoof, 2010). The sensation of thermal comfort is defined as the conscious state that expresses satisfaction to the existing thermal environment and whose evaluation is realized with the help of the subjective level of comfort with seven levels: +3 (very warm); +2 (warm); +1 (slightly warm); 0 (neutral); -1 (cool); -2 (cold); -3 (cold). For an ambiance, it is possible to define 4 zones of microclimate for the human body in terms of indoor air temperature: cold area; indifferent area - corresponding to the thermal comfort; warm area - increased sweating phenomenon; overheated area - physical work is not recommended for a long time.

From a physiological point of view, for average values of outside temperature (t_e), current legislation and literature recommend: $t_i = +22°C$ during the winter and $22 \div 26°C$ during the summer – for a seated person, normally dressed and without physical activity; $t_i = +28°C$ - for the naked body and $t_i = (20 + t_e)/2$-on warm summer days and for a short stay in air-conditioned environments. (Domniţa, 2011)

Relatively low variations in indoor air temperature are immediately detected by the human body, which has to quickly cope with the new changes to keep the heat exchange constantly with the environment. Deviation from the thermal comfort threshold for buildings' indoor environment (Epstein, 2006) may result in occurrence or increase of SBS symptoms.

Humidity

The forms under which water acts on buildings and building materials can be mentioned as follows:

- Water in the ground acts by pressure or by capillary climb on the building envelope.
- Meteorological water, in the form of weathering (rain, snow), acts on building closures (roof, exterior walls).
- The initial construction water resulting from the process of putting the building into operation (concrete pouring, wet plastering, etc.) acts on the mass of the building elements.
- Operating water results from the functional-technological process that takes place in buildings (bathrooms, kitchens, etc.).
- Hygroscopic water is derived from indoor and outdoor air humidity, depending on the physical structure of the building material.
- Condensate water results from condensation of water vapor on the surface and mass of the exterior construction elements (closure walls, terraces, bridge floors, etc.).
- Water of biological nature results from the processes of breathing and evaporation of water from the surface of the human skin.

From a physiological point of view, the values of relative humidity of the indoor air it is recommended: $_i$= 45% ÷60% in climatic environments and $_i$ = 30% ÷ 70% - in unclimated enclosures.

The existence of humidity can potentially assure a hospitable environment for microscopic organisms such as mildews and moulds (Ghaffarianhoseini, 2018) to be grown up indoor and represent a reliable danger upon the health status of occupants. Also, the low or very high humidity parameters cause human discomfort, may cause or worsen some symptoms of diseases of building residents. (Mishra, 2013)

Indoor Air Speed

The air moving velocity at the level of the occupants influences the exchange of heat between the human body and the environment, in conclusion, it determines the thermal comfort. It is recommended that the speed of air movement in the room is not accentuate, because the sensation of current appears if the moving air has a lower temperature than the indoor air and the jet is directed to parts of the human body. In the area from floor to 2 m high in living rooms and offices with a temperature of 20÷22 ° C, the air movement speed of 0.1 ÷0.15 m / s is considered admissible. Discomfort sensation is felt even more so as the temperature of the moving air is lower than the ambient temperature. Average air velocity, according to the standards in force, Normative for thé Design, Execution an Operation of Ventilation and Air-conditioning Systems - Indicative I 5 - 2010, for a building occupant in a seated and dressed type activity is shown in Table 4. (Popovici, 2010; Domniţa, 2011)

According to Mayer and Fanger (Fanger, 1967), the air flow discomfort (ICA) can be calculated according to the degree of turbulence of the indoor air (T_u), the air velocity (v_i) and the felt temperature (t_c), with the formula (1.1):

$$ICA = \left(34 - t_c\right)\cdot\left(v_i - 0.05\right)^{0.62}\cdot\left(0.37\cdot v_i\cdot T_u + 3.14\right)\left[\%\right] \quad (1)$$

These conditions may lead to adverse health effects, the air flow discomfort may be a strong predictor of SBS symptoms.

Average Radiation Temperature of the Room Delimitation Perimeter Elements

The thermal exchanges of the human body are based on the average temperature of the enclosure walls and radiators called the average radiation temperature θ_{mr}, that can be calculated with the relation (1.2):

Table 4. Values of average air velocity

Indoor air temperature [°C]	Indoor Air Velocity [m/s]
20	0.1 ÷0.16
21	0.1÷0.17
22	0.1÷0.18
24	0.13÷0.21
26	0.15÷0.25

$$_{mr} = \frac{\sum S \cdot \theta}{\sum S} \left[^{\circ}C\right] \qquad (2)$$

where: S is the surface of each delimiting area (wall, window, radiators) and - the temperature of each delimiting area. From a mathematical point of view, the average radiation temperature represents a weighted average of specific surface products and their specific temperatures, being in fact a radiating effect on a body inside the room, hot surfaces (heaters) and cold surfaces. (Domniţa, 2011)

The average room temperature of the wall must be very close to the indoor air temperature. If it is well below the temperature of the indoor air (e.g. winter), an ambient temperature of = 20° C will be felt as very cold and will have to be raised well above 20° C (2÷4° C) have comfort. The indoor temperature can be chosen between 19÷23° C, provided the temperature $_{mr}$ has equivalent values between 16÷25° C. Surface temperatures also have to be considered due to their large influence on perceived temperature.

Lower surface temperatures may result in local discomfort and water condensation. Also, low surface temperatures often result in thermally discomforted conditions and higher prevalence of SBS symptoms.

Clothing, Activity and Metabolism of the Human Body

The International Organization for Standardization (ISO-7730) has provided for the definition of thermal comfort six main factors, adding to the four parameters of the indoor microclimate the intensity of the work i_M, expressed in met[2] (human heat release) and the thermal resistance of clothing R_{cl}, expressed in clo[2]. It can feel the feeling of very good in a cooler room but clothed thicker and vice versa in a warmer room with lighter clothing.

The human body has the quality of maintaining a temperature of 37° C, regardless of ambient temperature and physical activity. The heat produced by the body during the activity is transmitted to the environment by: radiation, convection, thermal conductivity and water evaporation through breathing and sweat.

The increase in heat ceded by the body, when it is physically carried out, is due to the increase in heat yielded by evaporation (latent heat release). At air temperatures of 18° C and below, the heat loss of the human body to the environment is mainly due to radiation and convection; at temperatures above 30° C are mainly due to evaporation of sweat and at low temperatures, heat loss by evaporation of perspiration is constant. The importance of the evaporation phenomenon depends on the partial pressure difference of the vapor from the surface of the skin and the ambient air.

When the phenomenon of evaporation is not sufficient for total heat evacuation, the subject is exposed to illness (headaches, circulatory disorders), frequent phenomena in overcrowded or unventilated rooms, being relevant SBS symptoms.

Ventilation

The current ventilation standards and guidelines (I5, 2010) recommend a minimum ventilation rate of 8 l/s per person in order to extract the bio effluents (odors). The controlled introduction into a room of a treated air flow follows: ensuring thermal comfort and ensuring indoor air quality by refreshing the air and removing odors and toxic or even toxic odors. This is done by the ventilation and conditioning system that serves that room.

The role of ventilation and conditioning installations is represented by: the introduction of treated air into rooms, called replenished air or introduced air, which takes over excess harmful (heat, humidity, gases, vapors, dust); evacuation of vicious air (odors, toxic or noxious releases), called suction air or exhaust air; ensuring the required fresh air flow in the rooms and ensuring thermal comfort in the rooms.

The main causes for SBS symptoms related to building ventilation were inadequate functioning, old and unmaintained HVAC system, decreased number of air changes and decreased volume of clean air. Poor ventilation can potentially trigger the emergence of SBS symptoms.

Olfactory Comfort

Perception of odors is not a spontaneous process; it takes place only when the air is breathing intensely when its speed is 2÷4 times higher than normal breathing. The condition of achieving human body metabolism in a closed space is oxygen (O_2) uptake and carbon dioxide (CO_2) release (Table 5).

The CO_2 concentration in the outdoor air is 300÷400 ppm (cm^3/m^3) and in the homes air is approx. 900 ppm. The maximum permissible limit of the CO_2 concentration in the inspired air is 1000 ppm.

The level of excitation for some odors is very low, the olfactory organ having an important characteristic, namely, the adaptation. After a certain amount of time, due to continuous perception, the intensity of the scent smell decreases. However, odor can lead to excitement of the olfactory system producing disagreeable and uncomfortable sensations, it can also cause physiological reactions. Although there is a means of measuring the concentration of an odoriferous substance, it is not possible to measure the intensity of the sensation they cause. (Popovici, 2010; Domniţa, 2011; Kapalo et al., 2018)

Visual Comfort

Human life is closely related to the visual environment, because information is perceived in a 90% visual way, and activity is also related to seeing in most cases. Visual comfort is the conscious state, which arises due to physiological and psychological actions that express the satisfaction of the environment. The visual environment in a closed space occurs if it is illuminated and has two components: the room, bounded by opaque or transparent surfaces (the passive component); light, which makes the room visible (active component) (Beu, 2001) The human perceives only the light that reaches the eye, the vision exactly meaning the correct perception of colors. These are considered to be appropriate if they coincide with the colors perceived by natural light, resulting in the need for natural lighting through windows, skylights, etc. The poor quality of light can directly influence human health by affecting the occupants' visual comforts.

Table 5. The expired CO_2 flow

Activity	M [W/pers]	Inspired air [m^3/hour]	Expired CO_2 [l/hour]	Consumed O_2 [l/hour]
Resting condition	-	0,3	12	14
Intellectual work	120	0,375	15	18
Physical work very easy	150	0,575	23	27
Light physical work	190	0,75	30	35
Heavy physical work	>270	> 0,75	> 30	> 35

Acoustic Comfort

There must be a difference between sound and noise. Sound can be defined according to the meaning of interpretation, as follows:

- If it is considered a physical phenomenon, then it represents the elemental vibration of the elastic matter that propagates in the waveform transport environment.
- Can be defined as the external excitation of the human hearing organ, which leads to various reactions from them.
- If the aesthetic and meaningful effect is considered, the sound has encoded information content that the brain decodes, and sound perception of sound influences human comfort.

The main element of the relationship between people is speaking, and any moment that disturbs the understanding of speech creates uncertainty and, together with it, uncomfortable.

Noise can be defined as those effects of sound, perceived negatively by each person, the difference between noise and sound is made by human through the interpretation given. Noise is an overlap of sounds, with variable frequencies and amplitudes, producing an auditory feeling that is disagreeable or embarrassing. (Popovici, 2010; Domniţa, 2011) There are rules that refer to the quality of the desired acoustic environment, expressed by an acoustic intensity level for the considered situation and a determined reference period. For example, recommended maximum values: inside the stores - 75dB (A), ambient music - 65dB (A), residential building (resting areas) - 35 dB (A), (kitchens) - 38 dB (A).

Electromagnetic Radiation

The electromagnetic spectrum refers to a group of distinct forms of energies emanating from multiple sources with various frequencies. The high frequencies are gamma rays, X-rays and ultraviolet light; the lower frequencies of the spectrum include microwaves; medium frequencies include radio waves and light wave emission that provide the possibility to perceive the vision and the light and infrared energy which allows perceiving the heat. Exposures to higher levels that might be harmful are restricted by national and international guidelines (Ghaffarianhoseini, 2018).

Degree of Ionization

The existence of ions is given by the radiation of the natural radioactive materials present in the earth and in the air, and especially cosmic radiation and storms. Inside the rooms, the ion content is higher than in the outdoor air and depends on: ventilation, the particulate content of the air, building materials used, the objects in the room, etc. The average amount of continuous ions in the air: at altitude – 500 ÷ 1000/cm^3; at lowland -1,000 ÷ 5,000/cm^3; on valleys-5,000 ÷ 50,000/cm^3, in the rooms - 50,000 ÷ 100,000/cm^3. In general, air contains negative and positive ions that can be produced naturally or artificially. Researchers support the view that negative ions have a net positive effect on health, including improved mood, stabilized catecholamine regulation and circadian rhythm, enhanced recovery from physical exertion and protection from positive ion-related stress and exhaustion disorders. The acceptable minimum concentration of negative ions for indoor air is 200–300 ions per cm^3. The optimal level is 1000–1500 negative ions per cm^3. The lack of negative ions in the air may be responsible for SBS (Kukec, 2014a, 2014b).

Biological Contributors

Several contributors can be mentioned with the association to the growth of biological factors inside the buildings namely, building's structural failure, poor air ventilation or improper maintenance of building (Ghaffarianhoseini, 2018). Biological contaminants present in indoor air include bacteria, moulds, mildew, viruses, animal dander and cat saliva, house dust, mites, cockroaches, and pollen. There are many indoor or outdoor sources of these pollutants (people, animals and plant debris) (Kukec, 2014a, 2014b).

Chemical Contributors

The most important factors for SBS among chemical contributors are used constructional and household products and emitted pollutants, especially formaldehyde, phthalates, volatile organic compounds, odors, environmental tobacco smoke, biocides, and others. Many construction products used for waterproofing, insulating, fireproofing, roofing, painting, plastering, building and treating of floors, as well as surface coating contain toxic chemicals. Constructional products may emit harmful substances in the surrounding environment during their whole life cycle. In addition to constructional elements, also household products have to be considered from the aspects of indoor environment quality. Moreover, due to low air humidity in buildings, humidifiers are often used. Humidifiers in the ventilation circuit provide a place for microbes to flourish, and also provide a reason for adding biocides to humidified water (Kukec, 2014a, 2014b).

Psychosocial Contributors

Sick buildings can cause the appearance of psychological symptoms as follow: anxiety, stress and aggression of inhabitants (Kamaruzzaman, 2011). These negative psychological issues of being in sick buildings can aggravate in further issues for human such as growing the possibility for occurring the dangerous events in living place (Gomzi et al., 2007). Literature presented the ambient noise level, poor illumination levels, the indoor climate, the air quality as the important risk factors contributors conducting to dissatisfaction of occupants (Newsham, 2009). It was studied the repercussions of physical measurements such as thermal, lighting, and acoustic parameters, furniture dimensions on the state of well-being and the degree of psychological satisfaction of occupants. Also, studies were being carried out regarding the potential exterior view on the building occupants' performances, highlighted the important role of window in psychological satisfaction with lighting, especially through its effect on contentment with outside view (Ghaffarianhoseini, 2018).

Individual Contributors

There are special individual and personal factors that have proven to growing the risk of developing SBS. Women have been revealed to be more probable than men to be affected by SBS. The literature suggests that the age is a risk factor for SBS. According to researchers, children are more susceptible to respiratory diseases due to indoor air pollutants. Those persons who have already been diagnosed with atopic skin and allergy and asthma are more prone to be affected by SBS (Sundin, 2012).

Other Contributors

The attributes causing SBS will not be limited to the mentioned factor contributors, other causes such as construction materials, weak sanitation, ozone availability, organic solvents or formaldehyde in the atmosphere, gadgets and office equipment, furnishings and other materials/products placed or used in the building with possibility to produce fume, airborne chemical fumes and gases from any parts of the building and vermin (i.e. mice, rats and cockroaches) infestation can also have a significant role towards the sick status of a building (Ghaffarianhoseini, 2018).

The basic symptoms of SBS refer to the health status of the occupants. These symptoms can cause many negative effects on the occupants, most of these effects being interdependent.

EVOLUTION OF BUILDING DEVELOPMENT FROM SBS PERSPECTIVE

Human well-being, including health and comfort, is a critical characteristic for defining the quality of life of a residential building occupant. In late 1980s and during the 1990s, WHO concept of health, became important for recognizing of the 'healthy building' concept in terms of building performances. In last years, the term 'well-being' has become a problem of social and cultural preoccupation. Residential building market personnel have become more conscious of well-being. Various media have meaningfully discussed the impact of residential environment on human health, conducting to the appearance of healthy housing as a social problem, thus boosting and widening the humans' interest in "well-being" residential buildings. (Lee J., 2011) Also, a healthy building should be prepared and capable to answer to future trends, adaptable to 'new direction' such as climate change, energetic efficiency with low pollutant emissions and responsive to the changes towards a multifunctional and various society, the growing individualizing and the ever-changing demands and preferences of resident occupants.

The understanding, measurement, and improvement of human experience have been major goals of individuals, researchers, communities and governments (Costanza, et al., 2007). To meet the occupant's comfort requirement as the top priority (Allen et al., 2017), but also the need to develop a healthy living environment in residential buildings, major changes were imposed, best practices in design and construction with proof based human health and wellness applications. The literature highlighting the evolution of building design and development from sustainable, intelligent and healthy perspectives, a special role having the deficiencies found in the existing buildings and the SBS symptoms (Table 6.).

Current societies make use of resources and energy to supply the needed for daily life, and these assets have a consequence on the level of human well-being. The method to mark the relationship between these well-being parameter and energy consumption is complex (Cravioto et al., 2011; Roulet, 2018).

Buildings are at the crucial centre of our lives. The characteristics of a building, its design, its aspect and feel, and its technical norm not only affect our productivity, our well-being, our disposition and our interactions with others, they also determine how much energy is consumed in and by a building, and how much heating, ventilation and cooling energy is needed to realize an agreeable environment (BPIE, 2011).

Table 6. Evolution of building design and development from sustainable, intelligent and healthy design perspectives (Ghaffarianhoseini et al., 2018)

Sustainable Buildings 1990-2000	**Intelligent Buildings 2000-2010**	**Sustainable & Intelligent Buildings 2010-2015**	**Healthy Buildings 2015-2018**	**Digitally Sensible & Fully Responsive Healthy Buildings 2018-Future**
Gradual attention towards sustainable buildings. 1993 - USGBC Establishment. Concept of the economic efficiency use of energy, waste, water and pollution. Global environmental actions & sustainable development trends.	Increasing concerns towards green buildings. Building automation as a core focus. Extensive development directions of intelligent buildings. Adapting buildings to technological dimensions.	Green buildings is developed widely. Low - energy demand & BIM application. Intelligent buildings provided with sustainable design. Gradual change to the social dimension of buildings.	Increasing attention to health & well-being in buildings. Sick Building Syndrome becomes a critical issue. Healthy buildings as part of smart city. Emphasis on design of healthy buildings.	Buildings to become fully sensible and responsive. Digital design for enhancing health and productivity. Health as the central point of future buildings.

WELL-BEING CONCEPT AND STANDARD

As human lead longer and generally healthier lives, aspirations, ambitions and expectations of health care extend to include well-being and increasing quality of life, and the space in which they live being a critical point. Overall well-being is expressed as the balance between two compensatory forces: positive experiences could compensate for negative feelings (Danna, 1999). Various scientific, research, public health communities have provided a series of definitions for well-being concept, which are found in the literature One of the most relevant is formulated by Stanford Encyclopedia of Philosophy, as follows: "A person's well-being is what is 'good for' them..." Oxford dictionary (2015), defines it "...as such human beings' state of comfort, health and happiness, referring to human beings' environments, bodies, minds, and all other situations being in the most harmonious and satisfactory state."

The worldwide legal framework regulates issues related to energy efficiency, human health, safety and well-being building. Presented in a synthetic but not limiting manner, at the EU level these include: European Energy Performance of Buildings Directive 2010/31/EU, Energy Efficiency Directive 2012/27/EU, Romanian Energy Strategy 2015-2035, Methodology for calculating the energy performance of buildings Mc001/1,2,3-2006, Mc001/4,5-2009, Mc001/6-2013 updated in 2017, Certification of Energy Performance of Buildings, OUG 18/2009 on increasing the energy performance of housing blocks.

The most relevant international standards and assessment tools for the well-being building are the U.S. Green Building Council's Leadership in Energy and Environmental Design (LEED) certification, Building Research Establishment Environmental Assessment Method (BREEM) certification and Well-Being Building Standard developed by International WELL Building Institute.

One source to view for advice and improvement in indoor environment is the U.S. Green Building Council's Leadership in Energy and Environmental Design (LEED) certification. It would the most widely used green building evaluation system in the world. The LEED credit system has standards for all types of sustainable and environmental practices, from water efficiency to innovations during the design process. This includes a very thorough standard for indoor environmental quality with two prerequisites and 17 credits available for buildings that show strong sustainable characteristics. (USGBC, 2018) LEED indoor environmental quality credits, presented in Table 7, are the following:

Table 7. LEED indoor environmental quality credits (USGBC, 2018; Hamilton, 2010)

No.	EQ Credits	EQ Credits Explanation
1	EQ-Credit 1.– Outdoor Air Delivery Monitoring	Monitoring the ventilation system to ensure compliance with minimum requirements and installation of CO_2 sensors.
2	EQ Credit 2.– Increased Ventilation	Increase breathing zone outdoor air ventilation rates by at least 30% above minimum rates set by SR EN Standards.
3	EQ Credit 3.1.– Construction IAQ Management Plan, During Construction	Quality Management in Construction by elaborating a plan for the implementation of the ventilation system.
4	EQ Credit 3.2.– Construction IAQ Management Plan, Before Occupancy	Designing the ventilation system to completely eliminate the air of the building after construction and before residents occupy the space.
5	EQ Credit 4.1.– Low-Emitting Materials, Adhesives and Sealants	Materials used must not exceed requirements set by legislation.
6	EQ Credit 4.2.– Low-Emitting Materials, Paints and Coatings	Materials used must adhere to standards and laws.
7	EQ Credit 4.3.– Low-Emitting Materials, Carpet Systems	Materials used must adhere to standards and laws.
8	EQ Credit 4.4.– Low-Emitting Materials, Composite Wood and Laminate Adhesives	Materials used must adhere to standards and laws. The products should not contain urea-formaldehyde resins added.
9	EQ Credit 4.5.– Low-Emitting Materials, Systems Furniture and Seating	All furniture, aged less than 12 months, must be greenguard air quality indoors requirements
10	EQ Credit 5.– Indoor Chemical and Pollutant Source Control	Special measures are required to design the ventilation system such as special filters or medium to capture pollutant particulates.
11	EQ Credit 6.1.– Controllability of Systems, Lighting	Provide lighting controls for resident occupants.
12	EQ Credit 6.2.– Controllability of Systems, Temperature and Ventilation	Provide thermal and ventilation controls for resident occupants.
13	EQ Credit 7.1.– Thermal Comfort, Compliance	In accordance with the standards in force.
14	EQ Credit 7.2.– Thermal Comfort, Monitoring	Provide monitoring system and process for corrective action to ensure system performance set forth in EQ Credit 7.1
15	EQ Credit 8.1.– Daylight and Views, Daylight 75% of Spaces	For 75% of the areas inhabited, reach a minimum Daylight Factor of 2%.
16	EQ Credit 8.2.– Daylight and Views, Daylight 90% of Spaces	For 90% of the areas inhabited, reach a minimum Daylight Factor of 2%.
17	EQ Credit 8.3.–Views for 90% of Seated Spaces.	Provide visibility to the outdoor environment for 90% of the areas occupied on a regular basis.

The vision at the U.S. The Green Building Council (USGBC) is that buildings and communities will regenerate and sustain the health and vitality of all life within a generation. The mission is to transform the way buildings and communities are designed, constructed and operated, enabling a healthy and prosperous environment that improves the quality of life. (USGBC, 2018)

Influence Factors on Human Health and Well - Being Building Standard

Over the last decade, green building standards and standard-setting organizations have made significant strides towards the market transformation of the building industry, resulting in a rapid expansion of green buildings and environmentally conscious building practices throughout the world. (WELL, 2018) Well-Being Building Standard is the leading tool for advancing health and well-being in buildings globally.

When analyzing the green retrofitting measures from a human well-being perspective is needed to be taken into account the main framework address issues that impact the health and comfort of occupants. Most actions require approaches to change behavior through education and create a culture for choosing a positive lifestyle.

Importance of the main contributors and risk factors of sick building syndrome can be emphasized from the perspective of influence factors on human health.

So that, in accordance with the literature and in particularly with Well-Being Building Standard, in Table 8 is suggestively presented the relationship of influence of these contributors on human body anatomical systems.

COMMON WAYS TO GREEN RETROFIT RESIDENTIAL BUILDINGS

Recent practices recommend that green retrofit has supported older existing buildings to increase energy efficiency, optimize building performance, growing occupants' satisfaction and boost economic return while decreasing greenhouse gas emission. Green retrofit of residential building could have different approaches. Between the prevailing practical views is the one certified by the U.S. Green Building Council (USGBC). It defining that green retrofit is "any kind of upgrade at an existing building that is wholly or partially occupied to improve energy and environmental performance, reduce water use, and improve the comfort and quality of the space in terms of natural light, air quality, and noise—all done in a way that it is financially beneficial to the owner" (USGBC, 2018).

A series of strategies and methodologies are presented and suggested in the literature to address the ways to renovate existing buildings and solve SBS-related issues. (Almeida et al., 2018) While there are many ways to retrofit a residential building, with the aim of achieving energy efficiency, healthy environment and well-being for the occupants, research identifies the most common interventions, as follows.

Envelope Insulation

The building envelope is made up of locking elements which, through their properties, play a decisive role in achieving the occupants' comfort and ensuring the human well-being. The closures of a building are characterized by finite geometric dimensions and by certain thermophysical characteristics such as thermal conductivity, air and vapor permeability, etc. From the point of view of thermal protection, these elements are important to ensure: creating a comfortable indoor climate (according to the purpose of the building) while reducing the heat loss to the outside; avoid condensation of water vapor both at the inner surface of the building elements and inside them; achieving the thermal stability necessary to limit the internal air temperature oscillations and on the inner surface of the building elements; achieving a resistance to air infiltration to reduce the heat losses corresponding to their heating.

The thermal rehabilitation solution of the building envelope is determined by a specialized technical expertise (Thermal, Waterproof and Energy Saving Requirement) depending on the destination of space and the envelope constituent elements (Abrudan et al., 2017). Adoption of the thermal rehabilitation measure of a building can be done either by thermal insulation of the wall on the inside or outside the building element.

Table 8. Influence factors on human body anatomical systems

		Body Anatomical System										
	Influence factors on human health	**Cardiovascular**	**Digestive**	**Endocrine**	**Immune**	**Integumentary**	**Muscular**	**Nervous**	**Reproductive**	**Respiratory**	**Skeletal**	**Urinary**
AIR	air quality standards	x		x	x	x		x		x		
	smoking ban	x	x	x	x	x	x	x	x	x	x	x
	ventilation effectiveness	x						x		x		
	volatile organic compounds (VOCs)	x		x	x	x		x		x		
	air filtration	x		x	x	x		x		x		
	microbe and mold				x	x				x		
	construction pollution management	x			x	x				x		
	cleaning				x	x			x	x		x
	pesticide	x	x	x	x			x	x	x		x
	fundamental material safety	x						x		x		
	moisture management				x	x				x		
	air flush	x			x					x		
	air infiltration management				x					x		
	increased ventilation	x			x					x		
	humidity				x	x				x		
	air quality monitoring and feedback	x				x		x		x		
	operable windows	x								x		
	outdoor air systems									x		
	displacement ventilation	x			x							
	advanced air purification	x			x					x		
	combustion minimization	x								x		x
	toxic material reduction			x					x			
	enhanced material safety	x	x	x	x	x	x	x		x		
	antimicrobial activity for surfaces		x	x	x	x			x	x		x
	cleanable environment		x	x	x	x			x			x
	cleaning equipment		x	x	x				x			x
WATER	fundamental water quality		x		x			x				x
	inorganic contaminants	x	x					x	x			x
	organic contaminants		x	x	x			x	x			x
	agricultural contaminants	x	x	x				x	x			x
	public water additives		x			x			x	x		
	water treatment		x		x							

continued on following page

Table 8. Continued

		Body Anatomical System										
	Influence factors on human health	**Cardiovascular**	**Digestive**	**Endocrine**	**Immune**	**Integumentary**	**Muscular**	**Nervous**	**Reproductive**	**Respiratory**	**Skeletal**	**Urinary**
NOURISHMENT	fruits and vegetables	x	x	x	x							
	processed foods	x	x	x					x			x
	food allergies		x	x	x	x				x		
	hand washing		x	x	x	x			x	x		
	food contamination		x		x							
	artificial ingredients		x					x				x
	nutritional information	x	x				x				x	
	food advertising	x	x	x	x	x	x				x	
	safe food preparation materials	x	x	x	x	x	x	x	x	x	x	x
	food storage		x		x							
	food production	x	x		x		x				x	
LIGHT	visual lighting design			x			x	x				
	circadian lighting design	x	x	x	x		x	x				
	electric light glare control						x	x			x	
	solar glare control						x	x			x	
	low-glare workstation design						x	x			x	
	color quality							x				
	surface design			x			x	x				
	automated shading and dimming controls			x	x		x	x	x		x	
	right to light	x	x	x	x		x	x	x		x	
	daylight modeling	x	x	x	x		x	x	x		x	
	daylighting fenestration	x	x	x	x		x	x	x		x	
FITNESS	interior fitness circulation	x					x				x	
	activity incentive programs						x				x	
	structured fitness opportunities	x	x				x				x	
	exterior active design	x					x	x		x	x	
	physical activity spaces	x					x				x	
	active transportation support	x						x				
	active furnishings	x					x				x	

continued on following page

Table 8. Continued

	Influence factors on human health	Body Anatomical System: Cardiovascular	Digestive	Endocrine	Immune	Integumentary	Muscular	Nervous	Reproductive	Respiratory	Skeletal	Urinary
CCOMFORT	ergonomics: visual and physical						x				x	
	exterior noise intrusion	x		x				x				
	accessible design						x				x	
	internally generated noise	x		x				x				
	thermal comfort	x			x	x		x		x		
	olfactory comfort		x	x	x	x		x		x		
	reverberation time	x		x				x				
	sound masking				x			x				
	sound reducing surfaces	x		x				x				
	sound barriers				x			x				
	individual thermal control	x			x	x		x		x		
	radiant thermal comfort	x			x	x		x		x		
MIND	health and wellness awareness	x	x	x	x	x	x	x	x	x	x	x
	integrative design	x	x	x	x	x	x	x	x	x	x	x
	post-occupancy surveys	x	x	x	x	x	x	x	x	x	x	x
	beauty and design							x				
	biophilia - qualitative							x				
	adaptable spaces	x			x							
	healthy sleep policy	x		x	x		x	x				
	building health policy	x	x	x	x	x	x	x	x	x	x	x
	workplace family support		x	x	x			x		x		
	stress and addiction treatment	x	x	x	x	x	x	x	x	x	x	x
	material transparency	x	x	x	x	x		x	x	x		
INNOVATION	innovation features	x	x	x	x	x	x	x	x	x	x	x

Installing an anti-vapor barrier on the inside of the wall is necessary in the following situations: when using mineral wool insulation materials on the inner insulation of a concrete wall in subsequent insulation of the basement walls to a degree high humidity in the rooms (i.e. kitchen, bathroom). A special case is the insulating material based on polyurethane foam because incorporates both the qualities of a vapor barrier and those of the insulators, hopping use without problems in any situation. The choice for inner isolation must be analyzed very carefully and critically. If there are already traces of moisture on the inside of the walls, or if stains or even molds have formed, it is necessary to find out the cause and treatment, disposal, in order to avoid further damage. If the ceilings are made of wood beams, the ends of the beams that are not sufficiently protected against the moisture entering the room will be analyzed.

The advantages of external insulation are the following: the solid wall can accumulate heat from the room and can have a balanced role; keep cool during the summer; the construction material of the wall is not affected by frost; reduction of temperature stresses between construction and support parts;

there is enough space to apply an insulating layer of any thickness; avoid thermal bridges through the enclosed insulating layer (Shahin, 2019). But it also presents disadvantages, such as: the rooms are used only sporadically, requiring a longer heating period, suppose a greater work effort, maybe we can put problems in the details or finishing, the outer layer is required by weather phenomena, especially temperature variations or the direct action of the sun, as well as various mechanical actions.

Generically, the materials used to isolate opaque elements of the building are as follows: wood fiber boards, classic thermal insulation systems, extruded polystyrene foam, rigid phenolic insulation, rigid sprayed polyurethane, cellulose fibers, aerogel isolation, transparent thermal insulation systems with polycarbonate capillary plates, vacuum insulated panels, insulate concrete foam, isolated structural panels, phase changing materials. (Moga, 2018)

External furniture of windows plays a particularly important role in the thermal rehabilitation of residential buildings because glazed surfaces make the largest heat transfer. Most residential buildings in Romania, built before 1990, are equipped with wooden windows with simple glass, being far from achieving energy performance and the comfort of a modern home. In the thermal rehabilitation programs, replacement of traditional windows with simple glazing, with windows with thermal insulating glass, the following advantages are obtained: decreasing thermal transfer through the glazed surface leads to thermal protection on summer time, provides special conditions for permanent ventilation of the rooms, increases the quality of life by providing modern comfort, sound insulation reduces noise, conservation of the environment by reducing CO_2 emissions in the atmosphere due to reduction the consumption of primary fuels used for heating the dwellings, lowering costs for heating. The elements from which a thermal insulating window is made must firstly give it functionality and durability as follows: minimum 3 isolation chambers, the recipe of the material from which these profiles are extruded must correspond to the severe climate in Romania, be reinforced with gallivanted steel profiles of 1.5 mm thickness, minimum profile width 58 mm. Thermoisolator glass with low-E film; its thickness must be at least 24Mm, window heat transfer coefficient - Uw $\leq$ 2 W/ m^2K. In addition, the window can be provided with a ventilation system to ensure continuous air circulation in the room and prevents condensation and mold formation under certain conditions. The following variants may be adopted: insulating glazing, tinted stained glass (protected), reflective coatings for glazing, glass with low - E layers, laminated glass. (Şoimoşan, et.al., 2017)

When determining the thermal rehabilitation solutions of the terraces and roofs, account is taken of the condition (degree of deterioration) of the existing waterproofing and thermoinsulating layers, and the following situations may exist:

- The state of waterproofing and thermal insulation is not appropriate; thermal rehabilitation involves: removing all existing layers to the upper slope of the slope concrete; restoring them under conditions of replacement of the existing thermal insulation layer with a new heat insulating layer of quality and thickness corresponding to the new requirements.
- The condition of the existing insulation is good, but the waterproofing is damaged and it needs to be restored; thermal rehabilitation consists of: removal of the existing layers to the existing waterproofing, as long as it retains its function as a vapor barrier and maintains the existing thermal insulation layer; installing an additional heat-insulating layer of quality and thickness corresponding to the new requirements; the installation of the other layers, including the waterproofing layers.
- The state of waterproofing is good and all existing layers are maintained, less the waterproofing layer, providing good mechanical and thermal protection (including the direct action of solar ra-

diation) of the waterproofing layers; thermal rehabilitation implies: removing the waterproofing layer; performing local repairs of waterproofing; installing, if necessary, an additional waterproofing layer; installing a thermal insulation layer over waterproofing.

- The state of waterproofing, thermal insulation and slope concrete are not appropriate; thermal rehabilitation implies: removing all layers, including slope concrete; complete restoration of the terrace, possibly with the installation of sloping concrete over the thermo-insulating layer.

To substantially reduce the unfavorable effect of thermal bridges on the floor plan above the last level, it is very important to take care of the attics with a thermal insulating layer. For attic that do not require conditioning, the thermal rehabilitation is achieved by the thermal insulation of the attic floor, as follows: cleaning and leveling the support surface; application of the vapor barrier; application of thermal insulation; application of a separation layer (polymeric foil, geotextile); floor installation. For attics chamber requiring air conditioning the thermal rehabilitation is achieved at the level of the roof structure, as follows: unlocking the internal enclosure structure; installing thermal insulation between or below rafters; applying the vapor barrier inwardly; mounting the ceiling. (Moga, 2017; Moga, 2018)

Ventilation and Air Conditioning (HVAC)

Ventilation and air conditioning for living spaces in residential buildings must comply with a few basic rules: ensuring sufficient hygiene by establishing minimal air flows introduced and discharged from the enclosure; conducting a general and permanent ventilation; the air circulation is to be made from the main rooms (bedrooms, living room) to the service rooms (bathrooms, kitchens), where the air vents are located, thus washing the whole house.

Natural ventilation of living spaces is the most widespread in the country, air refreshment is ensured by opening windows or through the permeability of building elements to the passage of air, the air flow is reduced to the minimum possible to eliminate the heat loss from enclosures through evacuated air.

Natural ventilation methods are as follows: ventilation by opening the glazing (the system is simple, but the air flow is random and intermittent); ventilation through faults in the façade associated with vertical evacuation piping (the system is quite uncomfortable, but conditions only improve if the air inlet is self-regulating); horizontal duct ventilation associated with vertical outlet piping (the solution is the best one, ensuring proper ventilation) (Popovici, 2010; Domniţa, 2011).

Mechanical ventilation of living quarters allows a general and permanent ventilation with a fresh air flow independent of atmospheric conditions.

Simple flow mechanical ventilation has the following features: fresh air is introduced exclusively into the main compartments (bedrooms, living room, etc.) through self-adjusting discharge nozzles which have the role of attenuating the effect of the wind in order to keep the input flow constant; under the effect of the depression of the extraction fan, the air passes through the main premises to the heavily polluted environments (kitchens, bathrooms, etc.) through door leaks or specially designed spaces for the purpose at the bottom of the doors overpressure). (Popovici, 2010; Domniţa, 2011)

Double-flow mechanical ventilation involves the existence of two fans, components of the air treatment plant, one that extracts the vicious air from the "dirty" rooms (the bathroom and the kitchen), while a second fan introduces fresh air into the "clean" "(rooms, living room, bedrooms, etc.). The advantages are the following: control of the inlet and outlet flows through the two fans; the possibility of complex treatment of the introduced air (filtration, heating, cooling, dehumidification); the possibility of preheating or precooling the fresh air introduced, using the thermal energy from the exhaust air, by means of an air-to-air heat recovery or air-to-air heat pump; the possibility of preheating or precooling the fresh air introduced using the thermal energy from the soil through a ground-air heat exchanger ("Canadian well"); good noise protection, the boiler is located in a separate or soundproofed room. (Popovici, 2010; Domniţa, 2011)

For human health and well-being, the air in residential buildings must comply with the standard values set by the regulations, norms, lows regarding: formaldehyde, volatile organic compounds, particulate matter and inorganic gases, radon, smoking ban. Through maintenance and upgrades, the performances of the HVAC system can be better resulting in improving occupants' comfort and diminishing the negative impact on the environment. In this sense, buildings' owners are encouraged to clean the air filters, ventilators, boiler tubes and make sure that all the HVAC equipment is isolated so that heat transfer take place only when and where it is demand.

Lighting

The most common retrofit type that building occupants achieve is lighting retrofits for many spaces do not have energy efficient lighting. Retrofit practice signify that upgrading lighting accessory can ensure in an increase in the lighting level while decreasing energy consumption up to 70%, rebounding significant cost saving. Also, lighting retrofits is one of the easiest to conduct involving short or no interruption to building's daily operation. There are some options for a lighting retrofit designing: indoor lighting will be ensured in all building environments, but also outdoor lighting of the building's courtyard; using natural light over a period of time in the day as much as possible and minimizing artificial lighting; carefully identify and select lighting solutions so as to integrate as few sources as possible and with efficient energy performance; replacing incandescent bulbs with fluorescent lighting and LEDs - more efficient and suitable solutions, fluorescent lighting is less efficient than LED, but produces a warm light suitable for interior space, LED illumination is more suitable for outside space; it is recommended to alternate between the two types of illumination, the solution that offers the highest efficiency and the best results; designing a sensor-based intelligent lighting system and reducing the intensity of artificial light when natural light is available, i.e. switching off luminaries when artificial lighting is no longer required; the number and characteristics of luminaries shall be determined taking into account architecture, furnishings, the degree of protection required by the premises, the minimum level of lighting prescribed, and the particular conditions imposed by the rules in force for each required lighting space. (Beu, 2001)

Water System

Water conservation is important because water is essential to human health and life and many places lack water, particularly drinkable water (Deac, 2017; Aşchilean, 2014). The overarching retrofitting strategy is to determine the building use less water whenever possible (Iacob et al., 2018). The following specific measures are helpful to conserve water: upgrade faucets, toilets, and shower heads accessories that were produced before 1992, for the upgrade will result in significant saving on water consumption; adding aerators and occupancy sensors on lavatory faucets for they decrease the rate of water flowing through the faucets by mixing water with air while obtaining the pressure of the water; consider decreasing water use by recycling it; rainwater can also be captured for irrigation or even to flush toilets and other application in the building; indoor sanitary items will be equipped with sensors to optimize household water consumption; the accumulation system and distribution pipes will be designed for hot water consumption at 45° C; provision should be made for recirculation of hot water; in order to minimize heat loss, or to avoid the occurrence of condensation on cold surfaces, the pipes for the domestic water distribution system will be provided with the appropriate insulation (Aşchilean, 2009; Aşchilean, 2014).

Building Automation System (BAS)

Building automation is the term that describes the functionality of a smart, computerized network of electronic devices designed to monitor and control mechanical and lighting systems in a building. The basic functionality of BAS is to maintain the building's climate between certain parameters, provide lighting based on a employment program and monitoring of system performance as well as device failures. The building automation system manages all the operation systems including the HVAC system, lighting and the appliances, and helps to reduce utility costs and maintenance while improving occupants comfort (Kheir Al-Kodmany, 2014).

For example, the BAS aims to provide a comfortable climate and adequate illumination, often related to the control area, so that users in a section of the building benefit from settings other than users in other sections; the zone temperature sensor provides feedback to the controller so that it can supply the coolant or heating agent as needed. Lighting can be turned on or off by the building automation system, based on daytime, on occupancy sensors or timers. A typical example is to turn off the lighting in a space half an hour after the last motion detection. Photocells placed outside the building sense the level of illumination and accordingly modulate the lights in the enclosed spaces and parking.

Most air treatment units mix the return air with the outside air so that the smallest temperature change is needed. This can save money by using a smaller amount of chilled or heated water (but not all air handling units use cold or hot water circuits). It is also necessary to use outside air to maintain the air quality inside the building. Analog or digital temperature sensors can be placed in rooms or chambers, in return and air supply pipes, and sometimes in outdoor air. Actuators are placed on hot or cold water valves, return air dampers or external air dampers. Supply fans (and return, if applicable) start and stop, either based on daytime or temperature, or on building pressures, or on a combination of them.

Many building automation systems have alarm capabilities. If an undesired event is detected, the system can be programmed to notify someone. Notification can be done through a computer, pager, cell phone, or an audible alarm. For example: differential pressure switches on the filter can be introduced to determine if it is dirty; state alarms - if a mechanical device, such as a pump, is required to start, and the status indicator indicates that it is not working, this may indicate a mechanical failure; some valve actuators have end switches to indicate whether the valve is open or not; carbon monoxide and carbon dioxide sensors can be used to alert if the concentration level is too high; coolant agent sensors may be used to indicate possible refrigerant leakage; current sensors can be used to detect the low current condition caused by slipping the fan drive belts or clogging the pumps from the pumps (Popa, 2011).

Renewable Energy Sources

The extent of the depletion of fossil fuels, such as oil, coal and natural gas, which are in limited quantities, many of the very profitable deposits elsewhere, starting to diminish production, brought the attention of modern society to the creation new energy resources, superior valorisation of materials, diversification of products, improvement of the quality of life through creation and implementation of new processes and technologies.

Against this background, alternatives for obtaining electricity and heat are currently being sought by using technologies that offer maximum efficiency, high reliability and minimal pollution. Officially recognized renewable energies originate from the sun's rays, the internal temperature of the earth, or the gravitational interactions of the sun and the moon with the oceans (Mârza, et al., 2013). In addition, today hydrogen (Badea, 2017; Felseghi, 2015) is recognized as a carrier of non-polluting energy because it does not contribute to global warming if it is produced from renewable sources.

Solar photovoltaic and solar thermal panels are increasingly popular, wind turbines are also becoming disposable and affordable, geothermal energy could be cheaper than the solar energy but are disposable in limited geographic areas, biomass is a growing area due to the increasing interest in alternative energy sources (Şoimoşan, 2015). All these alternative resources but not limited, can be incorporated in residential building retrofit, in order to generate on-site energy, either by their direct use or conversion through the conversion equipment used individually or in combination by hybridization.

Innovation

Technological, scientific and research advancement provides designer and architects with new ways to improve the environmental performance and energy efficiency of residential buildings. For example, aerogel isolation, phase changing materials, echo-skin materials, a lightweight, transparent fabric that can increase building insulation and produce electricity by wind or sunlight, 3M solar film that can be placed on windows to produce energy and to decrease energy necessary for cooling building by absorbing more than 90% of infrared light, easier rooftop installation for residential solar photovoltaic panels, software used to optimize a series of performance parameters of the residential building. Internet of Things (IoT) together with learning algorithms can be present in our homes, where smart meters coordinate and save energy consumption, improve indoor and outdoor comfort parameters, facilitates the entire daily human activity, ensures human well-being.

The strategy for mobilizing investment in the renovation of residential and commercial buildings, both public and private, existing at Romanian level provides that in all scenarios, the renovation work should be "technologically neutral". In other words, no assumptions were made as to the specific measures that would be applied to achieve a certain level of energy savings or a certain level of well-being for the occupants. The package includes a number of measures such as the following:

- Thermal insulation of building envelope - the opaque part.
- The modernization of windows and doors.
- Solar shading - in particular to reduce the air conditioning requirement of buildings.
- Reducing outdoor air infiltrations.
- Upgrading HVAC systems.
- Installing combined heat and power systems.
- Connection to centralized district heating systems.
- Installation of heat recovery systems (from exhaust air).
- Modernization of interior lighting systems.
- Improving energy control mechanisms used.

- The installation of energy equipment using renewable energy (such as solar water heating installations, photovoltaic panels, heat pumps, biomass heating boilers, wind turbines, micro-cogeneration systems, etc.).

An ideal approach would be to consider the best package of measures that could lead to maximum improvement in energy performance and human health and well-being for each type of building, also taking into account the particularities of location in the area's climate.

FUTURE RESEARCH DIRECTIONS

Discussing about green retrofitting of building from SBS and human well-being perspectives is an extremely complex approach. The general aspects dealt with in this chapter and the critical analysis of the above-mentioned concepts, but also the problems, technical limitations encountered allow to identify and establish future research and opportunities directions in the field of the topic approached:

- Discussion, analysis and approach the more specific strategies, performance targets and achieving steps required for green retrofitting of existing residential building so as to meet the Well-Being Building Standard requirements.
- Expanding research on the implementation of innovative green retrofit solutions for energy intensive building in the sense of turning them into passive buildings, nearly Zero Energy Buildings or active energy buildings.
- Developing a database in order to create the necessary premises for the elaboration of some procedures, norms and standards improved with regard to the design, execution and safe operation of residential buildings, which will contain the essential aspect of the disposition stipulated by Well-Being Building Standard.

DISCUSSIONS AND CONCLUDING REMARKS

This chapter attended as a review of major issues, aspects and concepts related to an emerging and important problem concerning green retrofitting residential buildings from human well-being perspectives.

At EU level, more than 80% of residential buildings are built before the 1990s, most of them being inadequate to current demands on the sustainability of the housing stock. The inventory, expertise and analysis of these buildings reveal a number of impediments to realize the comfort parameters inside the buildings, which transforms them into intensive energy buildings, favoring human health problems, with a mix of causative factors of SBS.

Sustainable buildings or "green" buildings are high performance, energy-efficient buildings designed to conserve water and energy, use space, materials and resources efficiently to minimize residue in their construction, and to create a healthy interior space.

Health, although not an obvious policy area for renovating buildings, is the fact that, not insuring the indoor climate conditions (insufficient heating or overheating, condensation on building elements, mold development and indoor air pollution, etc.), it can lead to major health problems that cause costs for the population (lost working days and impact on health services).

As in many other countries, issues related to the quality, comfort and financial accessibility of residential building are a national concern. Energy costs are a key component of maintenance costs, and the only sustainable long-term solution to ensure affordable financial heating is to improve the energy performance of the real estate fund. Green retrofitting of residential buildings is a major opportunity for sustainable upgrading of the existing built-up fund to bring benefits to multiple households, businesses and the public sector.

The comfort and availability of energy have in the meantime led to increased requirements along with technical progress and will certainly not give up this way of life in the future.

REFERENCES

Abdul-Wahab, S. A. (2011). *Sick Building Syndrome*. Berlin: Springer-Verlag. doi:10.1007/978-3-642-17919-8

Abrudan, A., Rus, T., & Mare, R. (2017). Thermal rehabilitation influence upon the comfort in hospitals. In International Conference on Advancements of Medicine and Health Care through Technology, Cluj-Napoca, Romania (pp. 155-158). Springer. 10.1007/978-3-319-52875-5_35

Al-Kodmany, K. (2014). Green Retrofitting Skyscrapers: A Review. *Buildings*, *4*(4), 683–710. doi:10.3390/buildings4040683

Allen, J. G., Bernstein, A., Xiadong, C., Eitland, E. S., Flanigan, S., & Gokhale, M. (2017). *The 9 Foundations of a Healthy Building*. Harvard T.H. Chan School of Public Health. Retrieved from http://forhealth.org/Harvard.Building_Evidence_for_Health_the_9_Foundations.pdf

Almeida, C. P., Ramos, A. F., & Silva, M. (2018). Sustainability assessment of building rehabilitation actions in old urban centres. *Sustainable Cities and Society*, *36*, 378-385.

Amin, N. D. M., Akasah, Z. A., & Razzaly, W. (2015). Architectural evaluation of thermal comfort: Sick building syndrome symptoms in engineering education laboratories. *Procedia: Social and Behavioral Sciences*, *204*, 19–28.

Araki, A., Kawai, T., Eitaki, Y., Kanazawa, A., Morimoto, K., Nakayama, K., ... Kishi, R. (2010). Relationship between selected indoor volatile organic compounds, so-called microbial VOC, and the prevalence of mucous membrane symptoms in single family homes. *The Science of the Total Environment*, *408*(10), 2208–2215. doi:10.1016/j.scitotenv.2010.02.012 PMID:20188399

Aşchilean, I. (2009). Revamping and upgrade methods for wells. *Acta Technica Napocensis, Section: Civil Engineering – Architecture*, *52*, 271-282.

Aşchilean, I. (2014). *Reabilitarea si modernizarea sistemelor de alimentare cu apa a localitatilor urbane*. Cluj-Napoca, Romania: Ed. RISOPRINT.

Badea, G., Felseghi, R. A., Aşchilean, I., Răboacă, S. M., & Şoimoşan, T. M. (2017). The role of hydrogen as a future solution to energetic and environmental problems for residential building. In AIP Conference Proceedings (Vol. 1917, No. 1, p. 030004). AIP Publishing. doi:10.1063/1.5018277

Beu, D., & Pop, F. (2001). *Tehnica iluminatului în spaţii industriale, birouri şi locuinţe*. Cluj-Napoca, Romania: Mediamira.

Bourbeau, J., Brisson, C., & Allaire, S. (1997). Prevalence of the sick building syndrome symptoms in office workers before and six months and three years after being exposed to a building with an improved ventilation system. *Occupational and Environmental Medicine*, *54*(1), 49–53. doi:10.1136/oem.54.1.49 PMID:9072034

Buildings Performance Institute Europe - BPIE. (2011, October). *Europe's buildings under the microscope. A country-by-country review of the energy performance of buildings.* Retrieved from http://bpie.eu/publication/europes-buildings-under-the-microscope/

Clements-Croome, D. (2018). *Creating the Productive Workplace: Places to Work Creatively* (D. Clements-Croome, Ed.). London: Routledge.

COM. (2011a). 112 final. *A Roadmap for moving to a competitive low carbon economy in 2050.* Retrieved from http://eur-lex.europa.eu/LexUriServ/LexUriServ.do?uri=CELEX:52011DC0112:EN:NOT

COM. (2011b). 885 final. *Energy Roadmap 2050.* Retrieved from http://eur-lex.europa.eu/LexUriServ/LexUriServ.do?uri=CELEX:52011DC0885:EN:NOT

COM. (2011c). 571 final. *Roadmap to a Resource Efficient Europe.* Retrieved from: http://ec.europa.eu/environment/resource_efficiency/pdf/com2011_571.pdf

Communication from the commission to the European Parliament. (2011). *A Roadmap for moving to a competitive low carbon economy in 2050.* Retrieved from https://eur-lex.europa.eu/legal-content/RO/TXT/?uri=celex%3A52011DC0112

Costanza, R., Fisher, B., Ali, S., Beer, C., Bond, L., Boumans, R., ... Snapp, R. (2007). Quality of life: An approach integrating opportunities, human needs, and subjective well-being. *Ecological Economics*, *61*(2-3), 267–276. doi:10.1016/j.ecolecon.2006.02.023

Cravioto, J., Yamasue, E., Okumura, H., & Ishihara, K. (2011). *Performance analysis between well-being, energy and environmental indicators using Data Envelopment Analysis.* Japan: Kyoto University; doi:10.1007/978-4-431-53910-0_6

Crook, B., & Burton, N. C. (2010). Indoor Moulds, Sick Building Syndrome and Building Related Illness. *Fungal Biology Reviews*, *24*(3), 106–113. doi:10.1016/j.fbr.2010.05.001

Danna, K., & Griffin, R. (1999). Health and Well-Being in the Workplace:A Review and Synthesis of the Literature. *Journal of Management*, *25*(3), 357–384. doi:10.1177/014920639902500305

Deac, A., & Iacob, C. A. (2017). Reverse osmosis, the optimal solution for water treatment with membranes technology. *Bulletin of the Transylvania - Series I: Engineering Sciences*, *10*(59).

Domniţa, F., Popovici, T., & Hoţupan, A. (2011). *Instalaţii de ventilare şi condiţionare* (Vol. 2). Cluj-Napoca, Romania: U.T. Press.

Engvall, K., Norrby, C., & Norback, D. (2001). Sick building syndrome in relation to building dampness in multi-family residential buildings in Stockholm. *International Archives of Occupational and Environmental Health, 74*(4), 270–278. doi:10.1007004200000218 PMID:11401019

Environmental Protection Agency (EPA). (1991). *Indoor Air Facts No. 4 (Revised) Sick Building Syndrome.* Retrieved from https://www.epa.gov/sites/production/files/201408/documents/sick_building_factsheet.pdf

Epstein, Y., & Moran, D. S. (2006). Thermal comfort and the heat stress indices. *Industrial Health, 44*(3), 388–398. doi:10.2486/indhealth.44.388 PMID:16922182

Evans, G. W., & McCoy, J. M. (1998). When buildings don't work: The role of architecture in human health. *Env Pschy, 18*(1), 85–94. doi:10.1006/jevp.1998.0089

Fanger, P. O. (1967). Calculations of thermal comfort: Introduction of a basic comfort equation. *ASHRAE Trans, 73*, 1–4.

Felseghi, R. A. (2015). *Contributions regarding the application of fuel cells in the passive house domain.* Unpublished doctoral dissertation, Technical University of Cluj-Napoca, Romania.

Ghaffarianhoseini, A., AlWaer, H., Omrany, H., Ghaffarianhoseini, A., Alalouch, C., Clements-Croome, D., & Tookey, J. (2018). Sick building syndrome: Are we doing enough? *Architectural Science Review, 61*(3), 99–121. doi:10.1080/00038628.2018.1461060

Goldman, L. (1996). How Sick is Your Building. *Utilities Law Review, 12*(3), 14–15.

Gomzi, M., & Bobić, J. (2009). Sick Building Syndrome. Do We Live and Work in Unhealthy Environment? *Periodicum Biologorum, 111*(1), 79–84.

Gomzi, M., Bobić, J., Radosevic-Vidacek, B., Macan, J., Varnai, V. M., Milkovic-Kraus, S., & Kanceljak-Macan, B. (2007). Sick Building Syndrome: Psychological, Somatic, and Environmental Determinants. *Archives of Environmental & Occupational Health, 62*(3), 147–155. doi:10.3200/AEOH.62.3.147-155 PMID:18400655

Greer, C. (2007). Something in the Air: A Critical Review of Literature on the Topic of Sick Building Syndrome. *World Safety Journal, 16*(1), 23–26.

Hamilton, L. (2010). *Sick Building Syndrome. When good air goes bad.* Lynchburg, VA: Strategic Sustainability Consulting.

van Hoof, J. (2010). Thermal comfort: Research and practice. *Frontiers in Bioscience, 15*(1), 765–788. doi:10.2741/3645 PMID:20036845

Iacob, C., Deac, A., Chira, T., & Bolboaca, A. (2018). *Study of performance indicators at a waste water treatment plant in Romania, 18th International Multidisciplinary Scientific GeoConference SGEM 2018.* Bulgaria: Abena.

Imai, N., & Imai, Y. (2011). Psychosocial Factors That Aggravate the Symptoms of Sick Building Syndrome and a Cure for Them. In S. A. Abdul-Wahab (Ed.), *Sick Building Syndrome* (pp. 105–111). Berlin: Springer. doi:10.1007/978-3-642-17919-8_6

International Well Building Institute. (n.d.). Well-Being Building Standard, v1 with Q3 2018 addenda.

Jansz, J. (2011a). Introduction to Sick Building Syndrome. In S. A. Abdul-Wahab (Ed.), *Sick Building Syndrome* (pp. 1–24). Berlin, Germany: Springer.

Jansz, J. (2011b). Theories and Knowledge About Sick Building Syndrome. In S. A. Abdul-Wahab (Ed.), *Sick Building Syndrome* (pp. 25–58). Berlin, Germany: Springer. doi:10.1007/978-3-642-17919-8_2

Kamaruzzaman, S., Sabrani, N.A. (2011). The Effect of Indoor Air Quality (IAQ) Towards Occupants' Psychological Performance in Office Buildings. *Jurnal Rekabentuk dan Binaan*, *4*, 49-61.

Kapalo, P., Domniţa, F., Bacoţiu, C., & Spodyniuk, N. (2018). The Impact of Carbon Dioxide Concentration on the Human Health-Case Study. *Journal of Applied Engineering Sciences*, *8*(1), 61–66. doi:10.2478/jaes-2018-0008

Kukec, A., & Dovjak, M. (2014a). *Prevention and control of Sick Building Syndrome (SBS). Part 1: Identification of risk factors, Slovenia n Public and Environmental Health Professionals*. Ljubljana, Slovenia: University of Ljubljana, Faculty of Medicine. Retrieved from https://journal.institut-isi.si/prevention-and-control-of-sick-building-syndrome-sbs-part-1-identification-of-risk-factors/

Kukec, A., & Dovjak, M. (2014b). *Prevention and control of Sick Building Syndrome (SBS). Part 2: Design of a preventive and control strategy to lower the occurrence of SBS, Slovenia n Public and Environmental Health Professionals*. Ljubljana, Slovenia: University of Ljubljana, Faculty of Medicine. Retrieved from https://journal.institut-isi.si/wp-content/uploads/2014/10/Prevention-and-control-of-sick-building-syndrom-SBS.-Part-2.pdf

Lee, J., Je, H., & Byun, J. (2011). Well-Being index of super tall residential buildings in Korea. *Building and Environment*, *46*(5), 1184–1194. doi:10.1016/j.buildenv.2010.12.010

Mârza, C., Hoţupan, A., Moldovan, R., & Corsiuc, G. (2013). *Surse neconvenţionale de energie*. Cluj-Napoca, Romania: U. T. Press.

Ministerul Dezvoltării Regionale. (2016). Administraţiei Publice şi Fondurilor Europene - MDRAP Strategia Energetică a României 2016-2030, cu perspectiva anului 2050. Retrieved from http://www.mmediu.gov.ro/app/webroot/uploads/files/2017-03-02_Strategia-Energetica-a-Romaniei-2016-2030.pdf

Ministerul Dezvoltării Regionale. Administraţiei Publice şi Fondurilor Europene - MDRAP (2017a, January). Strategia naţională a locuirii. Retrieved from http://www.mmediu.ro/app /webroot/uploads/files/2017-01-13_Strategia_Nationala_a_Locuirii_2016-2030.pdf

Ministerul Dezvoltării Regionale. Administraţiei Publice şi Fondurilor Europene - MDRAP (2017b, October). Strategia pentru mobilizarea investiţiilor în renovarea fondului de clădiri rezidenţiale şi comerciale, atât publice cât şi private, existente la nivel naţional. Retrieved from http://www.mdrap.ro/userfiles/Strategie_renovare_cladiri_2017%20final_23octombrie2017.pdf

Ministerul Dezvoltării Regionale, Administraţiei Publice şi Fondurilor Europene. (n.d.). Normative for the Design, Execution an Operation of Ventilation and Air-conditioning Systems Indicative I5.

Mishra, A. K., & Ramgopal, M. (2013). Field studies on human thermal comfort—An overview. *Building and Environment*, *64*, 94–106. doi:10.1016/j.buildenv.2013.02.015

Moga, L., & Bucur, A. (2018). Nano insulation materials for application in nZEB. *Procedia Manufacturing*, *22*, 309–316. doi:10.1016/j.promfg.2018.03.047

Moga, L. M., & Moga, I. (2017). *Punţi termice specifice planşeelor terasă, de pod, deasupra subsolului şi plăcilor pe sol la clădiri cu pereţi din zidărie*. Cluj-Napoca, Romania: U.T. Press.

Molina, C., Anthony, C., Pickering, C., Valbjorn, O., & Bortoli, M. (1990). *Sick Building Syndrome: A Practical Guide*. Luxembourg: Commission of the European Communities. Retrieved from http://www.buildingecology.com/publications/ECA_Report4.pdf

Newsham, G., Brand, J., Donnelly, C., Veitch, J., Aries, M., & Charles, K. (2009). Linking Indoor Environment Conditions to Job Satisfaction: A Field Study. *Building Research and Information*, *37*(2), 129–147. doi:10.1080/09613210802710298

Norhidayah, A., Chia-Kuanga, L., Azharb, M. K., & Nurulwahidab, S. (2013). Indoor Air Quality and Sick Building Syndrome in Three Selected Buildings. *Procedia Engineering*, *53*, 93–98. doi:10.1016/j.proeng.2013.02.014

Passarelli, G. R. (2009). Sick Building Syndrome: An Overview to Raise Awareness. *Journal of Building Appraisal*, *5*(1), 55–66. doi:10.1057/jba.2009.20

Popa, E. S. (2011). The buildings automation. In *The 11th National Multidisciplinary Conference with International Participation "Professor Dorin Pavel - founder of the Romanian energy,"* Sebes, Romania.

Popovici, T., Domniţa, F., & Hoţupan, A. (2010). *Instalaţii de ventilare şi condiţionare* (Vol. 1). Cluj-Napoca, Romania: U.T. Press.

Rostron, J. (2008). Sick Building Syndrome: A Review of Causes, Consequences and Remedies. *Journal of Retail & Leisure Property*, *7*(4), 291–303. doi:10.1057/rlp.2008.20

Roulet, C. A. (2018). *Well being and energy in buildings, an architectural issue*. France: École Polytechnique Fédérale de Lausanne.

Sahlberg, B., Gunnbjörnsdottir, M., Soon, A., Jogi, R., Gislason, T., Wieslander, G., ... Norback, D. (2013). Airborne molds and bacteria, microbial volatile organic compounds (MVOC), plasticizers and formaldehyde in dwellings in three North European cities in relation to sick building syndrome (SBS). *The Science of the Total Environment*, *444*(1), 433–440. doi:10.1016/j.scitotenv.2012.10.114 PMID:23280302

Scheel, C. M., Rosing, W. C., & Farone, A. L. (2001). Possible sources of sick building syndrome in a Tennessee middle school. *Archives of Environmental Health*, *56*(5), 413–417. doi:10.1080/00039890109604476 PMID:11777022

Shahin, H. S. M. (2019). Adaptive building envelopes of multistory buildings as an example of high performance building skins. *Alexandria Engineering Journal*, *58*(1), 345–352. doi:10.1016/j.aej.2018.11.013

Shan, X., Zhou, J., Chang, V. W. C., & Yang, E. H. (2016). Comparing Mixing and Displacement Ventilation in Tutorial Rooms: Students' Thermal Comfort, Sick Building Syndromes, and Short-Term Performance. *Building and Environment*, *102*, 128–137. doi:10.1016/j.buildenv.2016.03.025

Burge, S., Hedge, A., Wilson, S., Bass, J. H., & Robertson, A. (1987). Sick Building Syndrome: A Study Of 4373 Office Workers. *The Annals of Occupational Hygiene, 31*(4a), 493–504. PMID:3439759

Şoimoşan, T.-M. (2015), *Contributions regarding the optimization of the heating systems in urban localities.* Unpublished doctoral dissertation, Technical University of Cluj-Napoca, Romania.

Şoimoşan, T.-M., Danku, G., & Felseghi, R. A. (2017). *Termotehnica construcţiilor.* Cluj-Napoca, Romania: U.T. Press.

Stanford Encyclopedia of Philosophy. (2005). *Well-being* (R. Crisp, Ed.). Stanford University; Retrieved from http://plato.stanford.edu/entries

Sundin, J. (2012). *The experience of living with Sick Building Syndrome.* Unpublished Thesis, NOVIA University of Applied Sciences, Vaasa, Finland.

Takigawa, T., Saijo, Y., Morimoto, K., Nakayama, K., Shibata, E., Tanaka, M., ... Kishi, R. (2012). A Longitudinal Study of Aldehydes and Volatile Organic Compounds Associated with Subjective Symptoms Related to Sick Building Syndrome in new Dwellings in Japan. *The Science of the Total Environment, 417–418*, 61–67. doi:10.1016/j.scitotenv.2011.12.060 PMID:22265600

U.S. Green Building Council's Leadership in Energy and Environmental Design. (n.d.). Retrieved from https://www.usgbc.org/resources

Woods, J. E., Drewry, G. M., & Morey, P. (1987). Office worker perceptions of indoor air quality effects on discomfort and performance. In *Proceedings of the 4th International Conference on Indoor Air and Climate*, Berlin, Germany (pp. 464-408).

World Health Organization - WHO. (1983). World Health Organization. Indoor Air Pollutants: Exposure and Health Effects. *EURO Reports and Studies, 78*, 1–42.

World Health Organization - WHO. (1986). Indoor Air Quality Research – EURO Reports and Studies. WHO Regional Office for Europe, Copenhagen.

World Health Organization - WHO. (1999). *Strategic approaches to indoor air policy-making.* Bilthoven: WHO European Center for Environment and Health, Retrieved from http://www.paho.org/tierra/index.php?option=com_multicategories&view=article&id=75%3Astrategic-approaches-to-indoor-airpolicy making&catid=9%3A19962000&Itemid=136& lang=en

Yang, C. H., & Heinsohn, P. A. (2007). Sampling and analysis of indoor microorganism (1st ed.). Wiley-Interscience. doi:10.1002/0470112433

ADDITIONAL READING

Anielski, M. (2018). *An Economy of Well-Being: Common-sense tools for building genuine wealth and happiness.* Canada: New Society Publishers.

Aşchilean, I., & Băncilă, Ş. (2019). *Calitatea în construcţii.* Cluj-Napoca, Romania: Risoprint.

Johnston, D., & Master, K. (2009). *Green Remodeling:Changing the World One Room at a Time*. Canada: New Society Publishers.

Krarti, M. (2018). Optimal Design and Retrofit of Energy Efficient Buildings, Communities, and Urban Centers (1st ed.). imprint: Butterworth-Heinemann.

Lohse, R., & Zhivov, A. (2019), Deep Energy Retrofit Guide for Public Buildings. Springer.

Racusin, J. D. (2016). *Essential Building Science: Understanding Energy and Moisture in High Performance House Design*. Canada: New Society Publishers.

Rynska, E., Kozminska, U., Zinowiec-Cieplik, K., Rucinska, J., & Szybinska-Matusiak, B. (2018). Design Solutions for nZEB Retrofit Buildings. Hershey, PA: IGI Global. doi:10.4018/978-1-5225-4105-9

Şoimoşan, T.-M., Danku, G., & Felseghi, R. A. (2017). *Termotehnica construcţiilor*. Cluj-Napoca, Romania: U.T. Press.

Tobias, L. (2009). *Retrofitting Office Buildings to Be Green and Energy-Efficient: Optimizing Building Performance, Tenant Satisfaction, and Financial Return, Urban Land Institute*.

Yudelson, J., & Meyer, U. (2013). *The World's Greenest Buildings: Promise Versus Performance in Sustainable Design*. London: Routledge. doi:10.4324/9780203082164

KEY TERMS AND DEFINITIONS

Building's Performance: Qualitative indicators of operation and yield obtained in the field of residential buildings.

Energy Active Building: The energy produced by the renewable energy conversion systems incorporated in the building is higher than the energy consumed by the building.

Energy Intensive Building: Energy inefficient buildings, big energy consumers.

Green Retrofit: Retrofits that ensure the maintenance and conservation of existing buildings and improve residential building energy efficiency and human well-being comfort.

Indoor Comfort: The quality of the living environment and the parameters of the building's internal environment.

Occupant: Resident of building.

Residential Building: A building made up of one or more rooms used for housing, with the necessary facilities and utilities that satisfy the living requirements of a person or family.

Sick Building Syndrome (SBS): Situation where multiple inhabitants of the same indoor space present acute disorder due to poor indoor air quality.

Sustainable Design: Principles of designing the building and services to submit with the principles of sustainability.

Well-Being Building: Positive effects of building environments on occupant health and safety.

Chapter 8
Building Information Modeling for Energy Retrofitting

Laurentiu Catalin Fratila
The Bucharest University of Economic Studies, Romania

ABSTRACT

In the era of strong IT development, superior processing of information is a desideratum in any business. Used in modeling and simulation, information has better capitalization, having a cognitive role; introduced in various algorithms and models can generate knowledge and wisdom. The processing of information is realized within the information systems (IS). A sub-domain of the energy sector that has developed in recent years is energy retrofitting, seen as a border subdomain, including specific elements from building construction and the energy field, for building analysis and construction process to achieve much larger energy savings. The development of retrofitting can be organized, through using BEM (Building Energy Model) and Building Information Modeling (BIM) models.

INTRODUCTION

The chapter treat with the importance of information, how it is present and how it is processed in energy systems. In the context of important upgrading projects, it is a decisive factor in choosing the decision and method of retrofitting. Taking into account the very large data volume, their different sources and their complexity, the information has to be treated within the information systems specific to the field in which they are applied.

The process of retrofitting is seen in a unitary way, being developed as a process with well-defined phases, where appropriate methods and techniques are applied.

From retrofitting methods, a superiour step is to develop Building Information Modeling (BIM) through the creation of models for existing buildings and the analysis of retrofit measures. Supported by specialized software, these models can be used in all stages of the building's evolution, from the conception phase until the end of its existence.

DOI: 10.4018/978-1-5225-9104-7.ch008

INFORMATION SYSTEMS: A SUPERIOR WAY OF ORGANIZING AND PROCESSING DATA

In the era of strong IT development, superior processing of information is a desideratum in any business. Information has a primary, descriptive role, describing a phenomenon that will be subject to analysis. Used in modeling and simulation, information has better capitalization, having a cognitive role; introduced in various algorithms and models can generate knowledge and wisdom.

If we take into account their complexity, we can see an evolution of them, depending on their degree of complexity. Thus, we can group them into: data, information, knowledge and wisdom. This is also evident from the definitions of The American Heritage Dictionary of the English Language for the items of data, information, knowledge and wisdom.

Data is defined as "Facts that can be analyzed or used in an effort to gain knowledge or make decisions; information" (ahdictionary.com, n.d.). In a simplest way, data can be seen as "the recorded, uncontestable, observed facts about things, people and events around us" (Sætnan, Schneider & Green, 2018). Data is used and processed in any field of activity. It is used both in describing certain phenomena or processes, but also in making some outputs as processed in indicators or graphical representations.

Information can be defined as "Computers processed, stored, or transmitted data" (ahdictionary.com, n.d.). Information is used to communicate data to users in different forms (processed or not processed) and play a descriptive or cognitive role. Information can be seen as collection of data or facts capitalized only in a specific context.

Knowledge can be seen as "The sum or range of what has been perceived, discovered, or learned" (ahdictionary.com, n.d.). Knowledge occurs as result of experience or study.

Wisdom is the most complex form and can be defined as "The ability to discern or judge what is true, right, or lasting; insight" (ahdictionary.com, n.d.). Wisdom can have as result a wise outlook, but can also be capitalized through an action plan.

The most suggestive graphical representation of these terms taking into account the associativity between them is made by Pressman, as is shown in Figure 1. The complexity of the new entities, occurred as a result of the associations between data, makes the difference between data, information, knowledge and wisdom.

Data quality can be defined as "the totality of features and characteristics of a data set that bear on its ability to satisfy the needs that result from the intended use of the data" (Unite For Sight, 2018). These data characteristics are summarized in Table 1.

Given the huge amount of information and heterogeneous sources, these features are found in any kind of information, but with some particularities, in a more developed or narrower form. Also, some features can be fulfilled by redundant forms; for example, for critical processes, there may be redundant sources for the same information, which ensures continuity and security in providing information. In some situations, even legislation (rules and regulations) impose the characteristics that data used by information systems related to specific activities should have.

We can only talk about Information System (IS) as a component of the management system where information is processed. IS is one of the four components of a traditional management system. It contains four subsystems: organizational subsystem, information subsystem, decision subsystem, and management methods and techniques subsystem.

Figure 1. A graphical comparison between data, information, knowledge, wisdom
Source: Pressman, 2005

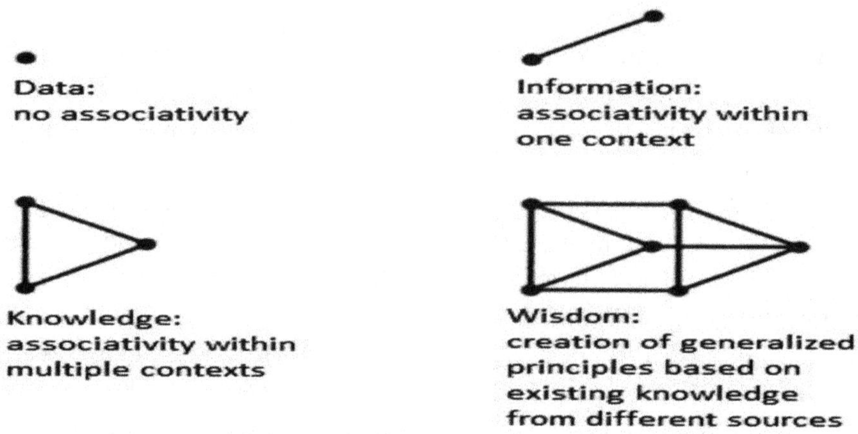

Table 1. Data quality

Characteristics of data	Explanation
Access security	Data must be restricted and kept secure to ensure confidentiality.
Accessibility	Data must be available or easily retrievable.
Accuracy	Data must be correct and free of errors.
Appropriate amount of data	The quantity of data must be appropriate.
Believability	Data must be regarded as true and credible.
Completeness	Data must be sufficient in breadth, depth, and scope for its desired use.
Concise representation	Data must be represented without being overwhelming.
Ease of Understanding	Data must be clear.
Interpretability	Data must be in appropriate language and units.
Objectivity	Data must be unbiased.
Relevancy	Data must be applicable to the task at hand.
Representational consistency	Data must be presented in a consistent format
Reputation	Data must come from a trusted source.
Timeliness	Data should be recorded as quickly as possible and used within a reasonable time period.
Value-added	Data must provide valuable insight.

Source: Unite For Sight, 2018

Based on the definitions of Henry Fayol and George R. Terry for management, the management system can be defined as an ensemble of elements and processes that act as a unitary entity in a coherent way to forecast, plan, organize, coordinate and control to achieve the objectives of an organization.

Nicolescu and Verboncu (2008) define the management system of organization as „all the elements regarding decision making, organizational, informational and motivational processes within the organization, through which it is exercised all activities and management relationship, in order to obtain an effectiveness and efficiency as large as possible". The management system is an organized system of elements (policies, procedures and processes) that create a well-defined basis for leadership in order to achieve the goals of an organization.

The organizational subsystem is defined by the totality of elements that act in a coherent and unitary way that ensures the framework, division, combining and functionality of the work processes necessary to achieve the company's goals.

The information subsystem is defined by a coherent mix of data (data, information, information flows) and means of processing them (procedures and equipment) that contribute to the company's performance and achievement of organizational goals.

The decision subsystem is the ensemble of all decisions discussed, adopted and implemented, taking into account the organization chart of the company, hierarchical structure and the settled objectives.

The management methods and techniques subsystem are the ensemble is a complex and coherent set of methods and techniques used to define managerial functions, relationships between them and the management elements that make them operational within an organization.

The traditional classification of Information Systems (IS) is: EIS (Executive Information System), MIS (Management Information System), DSS (Decision Support System), OSS (Operational Support System) and TPS (Transaction Processing System). Within them, different information systems, specific to the types of business, were developed.

PROCESSING INFORMATION IN A RETROFIT PROGRAMME

In the retrofitting process, information plays an essential role. The way data is acquired and information processed can make the difference between a successful project and a project with unsatisfactory results.

Ma et al. (2012) define five determinant phases for a retrofit system. In each phase the information has a well-defined position. The following features must be determinative: who is the issuer, who is the receiver, how it is processed and how it is capitalized (locally or transmitted to another receiver).

The implementation of a retrofit project can be done by a certain company (specialized in energy services), by a mixed team composed by the representatives of the specialized firm and the beneficiary, or, for the small projects, even by the beneficiary.

Phase 1: Project Setup and Pre-Retrofit Survey

In phase 1, it is defined the purpose of the retrofit project and the objectives by which this goal is achieved. The purpose of the project is set by the beneficiary and can be expressed by a qualitative or quantitative indicator.

The objectives can be established by the beneficiary in collaboration with the designer of the retrofit project and are expressed by quantitative or qualitative indicators, proposed by the company and validated by the beneficiary.

For setting the objectives, account must be taken of the available resources (financial, material, human), the available time horizon, but also the simultaneous existence of the activities in the building and the activities of implementation of the retrofitting project. Analysis of the available resources is an important element, which determines to continue or to cancel the retrofit project.

Figure 2. Key phases in a sustainable retrofit programme
(Source: Ma et al., 2012)

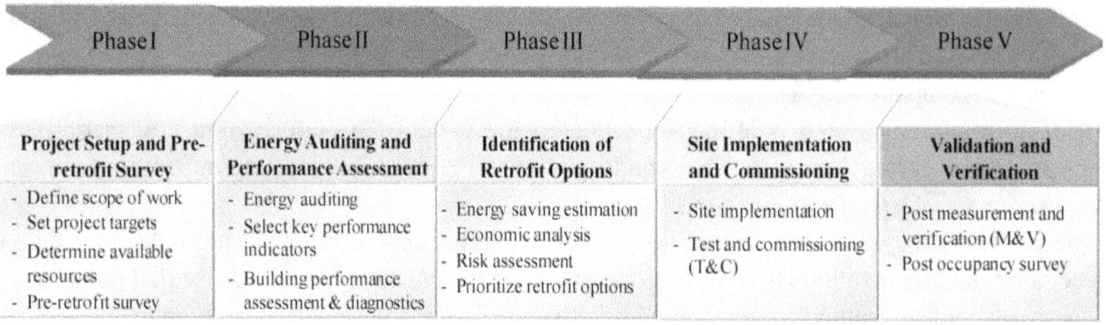

It is advisable for the company to carry out a pre-retrofit survey to have an overall energy view of the building. This survey identifies energy sources, consumers, energy flows, as well as energy issues of the building that the client claims. These inputs are provided by the beneficiary and are used by the company for various analyzes.

In the first phase we can talk about descriptive information. It has the role of describing the problems that will be managed by the retrofitting project that is to be implemented, starting from an existing energy situation.

Phase 2: Energy Auditing and Performance Assessment

The second phase represents the detailed analysis of the energy performance of the building. The first step is through an energy audit. Both the beneficiary and the energy auditor company have a well-defined role to play during the energy audit process.

According to The American Heritage Dictionary, energy audit is "An evaluation of energy consumption, as in a home or business, to determine ways in which energy can be conserved" (ahdictionary.com, 2018).

In Wikipedia, an energy audit is defined as "an inspection survey an analysis of energy flows, for energy conservation in a building, process or system to reduce the amount of energy input into the system without negatively affecting the output(s)" (wikipedia, n.d.).

The energy audit can be defined from the point of view of ISO 50002, a standard created for the regulation of the energy audit process. According this standard, an energy audit is "a systematic analysis of energy use and energy consumption within a defined energy audit scope, in order to identify, quantify and report on the opportunities for improved energy performance" (Leonardo energy, 2015).

According ISO 50002, an energy audit means to pass the following steps:

Starting from a general audit features, we can say that the energy audit provides an analysis of the energy consumption of the building.

It is analyzed the characteristics of the building, the constructive details, the materials used, but also their evolution taking into account the change of their properties over time.

It is explained the lighting system, the heating and cooling systems, the equipment used and their energy consumption. Monitoring systems, sensors, automation and control elements are taken into account, which make it possible to create automation for a part or for all of the electricity consumption control activity.

Figure 3. Energy audit steps
(Source: ISO, 2014)

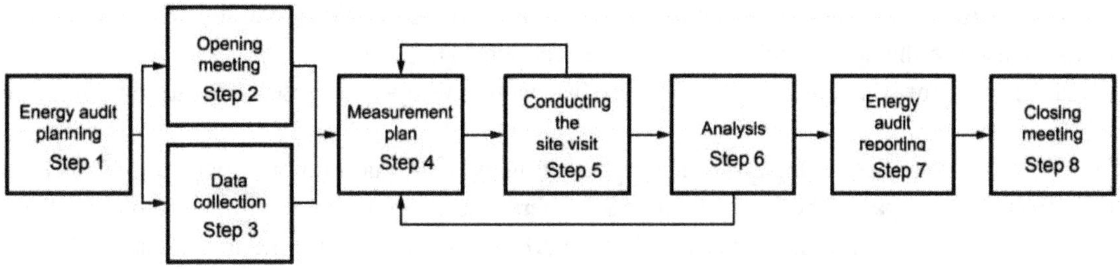

The human factor is an important factor treated from the perspective of occupant behavior, but strong connected with the organizational culture. The human factor is developed in a general frame created by the company's energy policies, internal rules and regulations.

Energy providers, tariffs and renewable energy sources are main elements that can help reduce energy consumption from suppliers and implicitly reduce costs.

The audit looks at ways of reducing energy consumption and possible uses of renewable energy, with positive effects on the company, but also on the environment.

The audit proposes ways of actions, changes to operations, improvement of staff behavior and prioritization of these measures. All will be done in a strictly financial budget and certain period of time.

Following the three stages of the audit process, the auditors in the first stage collect data and information, in the second phase they verify, analysis and validate all of this, and in the third phase they draw up the audit report using specialized software.

The audit involves data collection activities, processing them with specialized software and creating the audit report. A software support for the auditor may have the menu as in Figure 4.

In this diagram, each rectangle corresponds to a working procedure and has a screen associated. Green rectangles are information screens; do not allow data input or final outputs. Yellow rectangles are screens

Figure 4. Structure for Energy Audit Tool Manual
(Source: EATM, 2018)

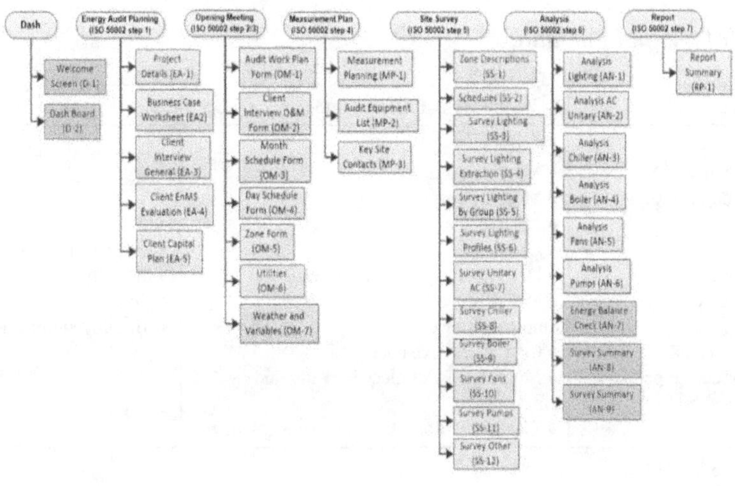

for input data. Using these screens has a mandatory character for data processing. Pink rectangles are associated with tools for analysis. Only AN1-AN6 rectangles are mandatories for analysis, for the rest of them is the auditor's decision whether or not use them. Blue rectangles are associated with forms for other inputs, used in the documentation and references section.

As can be seen from the diagram, the software has been developed in close association with the steps set in ISO 50002.

The software is varied, offering different outputs, along with the audit report. For example, the Energy Analysis from the company The Energy Audit Srl offers a software product organized in the following modules: Audit module, Modeling module, Benchmarking module and Budgeting module. The output list of the Audit module is detailed in table 2.

Also, in this stage, the performance indicators are set and the performance assessment and diagnostics are carried out.

Building performance assessment has as its starting point the selected performance indicators and aims to analyze the performance of the building through these indicators.

Diagnostics refers to the analysis of equipment in terms of malfunction or operation at parameters other than optimal operating parameters. Starting from Haves' observations on the diagnostic methods (Haves, 1999), we can affirm that "diagnostics" is not only the detection of operational problems, but the diagnosis of their cause.

Diagnostics is made using FDD (fault detection and diagnostics) tools, specialized software that treats input data and processes them according to well-defined algorithms.

FDD tools can be manual or automatic. Although the demarcation boundary between them is rather unclear, because each of them has automation elements in the three diagnostic stages, the involvement of the specialized human operator in the diagnostic phase, by interpreting the elements generated by the tool and identifying the problems in comparison with removing the human operator in automated tools determines whether a tool can be considered manual or automatic.

There are several classifications of diagnostic stages, and the most appropriate one can be proposed, starting from the classification of Friedman & Piette (2001): data acquisition, data management, and diagnostics.

Data acquisition stage involves the collection of raw data. The collection can be done: manual, (by introducing them by the human operator using a input screen - interface), random retrieval of data from

Table 2. Output list of the audit module

Consumption output	Bills output	Other energy vectors
• Loads and energy driver data • Data comparison between loads and energy drivers • carpet plot • specific energy indicator • Load consumption for aggregation type • Total consumption • Consumption profile from load • Consumption profile for each type • Consumption and involved profile	Electric energy: • Monthly request • Monthly cost per range • Load factor • Use factor • Daily electric average consumption • Monthly energy cost and cost of employed power • Cost of reactive energy • Monthly cost of electricity divided by entry • Electricity monthly summary break down	• Daily average consumption • Monthly cost • Cost divided by entry • Monthly summary break down

Source: the-energyaudit.com, 2018

an Energy Management Control System - EMCS (at the command of the human operator, by importing files with a certain format) or automatic retrieval of data from an EMCS (as a result of a complex communication protocol between the tool and EMCS that allows the tool to request and receive EMCS data in a default format).

Data Management stage aims to prepare data for processing by diagnostic analysis. This stage is realized, through three steps: timestamp synchronization, data validation and data archiving.

Tools that automatically receive data from EMCS can take up a very large volume of data, data that are heterogeneous and may not have synchronized timestamps. In order to be processed at the diagnostic stage, these data must have synchronized timestamps. The tool does this due to their implemented specific algorithms that process data and converge them into a single timestamp interval. Tools that allow data input by the human operator do not require timestamps.

Validation of data implies that data are analyzed in terms of classification within an acceptable range in order to process and correct them by eliminating the errors that may occur. Errors may arise from mistaken insertion of values by the human operator or even from mistaken communication of values by taking over from an EMCS system. The tool can also have interpolation, distribution, regression routines that allow data sets to be filled in where there are no input values. Validation is done automatically by the tool.

Most tools archive data. These can be viewed later or can be used for new analyzes. Archiving data involves saving them in a database. It can be archived raw data (all or part of it at user-defined interval or tool set), system alarm status, and diagnostic results.

Diagnostics is realized in two steps: detection and diagnosis.

Detection is the identification of problems by reporting the values of the indicators determined in the current system operation to the values provided for the optimal operation of the system. This is done by quantitative methods and qualitative methods.

Diagnosis means identifying and analyzing possible causes that have triggered problems and proposing solutions to control, mitigate, or eliminate them.

Phase 3: Entification of Retrofit Options

This stage aims at proposing alternative retrofitting measures. This is done in the following directions: energy, economy, risk.

From the energy point of view, it will create and use energy models for the system so that it can accurately estimate the performance indicators selected in the previous stage.

From an economic point of view, economic analysis will be based on these indicators to determine which alternatives provide the best (financial, environmental) results.

The risk factor must be taken into account because any process or activity involves a risk. In no business the risk cannot be totally eliminated, but only diminished and controlled. No business or business has all the necessary time, financial and human resources to eliminate all possible risks.

Economic analysis and risk assessment allow for the ordination of energy retrofit alternatives and prioritization of the measures to be taken under each alternative.

Simulation

Energy saving estimation follows the estimation of the energy performance parameters, those set in the previous step but not only those parameters that will be used in the economic analysis.

Determining these is done by applying one or more of the following methods: simulation packages, simulation models, and Building Information Modeling (BIM).

All simulation methods use the general format of an information system, defined by input, processing, output and reaction, as in Figure 5.

Reaction allows modification or adaptation of input parameters or even processing algorithms in those conditions where output is not required or not conclusive for analysis.

Energy simulation packages (BPS) are complex software, sold by specialized vendors. There are over 400 energy simulation packages. The most useful of these are: EnergyPlus, TRNSYS eQUEST, DOE-2, ESP-r, BLAST, HVAC-SIM +.

A comparison of a set of specialized software is done by Østergård (Østergård et al., 2016). They divide specialized software into four groups and analyze them based on the following criteria: users, design stage, interoperability, core complexity, objectives, and parametric simulation.

- **Users:** refers to the software user, architect (A) or engineer (E)
- **Design Stage:** refers to design stages to which software can be applied. The stages taken into account are: conceptual, preliminary, detailed and management
- **Interoperability:** refers to the possibility of software interconnection selected with other CAD software or simulation software

Figure 5. The general scheme of information system

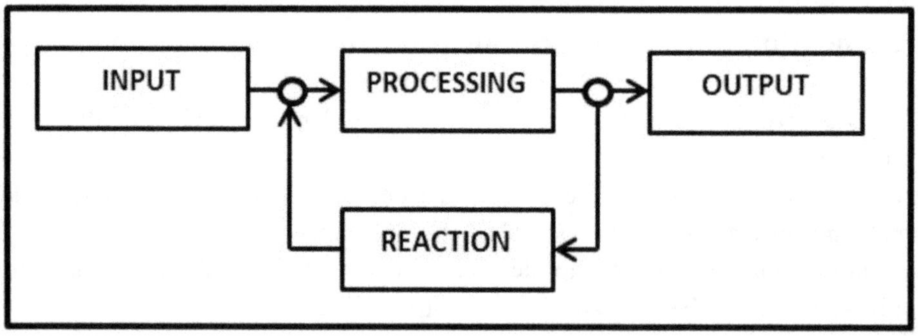

Table 3. Input and output for an information system of a building

Input	Output
• Parameters that make up or determine through calculation relationships the selected key performance indicators • General parameters that characterize the building • Parameters determined by data from various equipments that are not used by the selected simulation method	• Key performance indicators or energy performance parameters selected • Additional energy parameters (different from the selected energy performance parameters) resulting from the application of the selected method, but not used in the current analysis

Table 4. Software for analysis used in energy simulation

	Software	Users	Design stage				Interoperability	Core complexity	Objectives						Parametric sim.					
			Conceptual	Preliminary	Detailed	Management			Energy	Thermal	Daylight	Air Quality	LCA	LCC	Cloud	I/O Configurable	UA	SA	OAT	Optimisation
BPS (own engine)	Be10 (ISO 13790 monthly)[2]	(A)E		—			Standalone	Low	✓	(✓)						✓				
	Bsim	E		—	—		Standalone	High	✓	✓	✓	✓							(✓)	
	DOE2	E			—		Standalone	Medium	✓							✓				
	EnergyPlus (E+)	E			—		Standalone	High	✓	✓		✓				✓				
	EPC (ISO 13790 hourly)	(A)E	—	—			Standalone	Medium								✓				
	ESP-r	E			—		Standalone	High	✓	✓		✓				✓				
	IDA-ICE	E		—	—		File exchange	High	✓	✓		✓				✓			✓	
	iDbuild	E	—	—	—		Standalone	Medium	✓	✓	✓	✓							✓	
	IESVE	E		—	—		File exchange	High	✓	✓	✓	✓	✓	✓						
	Radiance	E		—	—		Standalone	High			✓					✓				
	VELUX Daylight Visualizer	AE	—	—	—		File exchange	High			✓									
BPS (external engine)	A+E3D	A	—	—			Integrated	Be10	✓	(✓)	(✓)									
	Daysim	E	—	—	—		Run-time	Radiance			✓									
	DesignBuilder	E		—	—		File exchange	E+, Radiance	✓	✓	✓	✓		✓						✓
	eQuest	E		—	—		Standalone	DOE2	✓											
	N++	E	—	—	—		Separated	E+, jE+, GenOpt	✓	✓		✓							✓	✓
	OpenStudio	E		—	—	—	File exchange	E+, Radiance	✓	✓	✓	✓		✓	✓	✓				
	Riuska	E		—	—		File exchange	DOE 2, own engine	✓	✓		✓					✓	✓		
	Sefaira	A	—	—			Run-time	E+, Radiance	✓	✓	✓				✓		(✓)	(✓)		✓
Plug-in	DIVA for Rhino	A(E)	—	—	—		Run-time	Radiance			✓									
	Green Building Studio	A(E)	—	—			File exchange	DOE2	✓						✓					
	HoneyBee (GH)	AE	—	—	—		File exchange	OpenStudio, E+, Radiance	✓	✓	✓			✓		✓				
	jEPlus (+ JESS)	E		—	—		Run-time	E+, DesignBuilder, N++	✓	✓	✓	✓		✓	✓		(✓)	(✓)	✓	
	Parametric Analysis Tool	E		—	—		File exchange	OpenStudio	✓	✓	✓	✓		✓	✓		✓	✓	✓	✓
	Solon	(A)E		—	—		File exchange	Green Building Studio	✓						✓			(✓)	✓	
Glue[1]	Dynamo	AE	—	—			-	-												
	Grasshopper (GH)	AE	—	—			-	-												

[1] Glue refers to software that enable links between BPS and geometrical modeling through graphical programming (also refered to as algorithmic modeling)

[2] Be10 is mandaroty to use for code compliance in Denmark

Source: Østergård et al., 2016

- **Core Complexity:** refers to the complexity of the parameters and the level of detail of the investigation.
- **Objectives**: refers to the energy characteristics of the system for which software is run. The analyzed characteristics are: energy, thermal, daylight, air quality, life cycle costs (LCC), life cycle analysis (LCA)
- **Parametric Simulation:** refers to uncertainty analysis (UA) and sensitivity analysis (SA) simulations, optimizations, facilities as one-factor-at-a-time (OFAT or OAT) approaches, cloud computing or facilities for input or output files.

Figure 6. Four Quadrants of simulation results
(Source: Long et al., 2010)

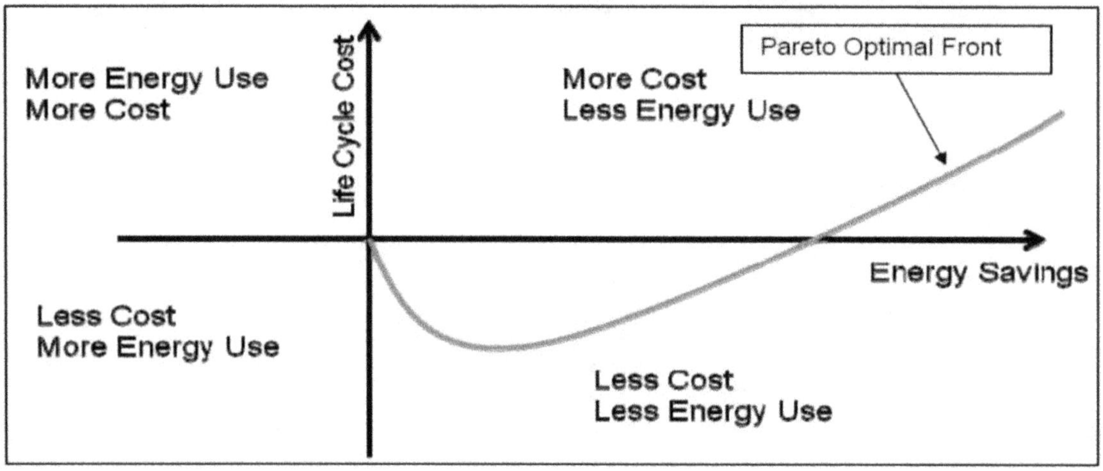

Simulation Models

Simulation models are typically mathematical models that simulate the energy behavior of the building before and after retrofitting, using different scenarios. There is no perfect or complete model or software that can take into account all the possible variables by applying the retrofit measures. It is advisable to use multi-criteria simulation models or to use several simulation models that take into account as many determinants as possible in the retrofit measures.

Modeling and simulation involve creating simulation models, verifying models, analyzing the results, assessing the economic impacts, and presenting all the results to the beneficiary. Base on this, the beneficiary can set the targets for the retrofit project.

The operation of such models involves the four phases (adapted after Long et al., 2010).

In the first phase it is created a baseline model associated with the buildings core characteristics (we mean geometric features, energy features and economic features such as location, floor area, program, loads).

In the second phase this model is checked from the point of view of input values, energy use, and economic results.

In the third phase a list of potential EDMs, which have a certain cost but can positively influence the characteristics of the building, is created.

In the fourth phase an analytical package is used which applies an iterative process. Each iteration analysis new possible values of EDMs and performs an energy simulation, resulting in the occurrence of a point in one of the four quadrants. If the Pareto curve for the new model is better than the base model, the solution is retained. The iterative process is repeated until Pareto optimal building design with associated EDMs is found. The number of possible combinations is very high, depending on the number and values of EDMs that can reach tens or hundreds of thousands, which makes desirable and requires the use of specialized software.

Table 5. Analysis of software used in energy simulation

Analysis methods for one economic indicator	Analysis methods for more economic indicators
• benefit-cost ratio (BCR) • net present value (NPV) • internal rate of return (IRR) • overall rate of return (ORR) • discounted payback period (DPP) • simple payback period (SPP)	• life cycle cost method • the levelized cost of energy • advanced analysis methods

Analyze

Retrofitting measures can be included in retrofitting alternatives - coherent packages of measures. These measures act independently or interdependently, can be studied economically individually or within the entire package. From the point of view of the retrofit measures analyzed and the economic indicators used, the analysis measures may be individual analysis methods or multiple analysis methods. In practice, it is preferable to apply a set of analysis methods in terms of several performance indicators.

Of the risk assessment methods, the most commonly used are: decision analysis, sensitivity analysis, expected value analysis, uncertainty analysis of building retrofits, Monte Carlo simulation, and risk-adjusted discount rate technique.

Building Information Modeling (BIM): Tool for Improvement Processing Information in Energy Retrofitting

A modern and complex tool for simulation and analysis used in energy retrofitting is Building Information Modeling (BIM). BIM is a top step in what concerns the creation of models for existing buildings and the analysis of retrofit measures.

BIM definitions are numerous and varied, each of which adds more particular information.

A general definition of BIM is the following: BIM is a process that involves data management that physically and functionally characterizes a building in order to create a digital model that will be used in all stages of a building's evolution at the conception phase until the end of its existence. BIM is supported by specialized software that helps the user in all stages of development and the existence of an asset (plan, design, build, operate).

Wikipedia highlights the existence of files or data from files in a certain format, data that can be "extracted, exchanged or networked to support decision-making on a building or other built asset" (Wikipedia, n.d.).

The US National Building Information Model Standard Project Committee propose the following definition: "Building Information Modeling (BIM) is a digital representation of physical and functional characteristics of a facility. A BIM is a shared knowledge resource for information about a facility forming a reliable basis for decisions during its life-cycle; defined as existing from earliest conception to demolition" (NIBS, 2014).

The specialists of the Royal Institute of British Architects define BIM in an evolution of its complexity, as follows: in a simplified form it is "the digital description of every aspect of the built asset", in a more structured formulation is "a process for creating and managing information on a construction project across the project lifecycle", and in a complete form is "a digital model [which] enables those who interact with the building to optimize their actions, resulting in a greater whole life value for the asset "(TheNBS, 2018).

In practice, software developers define BIM as "an intelligent 3D model-based process that gives architecture, engineering, and construction (AEC) professionals the insight and tools to more efficiently plan, design, construct, and manage buildings and infrastructure" (autodesk, 2018).

BIM Dimensions

BIM uses a lot of data as input parameters, data that refer to spatial, geographical, climatic characteristics that can be grouped in dimensions.

In the BIM reaction, the types of building information can be organized in "dimensions." An accepted organization of the information in the model is the following:

- **2D:** Two-dimensional graphical information; refers to the model involving the use of information that can characterize any point in a plane determined by two coordinates. This characterizes the primary, traditional development models, which represented the basic elements of the constructions (plans, sections, elevations, etc.)
- **3D:** Three-dimensional graphical information allowed spatial representation of the buildings using the three characteristics: width, height and depth;
- **4D:** Time and programme information; adds the time characteristic to spatial representation;
- **5D:** Cost information; is a representation that takes into account spatial, time and cost characteristics;
- **6D:** Facilities management information; is a representation that adds elements of building environmental and sustainability to the characteristics of the previous model.

Criticizing the exposed model, Richard Saxon (Saxon, 2018) has shown that the way in which information is grouped in 6D is wrong. Thus, this model groups information about the building, but also information regarding the commissioning and running of the asset. As a result, it proposed to reorganize the information specific to the 6D stage by splitting it into two dimensions:

- **6D:** Operational and maintenance information - is the last component which bring together the information about the current use (for which it was built) of the built asset; as a result, the first 6 models refer to building information;
- **7D:** information flowing from the building - refers to the information from the sensors and the information results from their analysis.

The model is created in the design phase by a complex team of architects, design engineers for building resistance, facilities engineers, surveyors, etc., and has the necessary and relevant information for all users. Each user can add in later stages information that enriches the model with discipline-specific data, but which allows for the uniqueness of the model, which allows keeping and completing the relevant information for each user. Software model involves creating objects, defining their attributes and the relationship between them.

As far as the data flow is concerned, their creation and use for the retrofitting process determines the following steps:

- Acquisition of geometric data.
- Acquisition of energy data.
- Creating a 3D model of BIM using an authoring tool and completing the BIM model by connecting with the thermal data.
- Verification the compliance between BIM and IFC or gbXML.
- Choosing a specialized software in the energy field.
- Simulation and analysis.
- Final decision on the selection of retrofitting methods and their prioritization.

The acquisition of geometric data implies the use of dedicated equipment for specific activities (e.g. laser scanning, photogrammetry) to measure the geometric dimensions of the building. This causes the creation of a point cloud to which 3D coordinates are attached.

Acquisition of energy data involves the use of specialized equipment for specific activities (e.g. thermography) that create a set of values that thermally characterize each building point.

Creating a 3D model of BIM is done by developing a project in an authoring tool. In this project, each geometric point will be encoded and automatically, using specialized software, a thermal value that will be taken from the energy data will be attached. Authoring tool selection and project development will be realized taken into account rules and regulations that standardize the field.

Thus, IFC is used, which is a standard in terms of defining and using objects within the BIM. Also, to ensure compatibility between BIM data and software applications, the gbXML requirements can be applied for BIM definition. Creating BIM taking into account the IFC and gbXML rules allows highlighting the evolution in time of the materials properties.

The choice of specialized software in the energy field is done with input / output, compliant with the BIM created and with the performance indicators selected for retrofitting. Several specialized software can be applied sequentially. Import or export of data is between BIM and software or even between different software.

Simulation and analysis involve running specialized software for different scenarios and analyzing results in terms of performance indicators. There is a correlation between the energy performance indicators and the economic analysis.

Final decision on the selection of retrofit methods and their prioritization will be done, depending on the energy and economic analysis and the available resources.

Note: Software applications can run without a BIM, but in this case is necessary to create a BEM (Building Energy Model) for each analysis software, through the variables it proposes, is the basis for running the software application.

Starting from the scheme presented by Sanhudo et al. (2018), different organizational chart can be developed. The author proposes a diagram for a BIM used for energy retrofitting in Figure 7.

Because multiple types of authoring tools and software products are used, problems may occur when importing data from the BIM to run in the power software applications. These problems can be of the following: location, geometry, material properties, building systems, building operation.

Figure 7. A BIM used for energy retrofitting

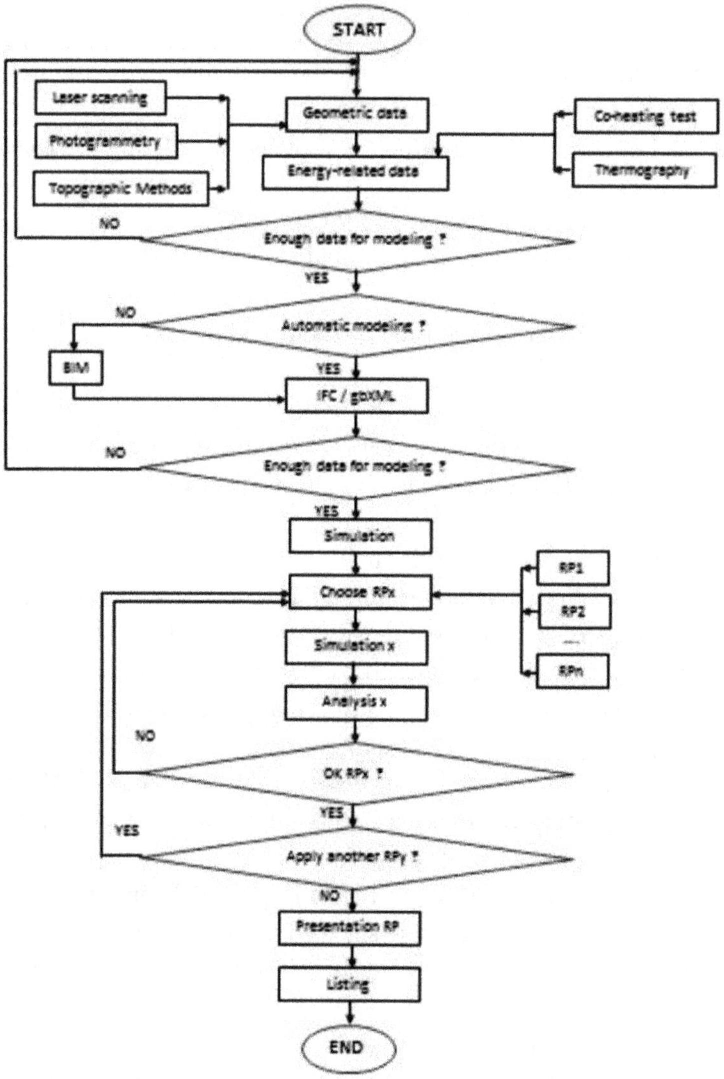

Thus, some software applications do not allow the input of data specific to the local weather conditions in which the upgrading project will be developed but use default data, despite the fact that the BIM can store this information.

Some buildings have a complex geometric shape with curved lines or architecture whose encoding in BIM is taken up incompletely or with errors in application software (e.g. surfaces or volumes), which results in erroneous results in simulation analysis.

The material properties have to be considered taking into account their evolution and behavior over time, and BIM or most application software does not offer coding or analysis capabilities. However, some software has created the possibility of introducing them manually.

Some software application cannot retrieve information about HVAC systems in the BIM because the IFC format does not allow them to be encoded. The problem can be solved by manually entering this data into the software application.

In terms of building operation, the problem is manifested in the form of the impossibility of taking over or encoding them in compatible software application data.

Phase 4: Site Implementation and Commissioning

Once the retrofitting measures have been established and prioritized, their implementation is proceeding. Implementation is done by specialized firms and involves the installation or upgrading of systems (equipment and software), testing and commissioning. For the installation of new equipment or upgrades made to existing systems, both the manufacturer's warranty and the post-warranty period must be established. In order to optimize these phases, rigorous planning is required if industrial or civil activities are carried out in the building. Some retrofitting measures allow them to be implemented without affecting the work carried out in the building, and others require reorganizing their activity or even temporarily stopping them.

Phase 5: Validation and Verification

The last step involves the validation of the retrofitting process thus implemented by reporting the values of the energy performance indicators selected in the 2nd step of the retrofitting process to the values obtained and measured after its implementation.

Subsequently, after the implementation of the retrofitting measures, periodic verification of the system operation will be carried out in desired parameters but also the energy savings achieved.

CONCLUSION

The energy sector has experienced a constant and strong development in recent years. Besides increasing and diversifying energy forms, as well as reducing pollution, an important role is played by increasing energy efficiency and reducing energy consumption of systems.

More, it was developed energy retrofitting, seen as a border subdomain in the energy field, that include specific elements from building construction and energy field, for building analysis and construction process to achieve much larger energy savings.

In this subdomain, the information is heterogeneous and is found in various forms. To be proper processed, the information must adhere to several essential features: access security, accessibility, accuracy, appropriate amount of data, believability, completeness, concise representation, ease of understanding, interpretability, objectivity, relevancy, representational consistency, reputation, timeliness, value-added. It is difficult to simultaneously fulfill these characteristics. In the same time, the variable volume of data, from a huge volume of data to an insufficient volume, and the different value of the data makes it difficult to process them.

The development of retrofitting can be done scientifically, organized, through using of Building Information Modeling (BIM) models.

BIM means modeling of building data to improve the decision-making process regarding energy retrofitting projects. BIM is used for new buildings in all phases, starting from the design phase, to building acceptance and commissioning. More, BIM can be used for old buildings to improve their energy efficiency. BIM allows identification of problems, but also offers solutions to solve them. The main objective of BIM authoring software is the development of an accurate 3D virtual representation of a building.

BIM can make an important contribution to achieving a BEM (Building Energy Modeling) performance. Both systems use to a large extent the same primary data, the functions provided by the BIM can be used within the BEM. An important role is played by the interoperability of information systems, being defined as the ability to communicate exchange and use data between two different software tools without an intermediate step of replication its content. The problem of interoperability has been solved by introducing and using IFC and gbXML data schemes for software development. IFC is an object-oriented data model for buildings, while gbXML represents a standard set of rules to define information using a customized markup language. By applying such schemes, the BIM model can be imported into various energy software tools without recreating the building geometry.

One direction of future development is to increase the degree of automation of BIM models. This can be done by automatically creating geometric 3D models in authoring tools from in-situ data, verifying them according to certain methods, and automatically realizing the linking of energy-related data from in-situ measurements. The result is a coherent, stable, and valid BIM model.

Another direction is the development of BEM products by improving its components: the graphical user interface (GUI) and simulation engine to provide as much information as support decision-making process. If GUI provides support for inputs and outputs, the simulation engine is based on performance data processing algorithms.

Last but not least, it is desirable to increase the interoperability between software products, especially if we are talking about products specific to BIM and BEM, which can be achieved by increasing standardization.

REFERENCES

Autodesk. (n.d.). Retrieved from https://www.autodesk.com

Designing Buildings Wiki. (n.d.). Retrieved from https://www.designingbuildings.co.uk/wiki/

Energy Audit Tool Manual (EATM), Energy Efficiency Management Office, Republic of Mauritius. Retrieved from http://eemo.govmu.org/English/Documents/Manual_Audit_Tool.pdf

Friedman, H., & Piette, M. A. (2001). Comparison of Emerging Diagnostic Tools for Large Commercial HVAC Systems. In 7th National Conference on Building Commissioning, Cherry Hill, NJ, May 9-11. Retrieved from https://eta.lbl.gov/sites/all/files/publications/lbnl-47698.pdf

Haves, P. (1999). Overview of Diagnostic Methods. In *Proceedings of the Workshop on Diagnostics for Commercial Buildings: From Research to Practice*, Pacific Energy Center, San Francisco, CA, June 16-17.

ISO. (2014). ISO 50002 Energy audits – Requirements with guidance for use. Retrieved from https://www.sis.se/api/document/preview/917504/

Leonardo energy. (2015). How many types of energy audit are there? Copper Alliance Initiative and partners. Retrieved from https://help.leonardo-energy.org/hc/en-us/articles/203598661-How-many-types-of-energy-audit-are-there-

Long, N., Hirsch, A., Lobato, C., & Daniel Macumber, D. (2010). Commercial building design pathways using optimization analysis, National Renewable Energy Laboratory. In *2010 ACEEE Summer Study on Energy Efficiency in Buildings*. Retrieved from https://aceee.org/files/proceedings/2010/data/papers/2007.pdf

Ma, Z., Cooper, P., Daly, D., & Ledo, L. (2012). Existing building retrofits: Methodology and state-of-the-art. *Energy and Buildings*, *55*. doi:10.1016/j.enbuild.2012.08.018

NIBS. (2014). National Institute of Building Sciences (the website). Retrieved from https://web.archive.org/web/20141016190503/http://www.nationalbimstandard.org/faq.php#faq1

Nicolescu, O., & Verboncu, I. (2008). *Fundaments of organization management (Fundamentele managementului organizaţiei)*. Universitara Publishing House.

Østergård, T., Jensen, R. L., & Maagaard, S. (2016). Building simulations supporting decision making in early design - A review. *Renewable & Sustainable Energy Reviews*, *61*, 187–201. doi:10.1016/j.rser.2016.03.045

Pressman, R. S. (2005). *Software Engineering: A Practitioner's Approach*. McGraw-Hill Companies Inc.

Sætnan, A. R., Schneider, I., & Green, N. (2018). *The Politics and Policies of Big Data: Big Data, Big Brother?* Routledge - Taylor & Francis. Retrieved from http://www.uniteforsight.org/global-health-university/quality-data

Sanhudo, L., Ramos, N.M.M., Martins, J.P., Almeida, R.M.S.F., Barreira, E., Simões, M.L. & Cardoso, V. (2018), Building information modeling for energy retrofitting – A review. *Renewable and Sustainable Energy Reviews*, *89*, 249-260. doi:10.1016/j.rser.2018.03.064

Saxon, R. (2018). Getting the dimensions of BIM into focus. *BIMplus*. Retrieved from http://www.bimplus.co.uk/people/getting-dimensions-bim-focus/

The American Heritage Dictionary of the English Language. (n.d.). Retrieved from https://ahdictionary.com/

The Energy Audit. (n.d.). Retrieved from www.the-energyaudit.com

theNBS. (n.d.). Retrieved from https://www.theNBS.com

Wanga, C., Yong, K., & Chob, Y. K. (2015). Performance Evaluation of Automatically Generated BIM from Laser Scanner Data for Sustainability Analyses. In *International Conference on Sustainable Design, Engineering and Construction*. Elsevier Ltd. doi: Retrieved from https://www.sciencedirect.com/science/article/pii/S1877705815021864 doi:10.1016/j.proeng.2015.08.531

Wikipedia. (n.d.). Retrieved from https://en.wikipedia.org/wiki/

KEY TERMS AND DEFINITIONS

Building Information Model (BIM): BIM is a process that involves data management that physically and functionally characterizes a building in order to create a digital model that will be used in all stages of a building's evolution at the conception phase until the end of its existence.

Data Quality: The totality of features and characteristics of a data set that bear on its ability to satisfy the needs that result from the intended use of the data

Diagnostics: The analysis of equipment in terms of malfunction or operation at parameters other than optimal operating parameters

Energy Audit: A systematic analysis of energy use and energy consumption within a defined energy audit scope, in order to identify, quantify and report on the opportunities for improved energy performance

FDD (Fault Detection and Diagnostics) Tools: Specialized software that treats input data and processes them according to well-defined algorithms.

Information System: A coherent mix of data (data, information, information flows) and means of processing them (procedures and equipment) that contribute to the company's performance and achievement of organizational goals.

Performance Assessment: Selection of a set of quantitative and qualitative performance indicators to analyze the performance of the building

Retrofit Project: Is a project focusing on increased building energy efficiency by enhanced heat recovery, taking into account technical aspects and economical constraints (available capital and payback times).

Section 3

Technical and Ecological Aspects of Green Retrofitting

Chapter 9
Retrofitting for Energy Management:
Using IoT Technologies

Michael Christian Krutwig
Krumedia GmbH, Germany

ABSTRACT

According to ISO 50001, energy management systems are increasingly being implemented in companies, with the continuous measurement of energy consumption playing an important role. At the same time, energy meters are increasingly becoming cyber-physical systems (CPS) in the IoT age. LPWAN radio networks and IoT clouds open up new possibilities for retrofitting meters and industrial components. By using these technologies, the measurement infrastructure of an energy monitoring system can be cost-effectively expanded. This section presents the technologies relevant for energy management and shows their application using selected use cases for LoRaWAN, OPC UA, and the combination of optical character recognition (OCR) and IoT Cloud.

INTRODUCTION

Systematic energy management is a common instrument for continuously improving energy efficiency in companies and organizations. Effective energy management saves money, strengthens the competitiveness of companies and lowers their carbon footprint. Therefore, energy efficiency is also an important policy instrument to achieve climate goals. In many European countries, the energy efficiency of companies has been demanded and promoted not only since the European Energy Efficiency Directive (EED) (European Parliament, 2012). A good energy management system provides information on where energy saving potential is in a company and influences organizational and technical processes. Hence, systematic energy management is used in many companies worldwide. From an economic viewpoint, improved energy efficiency not only reduces the company's overall energy consumption, but also the company's emissions and consumption of raw material. Energy improvements often bring additional advantages through non-energy benefits (Krutwig & Starosta, 2017). A systematic energy management

DOI: 10.4018/978-1-5225-9104-7.ch009

system based on the ISO 50001 standard or the Eco-Management and Audit Scheme (EMAS) enables companies to meet their energy policy commitments and continuously improve their energy performance through a systematic approach (Hillary, 1993). It comprises the necessary organizational and information structures, including the necessary tools. A central component of energy management is the monitoring of energy flows through temporary or permanent measurements, estimates, or calculations. With energy monitoring, potential savings are discovered and the savings already achieved through implemented efficiency measures can be checked.

The standard ISO 50001, published in June 2011, is the first established international standard for energy management systems. ISO 50001 is not sector-specific and can be applied by a variety of organizations, from small and medium-sized enterprises to large companies and public authorities. According to the standard, defined user groups and defined tasks with fixed deadlines are recorded and checked. The formal requirements for functions and tasks of systematic energy management must naturally cover a broad spectrum. In principle, the approach regarding energy balance limits and various energy sources must be pursued holistically. It is up to the respective company to tailor the various design options of the management system, as described in ISO 50001, to its own needs. Like most management systems that aim for continuous improvements, ISO 50001 is based on a plan-do-check-act (PDCA) cycle. It can be implemented independently of existing management systems or integrated into existing management systems, such as ISO 9001. Although energy monitoring via measurements is not mandatory according to the standard, temporary or permanent measuring equipment is used in many energy management systems today. Regular measurements are important for good transparency of energy flows. This is achieved by connecting the measuring devices and automatically transmitting the measured values to an energy monitoring system. In this process, the Internet of Things (IoT) opens up new possibilities.

The IoT can be described as a network consisting of numerous so-called cyber-physical systems (CPS) (Baheti & Gill, 2011; Monostori et al., 2016; Schlick, 2012; Wolf, 2009) that exchange information via M2M (machine-to-machine) communication. According to the CPS concept, objects become "intelligent". Therefore, a physical object has virtual representations that can be addressed on the Internet of Things. These self-describing cyber objects provide an interface to the data and services of the physical objects. Hereby, CPS is enabling so-called "smart" infrastructures: smart productions, smart cities, smart services, smart grids, smart healthcare, and many more. In this "4th Industrial Revolution," smart factories can reconfigure themselves quickly and flexibly and adapt to producing new and highly customizable products (ElMaraghy, 2005). In addition, IoT industrial applications can bring further added value to companies (Jeschke et al., 2017). In (Fantana et al., 2013), the prerequisites and possibilities of IoT use are described from the point of view of the industry sector. The concepts in this article can also be referred to as "smart retrofit" due to the use of CPS, a term that differs from *"traditional Retrofitting by not only replacing old components in the machine regarding their energy efficiency"* (Guerreiro et al., 2017).

The use of the IoT from the device to the application can be described via a technology stack (Figure 1). The device thus has software for providing the device-specific services and a communication interface. Communication usually takes place via the IP protocol and can also take place via various wired or wireless media. The cyber object is then globally available in the cloud platform. In our case, the cloud application shown in the stack is the software for energy monitoring or energy management. To achieve interoperability of all components, it is necessary to standardize the device software, the communication protocols, and the virtual representatives. Open Platform Communications Unified Architecture (OPC UA) (Mahnke, Leitner, & Damm, 2009), is a suitable and widely used key technology for this purpose.

Figure 1. The IoT technology stack
source: author

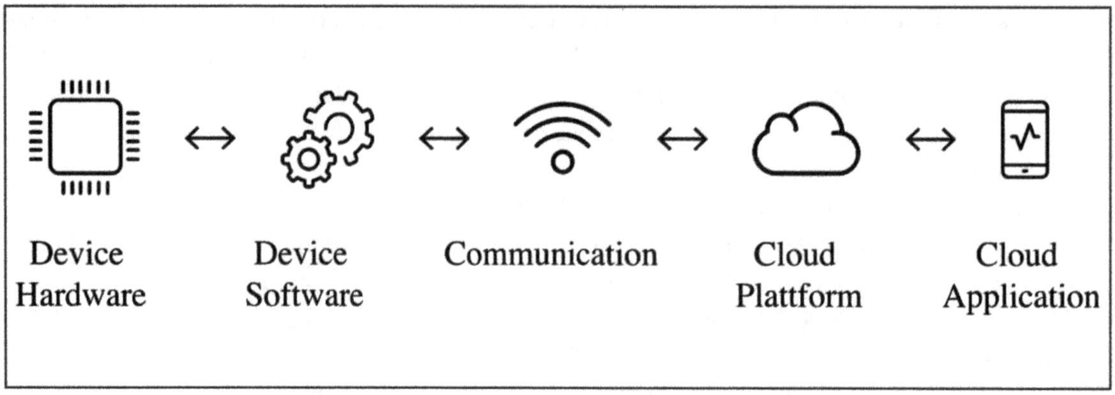

This chapter builds a bridge between energy monitoring and IoT. By modernizing or expanding existing plants and equipment, known as retrofitting, existing and previously non-connected energy meters and production plants are to be harnessed for continuous, networked energy monitoring. Considering a lifetime of production plants of up to 40 years and more, OPC UA becomes part of retrofitting in many companies. LoRaWAN[1], an important Low Power Wide Area Network (LPWAN) technology, is presented as a wireless communication technology particularly suitable for energy management. The retrofitting process is described and evaluated based on three concrete use cases.

BACKGROUND

The motivation for this contribution comes from energy monitoring, which is normally used for energy audits or for the implementation of an energy management system according to ISO 50001. This standard describes a management system that defines a continuous improvement cycle. It requires monitoring, measuring, and analyzing the energy use and energy consumption of a company. Within this process, balance limits must be defined that include areas with significant energy input, and variables must also be determined that have a significant influence on energy input. Based on the energy input and these variables, indicators (EnPI) are then defined, which are used as indicators of energy efficiency. Thus, the effectiveness of the plans and measures can be checked against the EnPIs. If possible and economically feasible, the data should be determined by means of measurements. For smaller companies this can be done by the electricity suppliers' meters. The further allocation of the consumptions could be calculated on the basis of consumer specifications or using measuring instruments used for a limited period without fixed installation. Within large infrastructures, permanent measuring equipment is installed. Calibrations and plausibility checks must be conducted to ensure that the data collected and the results of tests provide error-free and reproducible values. Ideally, time series with a high resolution in the minute range can be obtained via networked counters. Additional data for influencing factors, such as weather data, and non-energetic information, such as production data or sales figures, are added via automatic data connections, data imports, or manual entries to complete the database for energy management.

During the development and introduction of an energy management system according to ISO 50001, a measurement concept is created. This concept determines which media and which consumers are to be measured and how measurements are to be carried out taking into account the specified requirements. Records shall be kept of the planning and execution of the measurements and of the measuring and test equipment itself.

Figure 2 shows the setup and extension of a measurement infrastructure in the context of the PDCA cycle. After this, the energy manager can decide a new (PLAN) whether additional sub-measurements should be carried out for each run. For existing systems and meters, this extension (DO) is often the most economical way to gain new data. When deciding for or against a networked measuring system, economic efficiency is a factor that is difficult to calculate. While the costs for energy meters, installation, and networking are relatively easy to calculate, the monetary value can only be determined vaguely in advance owing to the additional knowledge gained and the resulting possible efficiency measures. Retrofitting gives us the opportunity to minimize costs here.

BASIC Technologies

In the text below, application examples based on IoT technologies are presented. It should be mentioned that there are numerous other IoT basic technologies that also make retrofitting possible. However, those are not relevant for understanding the examples shown. Especially in wireless transmission technologies, there are various sensor networks and mesh networks, based on technologies such as ZigBee, Z-Wave, Bluetooth, NFC, WLAN, 6LoWPAN and others, which save the costs for a wired network in the retrofit. An example of using ZigBee for energy management in the home is provided in the work of (Erol-Kantarci & Mouftah, 2011).

Figure 2. Measurement within the ISO 50001 PDCA cycle
(source: author)

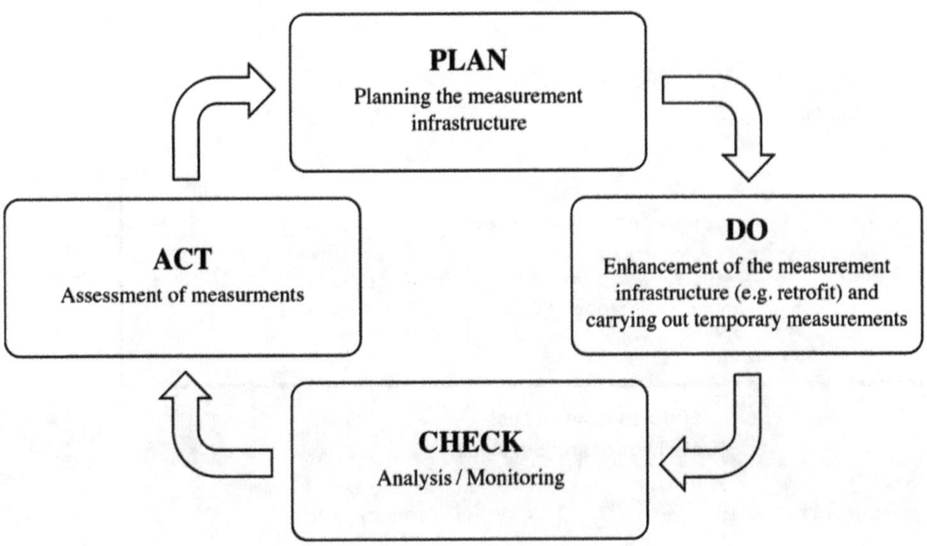

At the hardware level, non-invasive sensors, such as transducers, Rogowski coils, and bus sniffers are a good option for retrofitting. The work of (Nsiah, Schappacher, Rathfelder, Sikora, & Groza, 2018) describes a retrofit scenario using a Profibus sniffer and the Lightweight M2M protocol (OMA, 2017).

OPC UA for Energy Management

OLE for Process Control (OPC) (Rehbein & Pederson, 1996) is the original name for standardized software interfaces, which should enable data exchange between applications of different manufacturers in automation technology. In 1996, the independent OPC Foundation was established to maintain this standard, which now (as of 10/2018) comprises more than 620 companies. OPC's central idea is creating a platform-independent standard for the exchange of process data that abstracts from typical protocols such as Modbus, Profibus, etc. so visualizations (HMI) and similar systems can fall back on a common basis. However, the standard OPC is limited to Windows systems because it is based on Microsoft's OLE or COM/DCOM technology. This OPC protocol is also called "OPC COM," "OPC DCOM," or "OPC classic" to distinguish it from its successor OPC UA (Leitner & Mahnke, 2006).

In 2005, the OPC Foundation released the first draft of the OPC UA specification (OPCFoundation, 2017), which differed significantly from its predecessor. With this, the central problem areas of OPC were transformed into a modern and sustainable approach. OPC UA was adopted as IEC standard 62541 and is now widely used. OPC UA pursues a service-oriented, internet-enabled approach that is available for all platforms independently of Windows. It provides a communication model and a uniform data model for process data, alarms, and historical data that can be used across systems. This also includes a mature security concept. Figure 3 shows the OPC UA stack in a compact form. The companion specifications are cross-vendor information models for different industries and processes. For example, the Companion Specification OPC UA Robotics (VDMA, 2018) defines a standardized information model that can display all robot-related data uniformly, regardless of manufacturer and location.

In (Hannelius, Salmenpera, & Kuikka, 2008), several approaches describe how a migration from "OPC classic" to OPC UA can occur: OPC UA wrappers (for server) and proxies (for clients) translate the respective calls or via an OPC UA gateway, which implements the complete OPC UA stack and maps each UA call to an OPC COM call and vice versa. The use of adapter software is also possible,

Figure 3. OPC UA stack
(source: OPC UA Foundation[2])

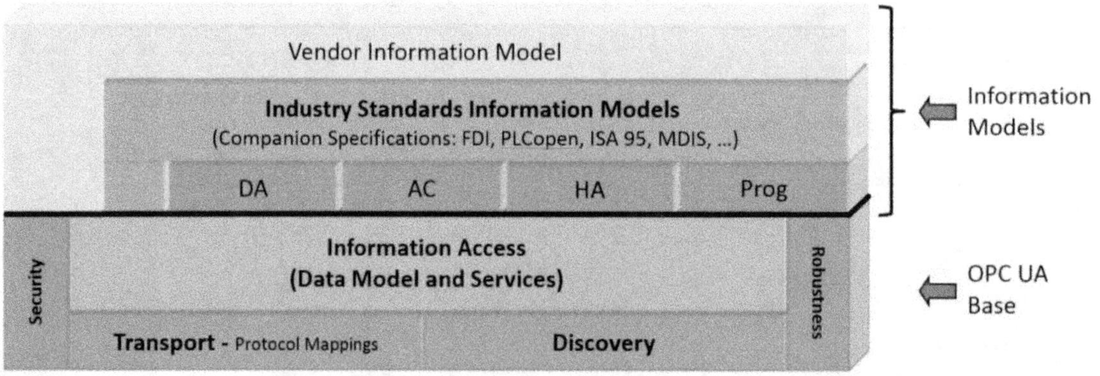

which forms a specialized and reduced gateway. In the presented use case, only the required services are implemented that directly query the OPC COM server. The migration to OPC UA is a retrofit, which converts production systems into cyber-physical production systems (CPPS). In (Schlechtendahl, Keinert, Kretschmer, Lechler, & Verl, 2015), this process is described as "CPPS enablers."

For energy management, OPC UA can also be used across domains. For this purpose, (Matzler Wollschlaeger, Fernbach, Kastner, & Huschke, 2013) have proposed an information model that is based on existing specifications from industrial automation and building management. This cross-domain data model is used as an example of monitoring and controlling heat recovery from production for an office building. The example impressively shows the significance of OPC UA goes far beyond its use in purely industrial plants. In fact, the standard is becoming increasingly widespread, for example, in building control systems, health care, canteen kitchens, and in many other domains. The properties of a "smart" production required in (ElMaraghy, 2005) are fulfilled by the automated engineering of production plants using OPC AU. In (Schleipen, 2008), the combination of OPC UA and the CAEX format (Ebel, Drath, & Sauer, 2008) shows how a plan for visualization and engineering can be described. This holistic description possibility is also of great importance for the use of holistic energy management.

In the application of energy monitoring, the OPC UA servers typically appear either in the form of machines, plants, or control systems that supply energy data, as well as many other production data and control functions. The OPC UA server can also consist of only one energy meter or one sensor. For this device class, there is only a lined option to use a host computer for the server. The OPC UA stack, available as open source for numerous programming languages, can be implemented on small single-board computers (Imtiaz & Jasperneite, 2013), so small and simple sensors will also be equipped with an OPC UA interface in the future.

While OPC UA has now become the leading standard, there are alternatives at the management level. KNX access based on web services offer, for example, BACnet/WS (Kastner & Szucsich, 2011) or oBIX (Neugschwandtner Neugschwandtner, & Kastner, 2007). Message brokers, such as MQTT (Hunkeler Truong, & Stanford-Clark, 2008) or CoAP (Shelby Hartke, & Bormann, 2014), which are optional components of the OPC UA stack, can also be named as lightweight alternatives to OPC UA. The Woopsa (woopsa.org) and Data Distribution Service (DDS) protocols (Pardo-Castellote, 2003) are relatively new and of reduced complexity.

LPWANs and IoT-Clouds

In energy management, energy meters and sensors are relatively small devices that make use of complex hardware for networking both uneconomical and technically problematic. The wired connection to power sources is also included in this problem. The solution is wireless connection between these devices via a class of networks, which are summarized under the term Low Power Wide Area Networks (LPWAN) (Bardyn Melly, Seller, & Sornin, 2016). These networks are specially designed for M2M communication and meet requirements that are not met by classic mobile radio networks. In M2M communications, data must be transmitted in the kilobyte or byte range. A few measurements per hour already provide a good data source for an energy management system. This concession provides the scope for meeting the following requirements:

- **Energy Consumption:** Communication must not require high energy consumption. Battery-powered devices must have a long service life (<1 year) or can be completely self-sufficiently supplied via energy harvesting.
- **Low Hardware Costs:** The hardware required for networking must be cost-effective, so an economic relation is maintained even with low-cost sensors. The chip costs for LPWAN hardware are in the price range of about €1 to €10. In the future, disposable objects should be available for a price of about 20 cents (source: Sigfox).
- **Area of Coverage:** According to the "wide Area," LPWANs should offer national or even international coverage with a corresponding number. Individual base stations thus cover a radius of up to several kilometers. If you can set up your own private infrastructure, a single district, a campus, or the property of a company will suffice as coverage area.
- **Reliability:** The accessibility of the devices must also be guaranteed inside buildings. It is not uncommon to find meters for electricity, water, and gas in a company's basement—thus, in areas reserved for normal mobile radio networks.
- **Security:** As part of the IoT stack, LPWANs must have a security architecture on board that protects data at the transport level and prevents unauthorized network nodes from participating.

In addition to these essential requirements, there are other requirements, such as a medium access method that does not require unnecessary wake-ups and the consideration of low-energy localization options, such as received signal strength indicator (RSSI) or time difference of arrival (TDOA) as an alternative to GPS, which has additional hardware and energy requirements. Especially, the economically motivated requirements for LPWANs offer an ideal prerequisite for retrofitting.

All LPWAN technologies are networked via a star-shaped topology with the base station connected to the Internet in the middle. If some of the requirements are waived, further radio networks for retrofitting exist, which play a role in home automation and are not discussed here. These include mesh networks such as Bluetooth LE 5, ZigBee, Z-Wave, 6LoWPAN, and the widespread WLAN, according to IEEE 802.11s. An interesting retrofit of air conditioning systems based on ZigBee is provided (Medina & Manera, 2017).

Several existing LPWANs fully meet these requirements. The best-known representatives available on the market are LoRaWAN (Sornin Luis, Eirich, Kramp, & Hersent, 2015), Sigfox (www.sigfox.com), NB-IoT, and LTE-M, which differ in technical parameters and marketing strategy.

Sigfox

Sigfox is a complete LPWAN infrastructure of the French company Sigfox, consisting of available base stations, a cloud portal, and device specifications for chips, which are certified by Sigfox for operation in its own infrastructure. Currently (August 2018), 12 countries are fully covered, including North America and Western Europe, 20 are under construction, and 28 are planned. Low-cost Sigfox modules are installed in over 300 devices for applications in smart homes, smart cities, agriculture, the environment, health and assisted living, and industry and fleet management.

Sigfox supports half-duplex communication. Only the node can initiate the communication with the server. This allows the node to retain control over its power consumption while still allowing the server to transmit parameter changes, commands, or updates to the node. For modulation, the DBPSK method is used for the uplink and the GFSK method for the downlink. If the node is a typical sensor node in energy management, such as an energy meter or temperature sensor, Sigfox can also be used in an operating mode with a unidirectional uplink so the nodes act exclusively as transmitters. Without a receiver, there are significant savings in energy and hardware costs.

The Sigfox data packets are small with a payload of 12 bytes. In the EU, up to 140 messages can be uplinked daily at a rate of 100 bits per second while, in countries where FCC approval is mandatory, the rate is 600 bits per second. Sigfox operates the uplink in the license-free ISM band (industrial, scientific and medical) at 868.00-868.60 MHz and the downlink at 869.40-869.65 MHz. For regulating competing access, there are restrictions on the assignment. In Europe, for example, the usage ratio must be less than 1% and the transmitted power must remain below 25 mW for the uplink and up to 500 mW for the downlink. The transmission range of a base station can be up to 50 km in flat rural areas, but only 3 to 10 km in an urban environment. Because more than one base station can receive a node's signal, there is no interruption in communication if a base station fails. For security, Sigfox uses a unique device ID and signed telegrams, as well as VPN and encryption on the application level.

For energy management, there are already several sensors with built-in Sigfox modules. In addition to complete energy meters, there are also camera- or pulse-controlled add-ons for existing meters so a retrofit can be carried out. The energy management software requires an interface to the Sigfox cloud API to retrieve the measured values. Therefore, the costs and effort of retrofitting are low, but the costs of using the infrastructure must also be considered. When using Sigfox in energy management, it should also be noted that the very low data rate does not allow energy monitoring in minutes or five-minute intervals; the service is sufficient for transmitting hourly values.

NB-IoT und LTE-M

With the Narrowband IoT (NB-IoT, also LTE Cat-NB1) (Ratasuk Vejlgaard, Mangalvedhe, & Ghosh, 2016) and LTE for Machines (LTE-M, also LTE Cat-M1 or eMTC) standardized by the 3rd Generation Partnership Project (3GPP), a complete cellular LPWAN infrastructure is also provided, so only the devices themselves have to be installed during retrofitting. As these networks are based on GSM/LT and, thus, operate in the globally licensed frequency bands, this infrastructure is offered by the mobile network operators, who can expand their existing radio systems with simple extensions for IoT services. By November 2017, over 30 of the world's largest mobile network operators NB-IoT, including Deutsche Telekom, Vodafone, AT&T, Telefónica, China Unicom, China Mobile, Verizon, and Telstra, will be in operation.

With LTE-M, a software update of the existing LTE radio systems is sufficient in many cases. With NB-IoT, the introduction requires a hardware upgrade of the base stations. To achieve the necessary range for IoT applications, NB-IoT offers an additional 20 dB power transmission balance. Thus, NB-IoT networks have up to ten times higher network coverage than GSM networks. Both technologies use the SC-FDMA method for uplink modulation and the OFDMA method for downlink modulation.

With its 180 kHz wide channels, NB-IoT fits into the specified GSM spectrum distribution and achieves a data rate in the uplink of about 150 to 250 kbit/s. Here, different sources vary. Further, NB-IoT fits into the protection gaps between LTE bands (guard band). The regular use of an LTE carrier (in band) is also possible. Owing to the low bandwidth, several 10,000 IoT endpoints can be supplied with one base station. Compared to broadband LTE, 3GPP has introduced some simplifications for NB-IoT. For example, a single antenna is sufficient for receivers. The technology works with half duplex in a paired spectrum and does not require hardware for full duplex mode. A transmission power control is not required for the terminals, as channel coding has also been simplified. This results in less complex radio modules and, therefore, LTE mobile devices that are cheaper than normal.

With up to 1 mbit/s in full duplex mode, LTE-M offers a significantly higher bandwidth than NB-IoT and is more oriented towards mobile use while also having a correspondingly higher energy requirement. This technology is, therefore, less predestined for use in energy management than NB-IoT. The results of (Lauridsen Kovacs, Mogensen, Sorensen, & Holst, 2016) show LTE-M coverage in a rural area of 99.9% of indoor (10 dB loss) and outdoor devices. However, for deep indoor uses, NB-IoT is required so the coverage remains at about 95% of all devices. The security mechanisms of both technologies are adapted from the LTE standard, where end-to-end encryption and the embedded NB-IoT SIM card protects against attacks.

LoRaWAN

LoRa, Semtech Corporation's patented radio technology, and the resulting communication protocol, LoRaWAN (Sornin et al., 2015), is of particular interest for retrofitting. This LPWAN technology enables users to construct their own infrastructure by setting up radio gateways in the IoT clouds, so they do not have to pass on their data to third-party service providers. There are no user fees for the infrastructure. Especially if use is limited to the real estate of a company, the technical construction of the radio network can be carried out cost-effectively. In many cases, using a radio gateway with a range of a few hundred meters to 15 kilometers per location is sufficient. In a suburban area, 3 km coverage could be achieved (Augustin, Yi, Clausen, & Townsley, 2016). The work (Bor, Vidler, & Roedig, 2016) presents an approach to extending the range over multi-hops with a such called LoRaBlink protocol. The penetration of buildings into basements and underground garages is problem-free. The data is transmitted encrypted with AES-128, and operator and application data can be securely separated by using a network key and an application key.

LoRaWAN defines three device classes (A, B and C), which differ in the possibility of communication (Sornin et al., 2015). This makes it possible to achieve simpler requirements with less complexity. The simplest and most economical Class A devices use the ALOHA procedure in the uplink and a downlink is only possible in a short time window immediately after the uplink and only serves receipts. For Class B devices, time windows can be defined for the data traffic in the downlink, in which the device is woken up via a time-synchronized beacon. Class C devices are permanently in receiving mode and can, therefore, not operate in battery mode. All three classes can be supplied in parallel in one network.

LoRa is a chirp spread spectrum 4 modulation technique for radio data transmission on the ISM band (SemtechCorp, 2015). It uses the license-free sub-gigahertz frequency bands 169 MHz, 433 MHz, 868 MHz (Europe) and 915 MHz (North America). The LoRa radio module is called the Concentrator. The LoRa modulation is based on the proprietary Chirp Spread Spectrum technology (Reynders & Pollin, 2016) with the possibility of controlling the data rate and signal strength for each individual node via a spreading factor (SF). This allows a minimum energy requirement to be found for each node depending on the signal attenuation and the amount of data to be transmitted. Another factor influencing the data rate is variable error correction, which is determined by a code rate (CR). The calculation of the data rate R_b also illustrates the relationship to the spreading factor:

$$R_b = \frac{SF \times BW \times 4}{2^{SF} \times \left(4 + CR\right)} \; bits/sec \tag{1}$$

where

SF: spreading factor (7 ... 12)
BW: modulation bandwidth (e.g. 125 KHz)
CR: code rate (1 .. 4)

Using this, the data rate can be adjusted from 0.3 to 50 kbit/s. For large-scale networks (Blenn & Kuipers, 2017), using a minimum SF in each case is recommended to minimize the collision rate by a short occupancy of the medium. For energy sensors in the corporate network, on the other hand, the sensor's removal and the minimization of the energy requirement will be the decisive influencing factors. Another limiting factor is the maximum duty cycle. This defines a maximum time in percent in which a node may actively occupy the air interface. In the ISM band in Europe, for example, this duty cycle is one percent.

In addition to own infrastructures, there are also LoRaWAN providers in some countries that offer nationwide coverage with radio networks or IoT clouds commercially. Such infrastructures are important for mobile LoRaWAN nodes. The community-based initiative The Things Network (TTN) (Blenn & Kuipers, 2017) aims to provide non-commercial, nationwide coverage with LoRaWAN. Every participant can expand the TTN-LoRaWAN with his own gateway and use all participating gateways as base stations for his LoRaWAN node.

LoRaWAN is well suited for metering and monitoring. Owing to technical limitations, it can only transmit measured value time series with a maximum resolution of five minutes when encoded efficiently. LoRaWAN is also successfully used for other fields of application, such as smart city applications. Applications in real-time monitoring, logistics, or wherever high data volumes are required are problematic (Adelantado et al., 2017). In the indoor area (Petäjäjärvi, 2017), a range of more than 300 meters through several buildings could be achieved, but no SF could achieve 100% loss-free transmission. In energy management, packet losses lead to irregular time series, for which the EnM software should be able to compensate.

Other Narrow Band Networks

There are numerous publications comparing different LPWAN technologies by technical characteristics and performance (Centenaro, Vangelista, Zanella, & Zorzi, 2016; Sinha, Wei, & Hwang, 2017; Vejlgaard et al., 2017). There are also other LPWANs, such as Weightless (weightless.org), Ingenu (ingenu.org), EC-GSM-IoT (3GPP) or the link labs based on LoRa (link-labs.com). Another application layer transport protocol to provide device management functionality over sensor or cellular networks is the Lightweight M2M (OMA, 2017) of the Open Mobile Alliance. In the current version 1.1 of LWM2M, objects for management of NB-IoT connectivity have already been defined and additional objects for LoRa will be added.

OCR Meter Reading

A retrofit of energy meters is intended to automatically deliver measured values at short intervals to an energy management system (EnMS) via a subsequently installed, cost-effective network. If available, an existing data interface can be used to transmit meter pulses or measured values for networking meters. If there is no data interface or if this cannot be used, a recognition-based solution can be used for a retrofit. The simplest retrofit approach is the detection and networking of measuring pulses, provided that the meter supplies pulse signals electrically or optically. More elegant is the recognition of the displayed meter reading. The meter reading is photographed and the image is converted into meter data using optical character recognition (OCR) algorithms. This data is then transmitted to the EnMS. In the case of a connection, the meter image can also be transmitted as such and an OCR process can only be applied on the EnMS side (offline recognition). A further motivation for OCR is the manual transmission of meter readings with a reading via mobile devices on which software performs the OCR and the transmission of the reading with fewer errors and faster than the classic way of paper and pen.

Machine-based character recognition is older than the internet. The first patent dates back to 1929 (Gustav, 1938) and provides a detailed description of the historical development of all technologies and products up to the 1990s (Mori, Suen, & Yamamoto, 1992), more modern approaches use neural networks (Jain Mao, & Mohiuddin, 1996), hidden Markov models (HMM) (Rabiner & Juang, 1986), a focus of research is context recognition. The recognition of counter readings is almost trivial to the general challenges of OCR: The values are not handwritten, there are only numbers instead of letters, the context is clearly defined, the image quality has a constant contrast, and the same quality and perfectly patterned templates are available. With these prerequisites, today's algorithms deliver a recognition rate of almost 100%. Nevertheless, some difficulties must also be overcome with energy meters, such as the observation of different digital and analog meters, confusion with meter numbers in images of the entire device, the correct recognition of decimal places and partially visible digits owing to the rotary movement of mechanical counters, and counter overflows in the plausibility check. With older counters, various characteristics must be considered, such as electromechanical meters with black or white meter registers, as well as digital meters with seven-segment displays in LED or LCD technology. Newer meters with matrix displays are not sufficiently "retro" for our purposes.

The automatic recognition of the digits occurs in several steps (Joachim Schenk, 2010):

Step 1: Image acquisition via a camera or CCD sensor. Here, a relatively low image resolution is sufficient for counter readings, colors can also be dispensed with. An LED is used for uniform illumination at all times.

Step 2: During pre-processing, the contrast of the image is optimized by filters, the image is converted into a monochrome image, and the line with the counter reading is extracted from the entire image and is normalized.

Step 3: The feature extraction divides the total counter reading into individual elements such as digits, commas, and spaces.

Step 4: In detection, the individual elements are identified as numbers by comparison with learned templates. If the HMM is often used for text, a much simpler matching algorithm can be used for pure number recognition. The composition of the recognized digits and separators results in the counter reading. After this process, the EnMS can additionally check the counter reading for plausibility and discard it in case of obvious reading errors.

Such a camera-based OCR system can be realized with low hardware resources. In (Castells-Rufas & Carrabina, 2006), a cost-minimal system is designed, which, with a VGA CMOS camera module and a low-cost FPGA, achieves a detection rate of 99.46% with changing counters. The approach of (Cortez, Felix, Girão, Frota, & Bessa, 2014) also develops an OCR for energy meters, using software components for Windows systems, and the authors examine and compare eight different preprocessing algorithms. Another prototype implementation for a mobile application for meter reading (Elrefaei Bajaber, Natheir, AbuSanab, & Bazi, 2015) is based on the free software OpenCV (Bradski & Kaehler, 2008), which achieves a recognition rate of 87.1%. For own software developments several open-source libraries exist, of which the Tesseract engine (Smith, 2007) should be mentioned exemplarily.

CASE STUDIES FOR RETROFITTING

This section presents several case studies that all have one thing in common: simple energy meters are retrofitted to cyber-physical systems. This allows the devices to be networked and to transmit energy measurements to the energy management system at short intervals. For all three examples, we use the *enerchart* software for energy monitoring, as all necessary interfaces to the IoT clouds and OPC UA servers are already implemented. This system also performs all conversions from line or meter reading to consumption.

Case Study: OCR Energy Cam

In this example, a three-phase meter of the energy supplier is retrofitted with the help of an OCR camera from *Q-loud*. The aim is automatic meter reading every five minutes with immediate transmission to the energy management system. The camera transmits the meter readings via a proprietary transmission protocol in the 868 MHz band to a local radio gateway. This protocol offers a transmission rate of 50 kbit/s and AES 128 encryption. The gateway then transmits the measured values via the internet to the camera manufacturer's IoT cloud service.

The first configuration of the camera is done locally using a Windows software. Here, you can set OCR parameters for the counter type, the digit size, the color (here: white digits on black rolls), the number of decimal places, and parameters for transferring the read counter readings. The camera is then glued to the counter without completely covering the counter reading. The camera is optimized for electricity meters; there is another version of this camera for gas meters. The camera is powered by a 3.6V 1600 mAh AA lithium ion battery, which is placed in separated case. An externally mounted antenna improves the transmission from a closed switch cabinet. Once the camera has been precisely positioned, it is automatically calibrated to the meter type with just a few button presses.

The wireless gateway connected to the internet is located in the same building four floors below the transmitter. The building fabric is concrete. Now the camera is registered with your device ID in the IoT cloud. Now all that is needed is to log in to the cloud, call up the functions for adding a sensor, and enter the camera's 20-digit hardware key. The read meter readings are now available for application servers in the IoT cloud and by the cloud's API.

The energy management system has to retrieve the data from the IoT cloud. For this, the system enerchart has to be extended by a device driver for the API of the IoT cloud. The camera data can then be queried in JSON format (Crockford, 2006). This custom driver in the EnMS is configured as follows:

Figure 4. Mounted camera module
(Source: author)

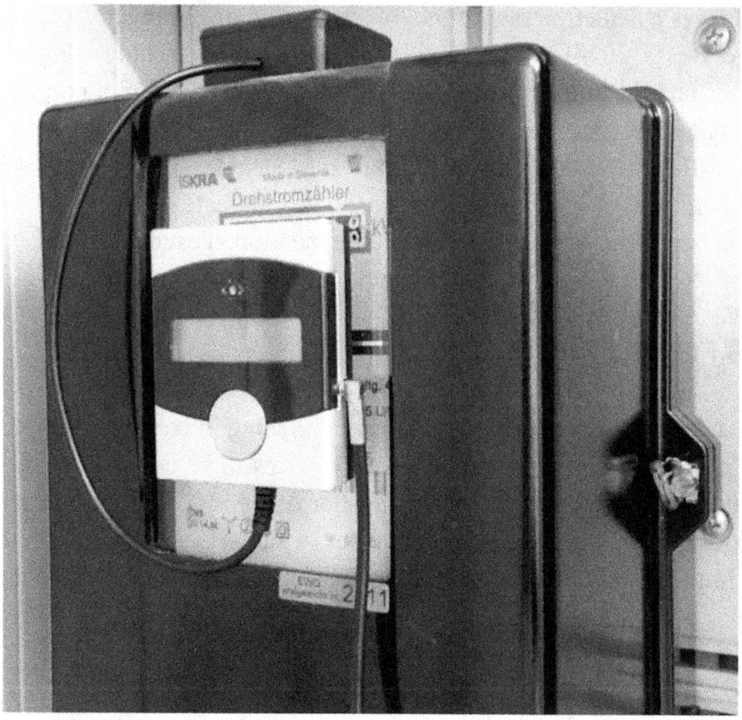

Step 1: Enter the cloud login data (username and password). This data remains stored in the EnMS.
Step 2: The API of the cloud returns all sensors stored there to the EnMS. The camera is now selectable as a sensor.
Step 3: In the EnMS, this sensor has to be mapped to an internal measuring point. This logical measuring point is used for the graphical visualization and for the monitoring. The smallest interval for the measurement data is minutes; we use a resolution of five minutes.

The conversion from counter reading to consumption occurs automatically in the energy management system. In a few cases, there is a loss during transmission, which is not critical for the overall evaluation. Despite the greater distance from the gateway, the wireless transmission works well. A dashboard was created in the energy management system for the meter data, which provides a clear overview of the meter reading, consumption, and energy costs.

Case Study: OPC UA Gateway for Industrial Plants

In this example, a simple 230V 1-phase AC meter with electrical pulse output (1,000 pulses/kWh) was extended by an OPC UA server using a modular industry-standard mini-computer. The counter with LCD display has a measuring range from 0.05 to 45A. The counter cannot be reset and keeps the counter reading even in case of power failure. This counter is to be integrated into an energy monitoring system for continuous measurement. The hardware platform of the mini-computer is an energy-related version of

Figure 5. Dashboard with charts (screenshot enerchart)

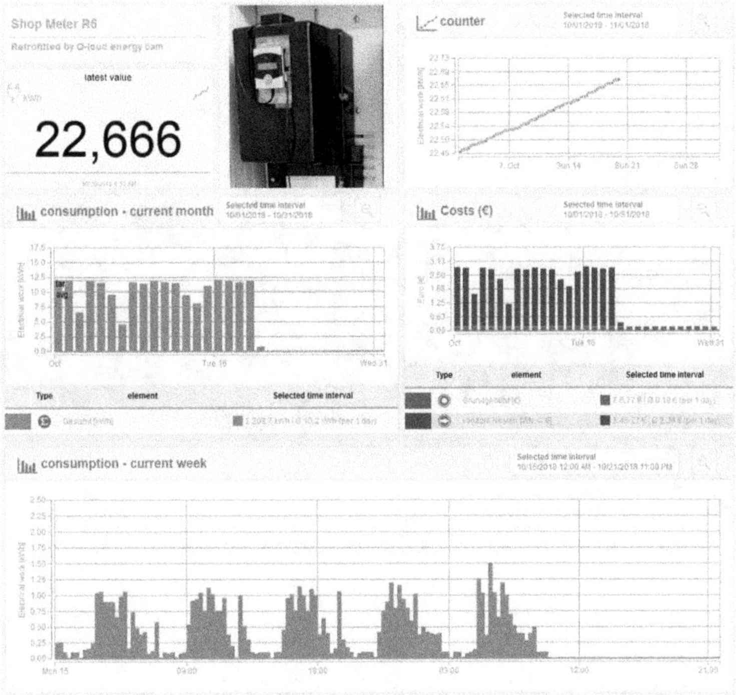

the industrial universal computer MICA from *HARTING*, which features an extra board with connections for a serial RS485 bus and up to eight analog pulse signals. On this board, each S0 channel appears as its own Modbus RTU device providing its virtual pulse counter and a slave ID. The platform is based on a 1 GHz ARM processor, 1 GB RAM. In addition, the memory was expanded by a 4GB flash memory of a micro SD card. On the network side, the computer was connected via a robust industrial Ethernet plug, which simultaneously supplies the device with power via PoE. Another GPIO connection with 8 x 12-24V inputs/outputs remains unused.

The Linux-based system software can be extended by different applications in the form of software containers. Standard containers are already available for numerous applications; an MQTT broker container is already installed in this version. For this case study, a custom container for the OPC UA server has been implemented. Although numerous more or less complete open-source implementations for OPC UA are available, the decision was taken in favour of a commercial OPC UA library because of the better developer support. The installed OPC UA server receives the virtual pulse counter over the state of the system variable, which is stored on the SD card and remains stored even in the case of a power failure. This counter reading is permanently integrated into the OPC UA information structure as an available variable. To synchronize the virtual counter with the physical counter, an OPC UA method for setting the counter reading has been implemented.

The existing OPC UA client in the energy management system must now be configured for the OPC UA server. A new data source of type "OPC UA" was created, regarding the network address of the server. Further settings include selecting the security profile, the authentication method, the polling interval, and some monitoring-related values. The last step was the mapping of the OPC UA variable with

Figure 6. Mini-computer with OPC UA server
(source: author)

a measuring point by browsing the address space of our OPC UA server (Figure 7). As a result, there now exists a measuring point which the energy management polled in minute intervals and a graphical analysis—analogous to our energy cam—can be performed to visualize the received time series.

Overall, this solution works without any problems and supplies the energy management system with the consumption values of the S0 meter every minute. The price for this retrofit is relatively high at around €800. Therefore, this investment is only economically reasonable if the system is fully wired with several meters. This approach is not robust against a computer failure, in which case the impulses would be lost and the consumption during the downtime would not be recorded. It would be better to use meters with Modbus. As an alternative to our self-made OPC UA server container, an already installed container with an MQTT broker or a container with a configurable Modbus TCP gateway can be used for the IP networking of the meter.

Case Study: LoRaWAN Modbus Gateway

In this use case, a three-phase power meter with the Modbus RTU interface has been retrofitted by a LoRaWAN. The *socomec* DIRIS A20 meter is a multi-purpose instrument for measuring currents, voltages, and frequency, as well as power and power factors. For readout via Modbus, a separately available add-on module is plugged into the device. This meter has to send the power of one phase via a LoRaWAN to an energy management system. A LoRa Modbus bridge from *comtac* is used as the transmission device, a LoRa Class C device for Modbus RTU and ASCII, and for a permanent power supply. Before we connect the bridge to the Modbus output of the device, it is connected to a notebook via a USB cable and configured to Modbus. Alternatively, an over-the-air (OTA) setup via a network server would be possible. With the USB configuration, the bridge appears as a data carrier. The configuration file can be adapted with a simple text editor. The Modbus parameters such as baud rate, register address, delays, and data

Figure 7. Configuration of the OPC UA measuring point (screenshot enerchart)

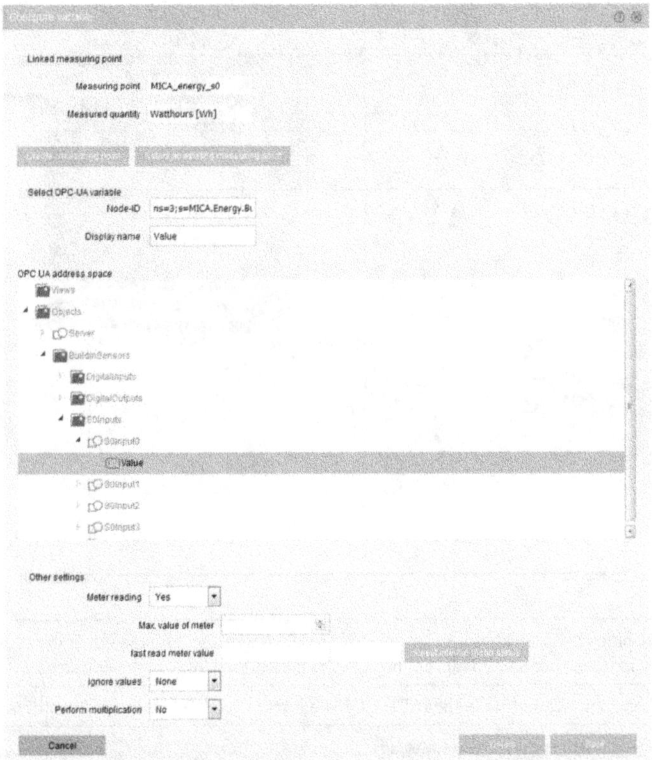

format of the connected meter are set here. For the LoRaWAN, the LoRaWAN keys of the device are stored here (Table 1) in addition to various transmission parameters.

The LoRaWAN gateway is a *comtac* LoRa base server that supports the simple service discovery protocol (SSDP) and can be configured via a web browser. The gateway is fixed to the LoRa frequency band 863-870 MHz of the EU. The base server supports two operating modes for packet forwarding: either as a pure LoRaWAN gateway for use with *Loriot*, a commercial, public, community-based network as a service (NaaS), or with internal forwarding to the built-in Lora Network Server. The use of Loriot makes sense when using several LoRaWAN gateways. The second option was chosen so the data is not forwarded to an external service. As the application server, an OPC UA server is included on the device, so the EnMS doesn't need a special driver for the specific base server. The Modbus parameters are entered in the bridge via a web interface. The identical bus parameters must now also be stored in the base server, so that meter readings are then available as OPC UA variables in the application server. The structure in the EnMS is then identical to the previous case study.

The visualization of the measured values is shown in Figure 9. The conversion of the meter readings into consumption values is carried out by the energy management system. The resolution of the time series is also carried out automatically depending on the zoom factor. Fifteen-minute values are displayed in the section shown. Although the Modbus counter and the bridge are installed in a shielded switch cabinet, there are no packet losses. The distance to the gateway is about 35 meters. There are no other LoRaWAN nodes in the test environment.

Figure 8. Device setup: (a) LoRaWAN base server (b) LoRa-Modbus bridge on a Modbus RTU meter

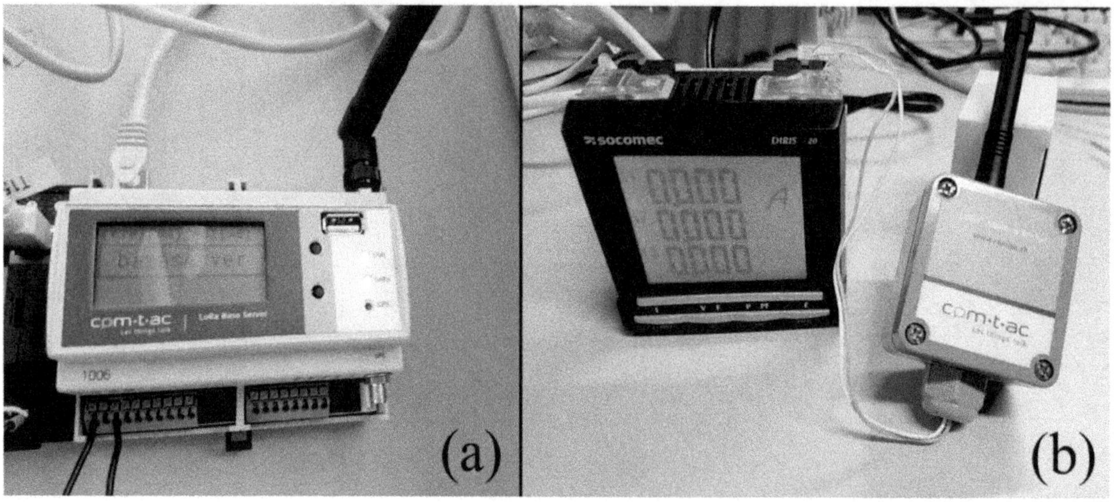

Table 1. LoRaWAN device keys

DevEUI	A 64-bit end-device identifier assigned to the device by the chip manufacturer. Seven bits are used as the network identifier and 25 bits are used as the network address of the end-device.
AppEUI	A global application ID in the IEEE EUI64 address space that uniquely identifies the owner of the end device.
AppKey	A 32-bit device address (non-unique)
DevAddr	A unique 32-bit unsigned integer identifier for the device in a particular network.
NetwSesKey	A key used by the network server and the end device to calculate and verify the message integrity code of all data messages to ensure data integrity.
AppSesKey	A key used by the network server and end device to encrypt and decrypt the payload field of data messages.

The LoRaWAN infrastructure created here can now integrate nodes that have no relation to energy management and which are used by third-party application servers. Concepts such as smart production, smart city, smart health care, and smart building are based on infrastructure that offers a high degree of flexibility and interoperability and where energy management is just one added value of many.

CONCLUSION

In all three case studies, a simple electricity meter was successfully retrofitted into a networked energy management system. The electricity meters became cyber-physical systems. Each case had a task with increased effort: for the OCR camera, a driver for the IoT cloud had to be developed in the energy management system. For the industrial mini PC, an OPC UA server container had to be programmed. With the LoRa solution, the high configuration effort in both the bridge and the LoRaWAN network server was noticeable. Both, the mini-PC and the LoRa base server, are expensive. Both devices cost around 800 EUR and are therefore only economically viable if used for a larger number of sensors. The MICA can coordinate up to eight S0 counters, as well as up to 247 devices on the Modbus. The LoRa base

Figure 9. Visualization in the EnMS (screenshot enerchart)

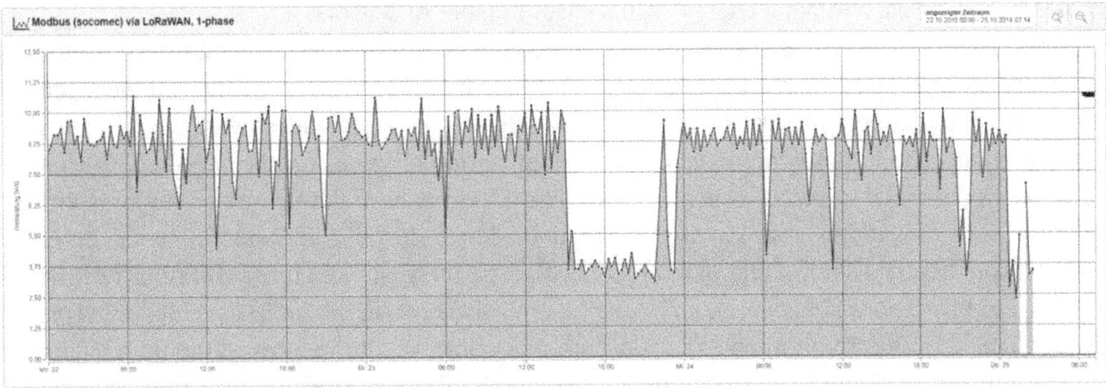

server can communicate with up to about 2,500 radio nodes, depending on the clock rate. If you only have a small number of radio nodes, the use of LPWAN infrastructures from third-party providers, such as NB-IoT or Sigfox, is a cheaper alternative. The price of the OCR camera is about 350 EUR per meter, so this form of retrofitting is only suitable for a small number of meters.

With the instrument of Companion Specifications, OPC UA pursues the goal of creating manufacturer-neutral standards for individual sectors. Energy management according to ISO 50001 is basically sector-neutral, but a Companion Specification especially for energy management could also be helpful here. In (VDMA, 2018), for example, a variable Total Energy Consumption in the object type ControllerType provides the cumulative energy consumption. In the future, software should at least be familiar with the most common companion specification and, thus, enable a simplified assignment between the data from the OPC UA server and the measuring points in the software. In addition, we see a need for research in developing specialized companion specifications for energy and resource measurements to create a standard for pure energy meters and to further simplify the client's configuration.

The LPWAN radio networks in the narrow band are an important factor in the decision for retrofitting. The technology is robust inside buildings and the costs for networking meters and sensors are relatively low. As energy meters are usually not moving objects and are spatially limited to one property, no area-wide WAN and no roaming techniques are required. If one wants to transfer the data to third-party service providers, LoRaWAN is without an alternative. If, as in the OCR camera example, an IoT cloud from third-party providers is integrated, further data protection regulations are relevant for this case. The interoperability of different cloud platforms must also be considered. One of the problems is, for example, the different use of device identifications, as each cloud uses a different device IS system (Koo & Kim, 2017). There are other OCR camera solutions on the market, including add-on cameras for water meters that are compatible with standardized wireless networks such as LoRaWAN, Sigfox, and NB-IoT.

For our energy management according to ISO 50001, the certainty now remains that retrofitting offers an interesting opportunity to further expand the measurement infrastructure and, thus, create more transparency about the energy flows. Particularly, technologies such as LoRaWAN can make the integration of wirelessly networked sensors economical. OPC UA has proven to be an important standard for interoperability. In fact, this standard is used in current industrial environments, so retrofitting makes the old world fully compatible even beyond energy management. Two of the three use cases have benefited from the fact that an OPC UA client is already implemented as a device driver in the energy management system. Only for the proprietary IoT cloud a driver had to be developed - an effort that is difficult to compensate by saving energy with a small number of sensors.

One approach for further research work lies in the area of cost analysis. For these case studies, the first solutions available on the market were used, alternatives were hardly available and therefore price comparisons were not possible. As the technologies will become more widespread, a larger number of different products and a significant reduction and stabilization of prices can be expected. Even the usage costs of commercial IoT clouds such as NB-IoT and Sigfox, have not yet settled through sufficient distribution and competition. As soon as this situation occurs, the costs for retrofitting can be reliably analyzed. More detailed studies can then be conducted on the costs and profitability of retrofit solutions compared with normal networking or complete replacement. In addition, usability concepts must be created that significantly minimize the configuration effort. In all three cases, the time required for configuration was too high. In everyday operations, the hurdle for retrofitting must be lowered even further by simplifying these processes.

ACKNOWLEDGMENT

Appreciations for their support of this work go to Christopher Bläsius, Dr. Dieter Braun, Michael Bryant, Prof. Dr. Bernhard Kölmel, Ivona Raguž and Prof. Dr. Adrian Tanţău.

REFERENCES

Adelantado, F., Vilajosana, X., Tuset-Peiro, P., Martinez, B., Melia-Segui, J., & Watteyne, T. (2017). Understanding the limits of LoRaWAN. *IEEE Communications Magazine*, *55*(9), 34–40. doi:10.1109/MCOM.2017.1600613

Augustin, A., Yi, J., Clausen, T., & Townsley, W. (2016). A Study of LoRa: Long Range & Low Power Networks for the Internet of Things. *Sensors (Basel)*, *16*(9), 1466. doi:10.339016091466 PMID:27618064

Baheti, R., & Gill, H. (2011). Cyber-physical systems. *The impact of control technology, 12*(1), 161–166.

Bardyn, J.-P., Melly, T., Seller, O., & Sornin, N. (2016). IoT: The era of LPWAN is starting now. In *Proceedings of the 42nd European Solid-State Circuits Conference, ESSCIRC Conference 2016* (pp. 25–30). IEEE. 10.1109/ESSCIRC.2016.7598235

Blenn, N., & Kuipers, F. (2017). LoRaWAN in the wild: Measurements from the things network. *arXiv preprint arXiv:1706.03086*. Retrieved from https://arxiv.org/abs/1706.03086v1

Bor, M., Vidler, J. E., & Roedig, U. (2016). LoRa for the Internet of Things. In *Proceedings of the 2016 International Conference on Embedded Wireless Systems and Networks.*

Bradski, G., & Kaehler, A. (2008). *Learning OpenCV: Computer vision with the OpenCV library.* O'Reilly Media, Inc.

Castells-Rufas, D., & Carrabina, J. (2006). Camera-based digit recognition system. *Networks*, *5*, 6.

Centenaro, M., Vangelista, L., Zanella, A., & Zorzi, M. (2016). Long-range communications in unlicensed bands: The rising stars in the IoT and smart city scenarios. *IEEE Wireless Communications*, *23*(5), 60–67. doi:10.1109/MWC.2016.7721743

Cortez, P., Felix, J., Girão, A., Frota, J., & Bessa, J. (2014). An OCR system for numerals applied to energy meters. *IEEE Latin America Transactions*, *12*(6), 957–964. doi:10.1109/TLA.2014.6893986

Crockford, D. (2006). *The application/json media type for javascript object notation (json)* (No. No. RFC 4627).

Ebel, M., Drath, R., & Sauer, O. (2008). Automatische Projektierung eines Produktionsleitsystems der Fertigungstechnik mit Hilfe des Datenaustauschformats CAEX. *Automatisierungstechnische Praxis*, *50*(5), 40.

ElMaraghy, H. A. (2005). Flexible and reconfigurable manufacturing systems paradigms. *International Journal of Flexible Manufacturing Systems*, *17*(4), 261–276. doi:10.100710696-006-9028-7

Elrefaei, L. A., Bajaber, A., Natheir, S., AbuSanab, N., & Bazi, M. (2015). Automatic electricity meter reading based on image processing. In *2015 IEEE Jordan Conference on Applied Electrical Engineering and Computing Technologies (AEECT)* (pp. 1–5). IEEE. doi:10.1109/AEECT.2015.7360571

Erol-Kantarci, M., & Mouftah, H. T. (2011). Wireless sensor networks for cost-efficient residential energy management in the smart grid. *IEEE Transactions on Smart Grid*, *2*(2), 314–325. doi:10.1109/TSG.2011.2114678

European Parliament. (2012). Directive 2012/27/UE of The European Parliament and of the Council of 25 October 2012 on energy efficiency, amending Directives 2009/125/EC and 2010/30/EU and repealing Directives 2004/8/EC and 2006/32/EC. *Official Journal of the European Union.*

Fantana, N. L., Riedel, T., Schlick, J., Ferber, S., Hupp, J., Miles, S., ... & Svensson, S. (2013). IoT applications—value creation for industry. Internet of things: Converging technologies for smart environments and integrated ecosystems, 153.

Guerreiro, B. V., Lins, R. G., Sun, J., & Schmitt, R. (2018). Definition of Smart Retrofitting: First steps for a company to deploy aspects of Industry 4.0. In *Advances in Manufacturing* (pp. 161–170). Cham: Springer.

Gustav, T. (1938). Reading machine. US Patent 2,115,563.

Hannelius, T., Salmenpera, M., & Kuikka, S. (2008). Roadmap to adopting OPC UA. In *6th IEEE International Conference on Industrial Informatics INDIN 2008* (pp. 756–761). IEEE.

Hillary, R. (1993). *The eco-management and audit scheme: a practical guide*. Hertfordshire, UK: Technical Communications Letchworth.

Hunkeler, U., Truong, H. L., & Stanford-Clark, A. (2008). MQTT-S—A publish/subscribe protocol for Wireless Sensor Networks. In *Proceedings of the 3rd international conference on Communication systems software and middleware and workshops COMSWARE 2008* (pp. 791–798). IEEE.

Imtiaz, J., & Jasperneite, J. (2013). Scalability of OPC-UA down to the chip level enables "Internet of Things. In *2013 11th IEEE International Conference on Industrial Informatics (INDIN)* (pp. 500–505). IEEE.

Jain, A. K., Mao, J., & Mohiuddin, K. M. (1996). Artificial neural networks: A tutorial. *Computer, 29*(3), 31–44. doi:10.1109/2.485891

Jeschke, S., Brecher, C., Song, H., & Rawat, D. B. (Eds.). (2017). *Industrial Internet of Things*. Springer International Publishing. doi:10.1007/978-3-319-42559-7

Joachim Schenk, G. R. (2010). *Mensch-Maschine-Kommunikation*. Springer-Verlag GmbH. Retrieved from https://www.ebook.de/de/product/9280122/joachim_schenk_gerhard_rigoll_mensch_maschine_kommunikation.html

Kastner, W., & Szucsich, S. (2011). Accessing KNX networks via BACnet/WS. In *Proceedings of the 2011 IEEE International Symposium on Industrial Electronics (ISIE)* (pp. 1315–1320).

Koo, J., & Kim, Y.-G. (2017). Interoperability of device identification in heterogeneous IoT platforms. In *Proceedings of the 2017 13th International Computer Engineering Conference (ICENCO)* (pp. 26–29). IEEE.

Krutwig, M., & Starosta, K. (2017). Characterization, Classification and Assessment of Non-Energy Benefits of Energy Efficiency Measures. In *Proceedings of 30th IBIMA Conference, Madrid 2017*. IBIMA.

Lauridsen, M., Kovacs, I. Z., Mogensen, P., Sorensen, M., & Holst, S. (2016). Coverage and Capacity Analysis of LTE-M and NB-IoT in a Rural Area. In *Proceedings of the 2016 IEEE 84th Vehicular Technology Conference (VTC-Fall)*. IEEE. doi:10.1109/vtcfall.2016.7880946

Leitner, S.-H., & Mahnke, W. (2006). *OPC UA–service-oriented architecture for industrial applications*. ABB Corporate Research Center.

Mahnke, W., Leitner, S.-H., & Damm, M. (2009). *OPC Unified Architecture*. Springer Berlin Heidelberg; doi:10.1007/978-3-540-68899-0

Matzler, S., Wollschlaeger, M., Fernbach, A., Kastner, W., & Huschke, M. (2013). An OPC UA cross-domain information model for energy management in automation systems. In *Industrial Electronics Society, IECON 2013-39th Annual Conference of the IEEE* (pp. 7513–7518). IEEE.

Medina, B. E., & Manera, L. T. (2017). Retrofit of air conditioning systems through an wireless sensor and actuator network: An IoT-based application for smart buildings. In *Proceedings of the 2017 IEEE 14th International Conference on Networking, Sensing and Control (ICNSC)* (pp. 49–53). IEEE.

Monostori, L., Kádár, B., Bauernhansl, T., Kondoh, S., Kumara, S., Reinhart, G., ... Ueda, K. (2016). Cyber-physical systems in manufacturing. *CIRP Annals, 65*(2), 621–641. doi:10.1016/j.cirp.2016.06.005

Mori, S., Suen, C. Y., & Yamamoto, K. (1992). Historical review of OCR research and development. *Proceedings of the IEEE, 80*(7), 1029–1058. doi:10.1109/5.156468

Neugschwandtner, M., Neugschwandtner, G., & Kastner, W. (2007). Web services in building automation: Mapping knx to obix. In *Proceedings of the 2007 5th IEEE International Conference on Industrial Informatics* (Vol. 1, pp. 87–92). IEEE.

Nsiah, K. A., Schappacher, M., Rathfelder, C., Sikora, A., & Groza, V. (2018). An open-source toolkit for retrofit industry 4.0 sensing and monitoring applications. In *Proceedings of the 2018 IEEE International Instrumentation and Measurement Technology Conference (I2MTC)* (pp. 1–6). IEEE. 10.1109/I2MTC.2018.8409633

OMA. (2017). Lightweight machine to machine technical specification. *Technical Specification OMA-TS-LightweightM2M-V1.0 approved.*

OPC Foundation. (2017). *OPC UA Specification: Part 1 – Concepts, Core Specification Parts, Version 1.04*. OPC Foundation. Retrieved from https://opcfoundation.org

Pardo-Castellote, G. (2003). Omg data-distribution service: Architectural overview. In *Proceedings. 23rd International Conference on Distributed Computing Systems Workshops, 2003.* (pp. 200–206). IEEE.

Petäjäjärvi, J., Mikhaylov, K., Yasmin, R., Hämäläinen, M., & Iinatti, J. (2017). Evaluation of LoRa LPWAN technology for indoor remote health and wellbeing monitoring. *International Journal of Wireless Information Networks, 24*(2), 153–165. doi:10.100710776-017-0341-8

Rabiner, L. R., & Juang, B.-H. (1986). An introduction to hidden Markov models. *IEEE ASSP Magazine, 3*(1), 4–16.

Ratasuk, R., Vejlgaard, B., Mangalvedhe, N., & Ghosh, A. (2016). NB-IoT system for M2M communication. In *Proceedings of the 2016 IEEE Wireless Communications and Networking Conference (WCNC)*, (pp. 1–5). IEEE. doi:10.1109/WCNC.2016.7564708

Rehbein, D. A., & Pederson, A. (1996). Impact of the OLE for Process Control (OPC) standard on the process industry. In *Proceedings of the Industrial Computing Conference* (Vol. 6, pp. 285–293).

Reynders, B., & Pollin, S. (2016). Chirp spread spectrum as a modulation technique for long range communication. In *2016 Symposium on Communications and Vehicular Technologies (SCVT)* (pp. 1–5). IEEE. doi:10.1109/SCVT.2016.7797659

Schlechtendahl, J., Keinert, M., Kretschmer, F., Lechler, A., & Verl, A. (2015). Making existing production systems Industry 4.0-ready. *Production Engineering, 9*(1), 143–148. doi:10.100711740-014-0586-3

Schleipen, M. (2008). OPC UA supporting the automated engineering of production monitoring and control systems. In *Proceedings of the IEEE International Conference on Emerging Technologies and Factory Automation ETFA 2008* (pp. 640–647). IEEE.

Schlick, J. (2012). Cyber-physical systems in factory automation-Towards the 4th industrial revolution. In *2012 9th IEEE International Workshop on Factory Communication Systems (WFCS)*, (pp. 55–55). IEEE.

SemtechCorp. (2015). *LoRa Modulation Basics, AN1200.22. LoRa Modulation Basics*. Semtech Corporation.

Shelby, Z., Hartke, K., & Bormann, C. (2014). *RFC7252: The constrained application protocol (CoAP). Internet Engineering Task Force*. IETF.

Sinha, R. S., Wei, Y., & Hwang, S.-H. (2017). A survey on LPWA technology: LoRa and NB-IoT. *Ict Express*, *3*(1), 14–21. doi:10.1016/j.icte.2017.03.004

Smith, R. (2007). An overview of the Tesseract OCR engine. In *Ninth International Conference on Document Analysis and Recognition ICDAR 2007* (Vol. 2, pp. 629-633). IEEE.

Sornin, N., Luis, M., Eirich, T., Kramp, T., & Hersent, O. (2015). *LoRaWAN specification, available from the LoRa Alliance (technical specification)*. LoRa Alliance.

VDMA. (2018). *OPC UA Companion Specification Robotics – Part 1: Condition monitoring, asset management, predictive maintenance, vertical integration*. VDMA.

Vejlgaard, B., Lauridsen, M., Nguyen, H., Kovács, I. Z., Mogensen, P., & Sorensen, M. (2017). Coverage and capacity analysis of sigfox, lora, gprs, and nb-iot. In *Proceedings of the 2017 IEEE 85th Vehicular Technology Conference (VTC Spring), Sydney, Australia* (pp. 4–7). 10.1109/VTCSpring.2017.8108666

Wolf, W. H. (2009). Cyber-physical systems. *IEEE Computer*, *42*(3), 88–89. doi:10.1109/MC.2009.81

ADDITIONAL READING

Chou, T. (2016). Precision: Principles, Practices and Solutions for the Internet of Things lulu.com. ISBN 1329843568

Howell, M. T. (2014). *Effective Implementation of an ISO 50001 Energy Management System (EnMS)*. ASQ Quality Press.

Lea, P. (2018). *Internet of Things for Architects: Architecting IoT solutions by implementing sensors, communication infrastructure, edge computing, analytics, and security*. Packt Publishing.

Rinaldi, J. (2016). *OPC UA - Unified Architecture: The Everyman's Guide to the Most Important Information Technology in Industrial Automation*. CreateSpace Independent Publishing Platform.

Veneri, G. (2018). *Hands-On Industrial Internet of Things: Create a powerful Industrial IoT infrastructure using Industry 4.0*. Packt Publishing.

ENDNOTES

1. LoRa™ and LoRaWAN™ are registered trademarks of Semtech Corporation. Patent No. US7791415, EP2763321. The abbreviations LoRa and LoRaWAN are used to make the text easier to read.
2. With permission of the OPC UA Foundation

Chapter 10
Green Retrofit Energy Efficiency Potential on Existing Building Envelope for Residential and Non-Residential Building

Robert Staiger
E[3]xpert, Germany

ABSTRACT

The chapter deals with the green energetic consideration of today's building envelopes for residential and non-residential buildings. It investigates the energetic effects the envelopes have on energy efficiency, energy consumption, material use, sustainable use of resources, lifetime considerations, economic and ecological impact. Today's it is estimated that approximately 30% of the annual primary energy demand for residential and non-residential buildings is needed. Energy resources for heat, electricity, air conditioning and cooling purposes, fossil fuels in form of gas and liquid are predominantly used.

INTRODUCTION

The aim of this chapter is to give a good understanding of possible renovations to building envelopes. The basics should be pointed out to show and estimate possible potentials in the renovation of building envelopes. First of all, the basic building structures are considered more closely for a better understanding. Then the different components for the building envelope are discussed. From this information, the current method of energy considerations of buildings is explained.

Based on two case studies of an outer wall, the physical and economic calculations are shown. The simulation shows different aspects of insulation, costs and savings.

DOI: 10.4018/978-1-5225-9104-7.ch010

GREEN ENERGETIC RENOVATION OF BUILDING ENVELOPES

The chapter deals with the green energetic consideration of today's building envelopes for residential and non-residential buildings. It investigates the energetic effects the envelopes have on energy efficiency, energy consumption, material use, sustainable use of resources, lifetime considerations, economic and ecological impact.

Today's thermal energy is estimated that approximately 30-35% of the annual primary energy demand for residential and non-residential buildings are needed. As energy resources for the transformation processes in heat, electricity, air conditioning and cooling purpose, fossil fuels in form of gas and oil are in present predominantly used (BP 2018).

The burning of fossil energy sources since the industrial revolution of the 1800s, CO2 pollution is steadily increasing in our atmosphere (Rosenzweig et al., 2018). This rising Concentrations of greenhouse gases are changing our climate at an ever-faster pace (Hutter, 2018). The consequences of the worldwide use of fossil fuels have been scientifically proven and the causative agent clearly transferred (Schönwiese, 2019).

The annual amount on fossil oil per annum are more than 14.000 Mill litre of crude oil per day and 100 mill tons on coal per day (BP 2018). Generating out of these fossil energy source different energy forms mostly through thermal burning processes more than 32.000 mill tons on CO_2/a are polluted in the air. CO_2 is the second most important greenhouse gas that drives climate warming process. In addition to that, a further large number of health-endangering gases and dusts are emitted into the environment.

The amount of Buildings in Germany in 2018 are roughly 42 Mill. The total surface area is roughly 3.900 Mill m^2 in Germany (DENA 2019). Worldwide estimation is 42.000 Mill houses and buildings. (Krieger, 2017).

The building envelope comprises all the components enclosing it to the outside, e.g. walls, windows, roofs and floors. For a sustainable construction, especially the thermal building envelope is of importance. It includes all components that separate heated rooms from outside air, from the ground and from unheated rooms. The thermal insulation of the thermal building envelope is crucial for the transmission heat demand of a building (Garzon, 2018; Swan, 2013; Wilson, 2013).

Buildings have a much longer service life compared for example to vehicles. While vehicles could theoretically be replaced by environmentally friendly new vehicles within a decade with appropriate regulations, this is not possible with buildings. And in terms of sustainability, it also makes no sense: the continued use of existing buildings, their renovation, conversion and further construction is usually connected with a significantly lower energy and resource consumption than a demolition and new construction, if the grey energy is included.

Against the background of climate change, resource-conserving energy conversion, efficient use of energy, a circular economy, minimization of waste materials is an existential necessity for humanity. With regard to a circular economy for green building elements, the construction industry is still largely at the beginning. A meaningful old building modernization should be preferred to a new building for economic reasons, because the core thoughts of the sustainability belong the recycling, the secondary use and extension of the life span in a reasonable frame.

With a green sustainable renovation, in addition to the known principles of planning and construction, higher-level aspects must also be noticed. They should ensure that the measure also proves to be ecologically compatible, economically convincing and energy efficient in the long term.

The rules that apply in principle to all modernization measures should be observed. These are: recyclable, energy efficient, eco-friendly, eco balance (LCA), build economically, Old building method of operation, Protection and reuse of existing components, prefer dry construction, using finished components

The goals of sustainable construction and renovation of building elements are to minimize the energy and material resources. All lifecycle phases of a building are considered. The aim is to optimize all factors influencing the life cycle: from the extraction of raw materials to the construction and dismantling. The average useful life of a house is about 50 - 100 years (Wilson, 2013).

The following factors have to be considered in sustainable rebuild construction: Reduction of energy demand, Reduction of consumption of resources, use of recyclable building materials and components, avoidance of unnecessary transport costs for the building materials, safe and efficient return of the materials used in the natural material cycle, reuse possibilities, protection of natural areas, space-saving building constructions

The energy and cost-conscious planning of buildings and renovation elements is usually done with the help of an energy balance method. Already in the planning phase, the main energy flows and thus also the energy demand of the building can be determined by calculations. Energy balance methods enable the optimization of building elements under the given economic and ecological conditions.

With energy efficient, modern refurbished buildings, owners can make their corporate image visible to the outside. These real estate investments also provide the long-term return on investment, an aspect that institutional investors attach great importance to. The investor having a major market advantage with a highly efficient building with less energy usage in the housing market with more stable rents, regardless of the price of energy (oil, gas) (Steiner et al., 2018).

Energy is a physical quantity and can therefore be precisely measured. Energy-efficient and sustainable buildings have become measurable, namely through the energy index and the life cycle calculations of different building materials, which quantify the operating energy requirements and the grey energy of the building. For many builder, planers and architects a terrible thought that the building can be measured, possibly still qualifiable. With today's energy pass, houses are already comparable in terms of energy per area, such as energy consumption per kilometer for automobiles. The planners and architects should make use of the advantages of sustainable construction and apply the changed conditions in their work in many ways.

Energy indicators are used for the energy assessment of a building and refer to different properties or values to be calculated. For example, the Association of German Engineers (VDI) distinguishes between energy parameters (specific energy consumption, efficiencies, utilization rates, rated power, etc.) and energy parameters for holistic assessment (cumulative energy consumption, harvest factor, energy payback period, etc.). A frequently used energy parameter is the energy index, which gives the annual heating demand in kWh in terms of floor area in square meters (kWh/m^{2}a). These area-related values allow the comparability of different buildings. This energy index now gives the opportunity to compare under the same frame conditions (reference building). The energy index can serve as a useful meaningful parameter for different decisions around the building. These decisions can be: sale of the building, energetic renovation of the building, estimates of energy usages.

REBUILD OF GREEN BUILDINGS ENVELOPES

The term "energetic renovation" essentially comprises the improvement or modernization of the thermal building envelope in terms of its heat transfer (Wang, 2012; Olander, 2007). The aim is to minimize energy usage for heating the premises and to reduce the ventilation heat losses. Modernization of building envelopes means as well more comfort for the homeowners and more eco-friendly material usage in the renovation processes.

The building envelope is a closed geometric structure and the physical separation between the internal and external environment of a building (Swan, 2013; Westermann et al., 2013). The outer envelope is a barrier against environmental hazards, precipitation, ambient air, ambient temperature, moisture/water barrier and noise radiation. The building envelope protects the internal facilities and installations along with the mechanical conditioning systems see standard EN 18599 heating and cooling, lighting, air conditioning, air recovery.

Energy cannot be lost in transformation processes. This is described by the basic laws of thermodynamics (Atkins, 2010). The amount of energy flows in the form of temperature differences between building elements depends on the insulating effect of the building components. all exterior elements of a building form the building envelope. The amount of energy flows is calculated using the energy balance method. this amount of energy depends on the different materials (transmission losses), the size of the surface, the ventilation losses caused by leaks in the casing. This necessary amount of energy to keep the temperature level in the building evenly, a heating system is needed.

Depending of the design of building there is a variation of considerations for the envelope of a building.

In Germany, buildings are divided into building classes depending of the state's regulation. The classification of a building into a building class depends on the height and the area of the building. Each federal state governs the division of the building classes differently in the respective state building regulations. The division of the buildings into different building classes depends on different component and building material requirements. The higher the building class, the higher the requirements for the fire resistance of the components. The state building regulations of the German federal states generally define as follows: "Buildings are independently usable, covered structures that can be entered by humans and are suitable or intended to serve the protection of humans, animals or things."

The Austrian building regulations, building laws and building technology laws of the countries define the buildings as "covered, all or predominantly enclosed structures that can be entered by persons." The building is thereby "a plant that is connected to the ground and their professional production technical knowledge is required."

Methods of Building

There are no exact building differentiations. The following possibilities for the differentiation of buildings are conceivable. The Table 1 shows the possible criteria for a differentiation of buildings.

For our investigations, we use residential and non-residential buildings as a basis. Residential buildings are buildings that serve residential purposes in any size. Non-residential buildings are buildings that are not used for residential purposes. This applies to public buildings, offices, industrial buildings and more.

Table 2 show the different examples of residential and none residential buildings.

The building envelope is made up of components that separate heated space from the outdoor air and the unheated rooms. The physical components of the building envelope consist of the foundation,

Table 1. Building types

Criteria	Details
Construction materials	Solid construction, skeleton, masonry, earth building, timber construction
By function	In distinguishing by function there is no uniform or mandatory categorization
Live	Residential and commercial buildings
Work	Workplace, office building, commercial, stores factories, workshops
Storage goods handling	Warehousing, warehouses, storage
Social live	Public buildings, town halls, government buildings
Education, science	Educational institutions, schools, colleges, universities, institutes
Health care	Hospital, sanatorium, nursing homes, spa building, penal institution
Culture	Libraries, museum, concert hall opera houses cultural centres, civic centres
Sport	Sport facilities, sport hall swimming pools
Religions	Sacred building, temples church
Traffic	Function of changing modes of transport in railway stations, airports, bus stations, reception buildings, parking spaces
Administration law politics	City hall district houses, courts parliaments

Source: Hess et al. 2018; Riley und Cotgrave 2018, own contribution

Table 2. distinctive features of residential and none residential buildings

Types	Examples
Residential building	• Dwellings • One- and two-family houses • more family building • Apartment blocks
Non-residential buildings	• Offices, office blocks • Industrial building • Public buildings like schools, city halls, • Manufacturing facilities
Mixture of both types	• Family homes with separate offices like a doctor office, lawyer office...

Source: Schröpfer 2016; Schröpfer und Menz 2019, own contribution

walls, doors, windows, ceilings and roof. The sizing, properties and compatibility of the materials, manufacturing processes and details, their joining and interrelation are the main factors that define the effectiveness and durability of the building seal. The envelope consists of several layers: support layer, heat-insulating layer, airtightness and vapor barrier layer, a moisture buffer layer and protective layer (Meyer, 2015; Hopfe, 2015).

First of all, all components of a building envelope of the heat-transferring surrounding surface are considered, such as outer walls, windows and outer doors, roof, upper story ceiling, basement ceiling or floor slab. Table 3 shows the envelope elements of a standard building and their special function and task of the elements.

Table 3. Building elements

Building Elements	Components function and tasks
Wall	Opaque element, Insulation surface, structural element, statically element
Roof	Opaque element, Insulation surface, weather protection, noise barrier
Ceiling	Structural element, snow weight, opaque and transparent (roof glass)
Floor	Separation
Windows	Transparent element, openings, fix glass
Doors	Openings, opaque, transparent, mix

Source: own contribution

For a successful green refurbishment strategy, not only the optimization of the insulation thicknesses is crucial, but the possible complete improvement of the entire thermal envelope with special "green" building elements and materials.

The exclusive consideration of individual components, e.g. insulation of the façade without consideration of windows and window reveals or the ventilation concept can lead to possible building damage. A holistic approach is indispensable from the numerous connections in renovation processes. This holistic approach is the basis for all renovation work on existing buildings

The decision as to whether the renovation of a building makes economic sense or whether the old building should be better replaced by a new building must be based on a cost-benefit analysis. A subdivision of the renovation work is absolutely necessary for the determination of exact investment costs. These can be better and more accurately analyzed under energetic aspects.

The overall consideration also includes the heat supply, as a better insulated house requires a smaller heating power means a smaller boiler or generator. Low temperature systems, such as Floor or wall heating systems are only in insulated buildings useful. When creating the renovation concept, the dimensioning of the building services plant considering the existing in the final state building envelope. Otherwise, the heat generators (boiler, etc.) may be over dimensioned, resulting in higher energy consumption and less efficiency. Another Chapter investigate green heating generators in details and will be discussed for the different building types later.

Sustainability in the Building Inventory

Buildings have a much longer service life compared to vehicles. While cars and trucks with appropriate regulations could theoretically be replaced within a decade by environmentally friendly new cars, this is not possible with buildings. And in terms of sustainability, it also makes no sense: the continued use of existing buildings, their renovation, conversion and further construction is usually associated with a significantly lower energy and resource consumption than a demolition and new construction, if the grey energy is included (Wind, 2017).

When it comes to sustainable planning and construction, it is not just about new buildings, but also a closer look at the inventory. This must be gradually upgraded energetically (Häfele et al., 2017). The variety of old buildings in need of renovation, however, is large: Which parts are worth protecting, which rather not, which building fabric can be preserved and which must be replaced, for this expertise and accurate inventories are necessary (Swan, 2013). General specifications are difficult, especially with architectural monuments and facades worth preserving. But this is only a very small proportion of buildings. Renewable energy systems can also be used to ideally heat such buildings, for example with biomass in the form of wood chips that is CO_2 neutral.

For an upgrade of the old buildings a state coordination is inevitable, the public over private interests. This approach has proved successful, for example in forestry, where nobody is allowed to cut down their private forest. No one will deny the necessity of measures that slow down and steer the urban sprawl, such as development plans and land use plans. Planners, authorities and politicians have to deal with the topic, it should also be analyzed at the universities and prepared for public discussion. The dogma "economic growth is equivalent to rising oil consumption" is outdated, on the contrary: Increasing the efficiency of machinery, vehicles and the construction stock boosts the domestic economy and reduces oil consumption.

By contrast, the rapid global warming caused by man-made climate change caused by the combustion of fossil energy sources must be controlled. This can be done with resource-efficient energy generation, efficient energy use, the introduction of a circular economy while minimizing production and transport energy. In the building stock, there is a huge potential for saving energy and reducing greenhouse gases.

In recent years, tremendous efforts have been made to use more efficient and greener technologies in the building sector. Due to the abundance of different building requirements, especially in the renovation area, there is a huge energy savings and cost potential that should be used. With regard to the introduction of a circular economy for building materials and components, the construction industry is still largely at the beginning. In particular, a comprehensive methodology or design and construction theory for recycling-compatible design is missing.

The research and development on recycling-compatible construction, which was made with the announcement of a take-back obligation in the automotive industry, and methodically well-structured cooperation, but also disassembly, can certainly be an inspiration and example for comparable developments that are urgently needed in the construction industry. If one were to create the basis for a perfect recyclability of the built environment, one would have to define the following goals (Sobek W., 2015).

These goals are formulated as follows:

- Build buildings that do not need energy for their operation in the annual sum ("zero energy consumption / zero energy").
- Build buildings that do not emit harmful emissions (zero emissions / zero emissions).
- Build buildings that are fully recyclable ("Zero Waste").

The demand for a triple zero: "Zero Energy / Zero Emission / Zero Waste" also stands for the so-called "Triple Zero Concept," which is currently being developed by the city and the Stuttgart Metropolitan Region in the form of a series of exemplary projects from the areas of old buildings, New construction and reconstruction is launched.

There is a global rethinking of the sustainability aspect of the built environment. Politics, science and industry are preparing to introduce the sustainability aspect in the built environment. This requires new concepts and ideas from architects and planners who need a new approach in our building stock (Sobek W., 2015).

The term sustainability was defined in the early 18th century against the background of an increasing national timber shortage. In 1713, Carl von Carlowitz (1645-1714), the Oberberghauptmann in Kursachsen, used it. However, even in the late Middle Ages, the capacities of forests in many regions of

Central Europe, especially in those with a strong mining and mining tradition, were exceeded and their limitations became clear. From this context, the actual forestry developed regionally and replaced the hitherto prevailing uncontrolled exploitation of the forests (Lüdtke, 2018; Ibisch, 2018; Baier, 2015).

The Helsinki Resolution of 1993 defines sustainable forest management in modern terms as: "The treatment and use of forests in a manner and to an extent that reflects their biodiversity, productivity, rejuvenation capacity, vitality and ability to absorb relevant environmental, economic and social functions at present and in the future at the local, national and global level, without damaging other ecosystems (Klenner, 2009). "This review is interesting and encouraging from a variety of perspectives. Because it was recognized that people by their action's nature, as they knew and loved irrevocably destroyed.

The goal of sustainability is to create a balance between use and regeneration of resources.

The goals of sustainable construction are to minimize the consumption of energy and resources. All lifecycle phases of a building are considered. The aim is to optimize all factors influencing the life cycle: from the extraction of raw materials to the construction and dismantling. The following factors have to be considered in sustainable construction:

- Reduction of energy demand
- Reduction of consumption of resources
- Use of recyclable building materials and components
- Avoidance of transport costs (of the building materials and components used)
- safe return of the materials used in the natural material cycle
- Re-use options
- Protection of natural areas
- Space-saving construction

Sustainable construction means reducing the long-term functionality and satisfaction of users with minimal environmental impact and optimized lifecycle costs. Consequently, especially construction products and components of the thermal building envelope are important in terms of their architectural quality and the fulfilment of the building physics requirements, such as in facades built-in building products due to their exposed location on the building. The increasing trend towards the integration of new technologies into the facade, such as Photovoltaics, solar thermal systems or building services components (so-called "active facades") often contradict a holistic life cycle analysis. Sustainability for facades from a non-energetic point of view means that on the one hand individual components are controllable, interchangeable and recyclable and on the other hand the Facade is maintenance and repair friendly. Using the example of these active facades, it becomes clear that interdisciplinary cooperation between engineers' various disciplines (construction, building technology, mechanical engineering, technical Physics, electrical engineering) and architects is required. Especially in the system development of component components leads a neglect of interdisciplinary Planning interfaces to a variety of negative consequences in technical, economic and environmental terms. While an increased installation effort or improvisations for connections or transitions are already visible in the execution phase, a number of effects occur only in the long term after many years of operation on, such as mold growth by condensation, component cracks by restraint stresses, a significantly increased heat demand or increased maintenance and disposal costs

In order to be able to implement the requirements for the sustainability of buildings, three categories were determined.

The Economic Evaluation

In addition to the acquisition and construction costs, the economic evaluation also includes the construction follow-up costs. These follow-up costs often have a negative impact on the life cycle cost, as e.g. high operating costs (for heating, hot water, electricity, etc.) or maintenance / maintenance costs are included. The dismantling costs are also considered.

The Ecological Rating

In addition to conserving resources, the ecological assessment also considers the optimal use of construction products and materials. The reduction in consumption of media (water, electricity, heating, etc.) plays just as important a role as the reduced burden on the environment. The assessment also includes the area occupied by the building, the primary energy consumption and the global warming potential.

The Socio-Cultural Evaluation

In addition to the aesthetic and design factors, sociocultural assessment is primarily about comfort and health protection. Thus, in addition to thermal (room temperature), acoustic (sound insulation) and visual (lighting) comfort, the use of low-emission building products also has an influence on the building or its inhabitants and users.

Building and Energy

Energy is a physical quantity and can therefore be precisely measured. This means that even energy-efficient and sustainable buildings have become measurable, namely through the energy index and the life cycle assessment of the building materials, which quantify the operating energy requirements and the grey energy of the building. Grey energy is the energy amount for manufacture, transport, store, sell and dispose of a product (Wind, 2017). In doing so, all preliminary products up to raw material extraction are taken into account and the energy input of all applied production processes is added.

Grey Energy

- Energy used to build
- Energy for maintenance
- Energy for demolish

The often-significant grey energy consumption arises without it being directly recognizable to us. This starts with the extraction of raw materials and the production of materials, goes through the production, processing and storage of components, includes the energy for the transport of people, machinery, components and materials to the construction site to the installation of components in the house Also included are the required maintenance, repairs, repairs and later disposal. The grey energy is also responsible for the harmful grey CO_2 emissions. The less energy our modern and economical houses consume, the greyer energy comes into play. In a low energy or passive house building the grey energy is a significant part of the total energy usage over a life cycle assessment. The loss of expensive heating is prevented by an increasingly well insulated building envelope. However, the increased use of insulating materials in particular requires more energy for the production and processing than it saves later. By using local materials and resource-saving construction, the grey energy built can be minimized.

According to the European Commission, the construction and maintenance of buildings, including heating, air conditioning, lighting and electrical equipment account for 40% of EU energy consumption (Oehler, 2018). This is where the responsibility of the architect and the builders begins - because residential and office buildings can today be realized with an energy balance over the year.

Different standards have been initiated, all aimed at minimizing the energy usage of buildings (for example, by measures such as thermal insulation, insulating glazing, energy-efficient heating technology) while at the same time allowing the house itself to generate energy. The Passive House standard serves as the conceptual basis, while in Switzerland this is the Minergie-P standard. It is complemented by the Minergie-Eco standard, which considers the life cycle assessment of building materials. The energy figures for these standards are well below the government-required levels. It should be noted that the values are used throughout the year to calculate the energy production / energy consumption ratio, and often "external" energy is still needed to cover the peak loads. The latest Energy Performance Building directives from the EU defined a new building standard "nearly zero energy building" which is in place since 2018 through the energy performance building directive (EBPD, 2018).

Passive House Standard

In a passive house, the majority of the heat demand is covered by passive energy sources such as solar energy, waste heat from people and technical equipment. According to the certification criteria of the Passive House Institute Darmstadt, a passive house must not exceed a yearly heating requirement of 15 kWh / m^2 (equivalent to approx. 1.5 litres of heating oil per m^2 per year). The primary energy figure for residual heating, water heating, ventilation and domestic electricity must not exceed 120 kWh / (m^2a). The maximum permissible heating load is 10 W / m^2 and must be able to be introduced via the supply air even in winter on unfavorable days.

Characteristics of a passive house:

- Highly insulated building envelope, U < 0.15 W / (m^2K)
- Avoidance of thermal bridges
- Compact structure
- Passive use of solar energy through south orientation and shading freedom
- Highly insulating glazing and window frames, Uw < 0.8 W / (m^2K); g-value by 50%
- Air tightness n50 < 0.6 / h
- Heat recovery from exhaust air, heat recovery rate > 75%
- Highly efficient power-saving appliances for the household
- DHW heating by e.g. Solar collectors or heat pump
- Passive air preheating by e.g. geothermal heat exchanger

Low-Energy House Standard

Low-energy buildings are buildings that clearly undercut the primary energy requirements permitted by EnEV, i.e. have an annual primary energy requirement of between 35 and 75 kWh / m^2a. The Passive

House standard is considered a further development of the low-energy house standard. In Switzerland, a low-energy house is a house built according to the Minergie standard.

Nearly Zero Energy Building

The Directive 2013/31 EU should take into account improvements in the overall efficiency of buildings, taking into account local conditions, indoor climate and cost-effectiveness (Article 1, first sentence). The Directive expressly deals with minimum requirements that Member States may go beyond (sentence 3).

Definition: Article 2, second sentence: "'nearly zero-energy building' means a building that has a very high energy performance, as determined in accordance with Annex I. The nearly zero or very low amount of energy required should be covered to a very significant extent by energy from renewable sources, including energy from renewable sources produced on-site or nearby". (Hermelink, 2013).

Thus, the balance sheet frame is already set: overall efficiency includes the entire energy needs. Energy production at the site will be included, as well as renewable energy generated in its vicinity.

Annex 1 (Common general framework for the calculation of the energy performance of buildings) states that building efficiency should be determined by the energy requirements for space conditioning (heating, cooling) and hot water production (Annex 1, first sentence) (Passer A., 2012). This is first in a certain contrast to the demand of the picture of the overall efficiency, since the energy requirement of highly efficient buildings is no longer dominated by heating, cooling and hot water. However, sentence 3 requires that (especially for non-residential buildings) built-in lighting be taken into account.

The required "cost-optimal level" (Article 2, sentence 14) includes the energy-related investment costs, operating (maintenance, energy) and possibly disposal costs over the lifetime. The lowest energy level will be required from 2019 for public, from 2021 for all other buildings. For rapid implementation, the member states set market incentives and remove market barriers (EBPD 2018).

Zero-Energy House Standard

For a zero-energy house, the amount of electricity produced (amount of energy) corresponds to the energy consumption of the house. It consumes as much electricity or energy as it produces itself and thus does not require any external energy from electricity, gas or oil. The energy for heating and hot water is gained through the use of renewable energy.

Characteristics of a zero-energy house:

- Large windows to the south
- A low A / V ratio (ratio of surface area to volume converted)
- Building external surfaces (facades, roof, windows, doors) with low heat transfer coefficient (= thermal insulation value, U value, early k value)
- A substantial airtightness

Plus-Energy House Standard and Efficiency House Plus

In a plus energy house, the amount of energy produced during the year is greater than the consumption. The energy required for the construction of the house is not taken into consideration for the production, transport, installation and disposal. This is also called grey energy.

With energy-efficient, modern buildings, corporations can make their corporate image visible to the outside world. These real estate investments also provide the long-term return on investment, an aspect that institutional investors attach great importance to. Also, in this point, "zero-energy projects" already offer great advantages with the same return.

The aim is to achieve stable gross rents in which the heating costs are already included. This benefits primarily the tenants, because they are immune from additional cost increases and so can plan their private budget better. At the same time rental apartments with a much higher living comfort arise. The investor has the advantage of having a major market advantage in the housing market with more stable rents, regardless of the price of oil.

With regard to the lifecycle costs of a building, it is noticeable that only 17% are required for construction and design costs, 40% for maintenance and renewal and 40% for heating energy, and 3% for deforestation. Compared to a modern zero-energy project, whose construction costs are 10% higher, equivalent to 1.5% of life cycle costs, 40% of the costs can be saved in the form of unneeded energy. In this analysis, the expected rising energy prices are not yet considered. Figure 1 shows the cost comparison of today's conventional buildings in Germany and an energy-optimized building like an efficiency house according to KfW criteria (Gabriel & Ladener, 2018).

Increased efforts in thermal insulation of buildings, in energy saving and in climate protection have led in recent decades to a stronger integration of building physics considerations in building design. Today, building physics knowledge is essential in the design, planning, practical implementation and use of buildings. Because in order to safeguard the functions and maintain a building, the indoor climate must be comfortable and tailored to its users, moisture damage avoided and the outside climate considered. Also necessary is the intelligent use of building physics, building materials, construction engineering, plant engineering and design resources.

Figure 1. Comparison conventional low energy building
(Source: (Gabriel und Ladener, 2018, own contribution)

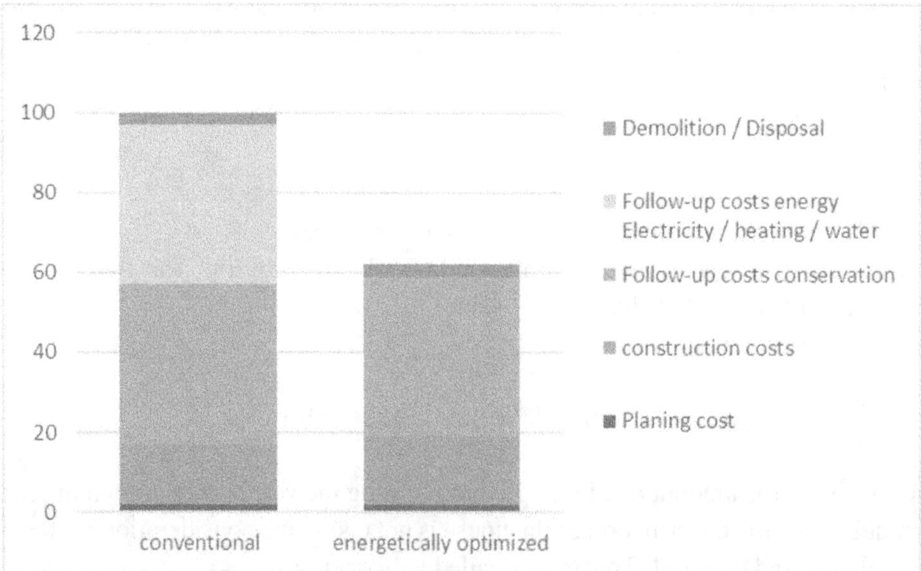

Building physics can be used to develop sustainable buildings with high thermal comfort, low energy consumption and the efficient use of residual energy supplied when needed. In addition to avoiding any condensation and mold problems, they also have a good supply of natural light and good room acoustics. The necessary building materials and constructions are examined in building physics especially with regard to the permeability of heat (heat conduction, heat radiation, thermal insulation), moisture (water vapor content, moisture transport, condensation of water, mold formation in apartments), acoustics and air.

Following aspect should be considered working on the renovations of the building envelope.

Airtightness of the building because of ventilation losses, type of the components which will be used, how the components are connected together, air pollutants through building materials, sound proofed materials in the building, summer heat protection through solar gain in summer times, diffusion of humidity in the components, dew point problems depending of temperature differences, different thermal bridges on the building, het convection on the components, thermal radiation and heat conduction are all points which should be care of.

Fundamentals About the Envelope Elements

There are a variety of factors that must be considered when planning changes to the building envelope. Factors which influence the envelope elements:

Planers engineers and architects must know how to design the different building elements and what characteristics/necessities are from importance. There are a numerous of amount of necessities for considerations the building envelope when rebuilding to a green and highly efficient building. The following Table 4 shows the different features around the building envelope. These are physical, static, economic, health, energy and aesthetic features.

Table 4. Feature for building envelope

Building features	Factors/importance's
Static considerations	The required evidence includes, among other things, a static calculation of the structure made by an expert. It serves as proof of stability, usability and durability
Construction, design	Look, design, regulations
Building physics	Noise, thermal, thermal heat bridges, thermal mass, thermal conductivity, surface resistance
Environmental	Air tightness, weather sealing, recyclability, sustainability, greenhouse gas emissions
Economics	Price, labour, building time, LCA live cycle assessment, recycling cost, investment cost
Materials	Natural, artificial
Health and safety	Health compatibility, health emissions
Safety	Fire protection
Energy relevant factors	Thickness, heat resistance, type of insulation, area of heated surface
General aspects	Availability, live time
aesthetic	Form, colours, surface

Source: Fouad 2015; Figl H. 2017; Garzon 2018; Hopfe, 2015, own contribution

Energy Balance Methods

To the thermodynamic law's energy will be not destroyed or used. Energy will only change in different energy forms or in other temperature level. According to the energy conservation laws, energy is never lost (Honig, 2013; Olander, 2007)

However, energy can be converted to a lower level by changing the temperature of a surface. These surfaces, for example, of a building are called the building envelope. The energy balance of a building is determined by the energy that emerges from the envelope through the various materials and surfaces, and the ventilation losses caused by the leaks of a building shell. The difference that needs to be applied to keep this balance the same is provided by an energy generator (heating system). Figure 2 shows a building with building envelope and thermal envelope of a standard building.

The energy and cost-conscious planning of buildings is generally carried out with the aid of an energy balance method. Even at the design stage, the main energy flows and thus also the energy requirements of the building can be determined by calculation. Energy balance methods enable the optimization of the building under the given economic conditions.

In conjunction with the draft Directive, CEN has developed various EN standards to enable the implementation of the Directives

Energy Consumption in Buildings

In order to calculate the total energy consumption in buildings, the standard CEN TR 15615 describes the structure of the standards. The EN 15603 standard defines the types of use of energy to be taken

Figure 2. Building envelope and energy flows
(own contribution)

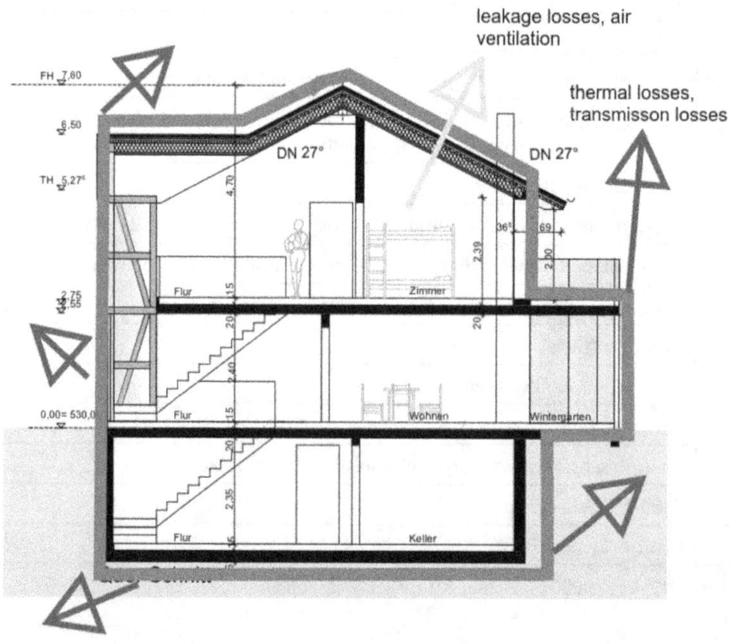

into account and provides procedures for the evaluation of energy efficiency and energy ratings for new and existing buildings. The EN 15217 standard explains the options for indicating energy efficiency in a certificate and setting out the requirements for energy efficiency. The standard EN 15459 provides a calculation method for the efficiency of heating and air conditioning systems, which refers to the energy consumption and energy consumption of a building (Simader, 2014)

Standards for Thermal Efficiency of the Building Envelope

For the calculation of the thermal efficiency of components, the standard EN ISO 13789 is used to determine the total heat transfer coefficient. The standards for U-values are divided into two groups. For simplified procedures:

- *EN ISO 6946, EN ISO 13370, EN ISO 10077-1, EN 13947*

For detailed procedures:

- *EN ISO 10211, EN ISO 10077-2,*

For thermal bridges:

- *EN ISO 10211, EN ISO 14683*

The EN ISO 10456 standard determines the thermal values of construction products.

How the Energy Requirement Is Calculated

In the case of demand-oriented identification, the information on insulation standard and heating technology is calculated using a computer program. The result shows both the final energy, i.e. the actual consumption, as well as the primary energy demand. This second value also includes the energy necessary for the production, transport and storage of the fuel and is therefore an important indicator of the ecological quality of the house.

Methodical Approach

This method is a standardize for defining the quality of a building envelope. This result gives a figure about how much energy is necessary to keep the building on a special temperature level over a period of time. The result will be a figure like comparing cars with liter/ gallons/ per mileage or kilometer.

In the monthly balance method, the annual heating demand of a building is determined from the heat balance of the individual months. The heat balance is composed of the monthly heat losses and the monthly heat gains taking into account the degree of utilization. The utilization rate is set, since usually not all profits can be used to 100%. The reason is that they do not always occur at all times as they are needed.

The building-specific heating time can be determined exactly with this calculation method. Conservatories, transparent thermal insulation, heat recovery systems or even passive solar energy utilization measures can be taken into account in the monthly balance sheet procedure.

The monthly balance sheet procedure records and accounts for both the building envelope and the plant technology.

The calculations according to the monthly balance sheet method are possible only with the help of computer simulation programs because of the very high computational effort.

Energy Index

Energy parameters are used for the energy assessment of a building and refer to different properties or values to be calculated. For example, the Association of German Engineers (VDI) distinguishes between energy parameters (specific energy consumption, efficiencies, utilization rates, rated power, etc.) and energy parameters for holistic assessment (cumulative energy consumption, harvest factor, energy payback period, etc.).

A frequently used energy parameter is the energy index, which indicates the annual heating demand in kWh in terms of floor area in square meters. These area-related values allow the comparability of different buildings. It is important that the characteristic values are based on the same reference values. It is important to pay attention to whether the areas are Gross Basic, Net Basic or Usable.

The relationship between building envelope and building surface has a major impact on the energy index. The more compact a building, the lower the transmission heat losses. Therefore, only comparing similar building types is possible. In the area of housing construction, for example, only the energy figures of detached single-family homes or of apartment buildings are comparable. The average value of the energy index per year for multi-family houses is 175 kWh / m_2 and for single-family homes rises to 300 kWh / m_2.

Other influencing factors on the energy index are the age of the building, the refurbishment status and - difficult to estimate - the energetic behavior of the users or the type of use.

Examples of the energy index annual heating requirement:

- Housing stock: 100-275 kWh / (m^2a).
- New construction according to WSVO 1995: 54-100 kWh / (m^2a).
- Low Energy House (NEH): 35-75 kWh / (m^2a).
- Passive house: <15 kWh / (m^2a).

Energy-efficient buildings are divided into four categories: low-energy, passive, low-energy and plus energy houses. Passive houses are often the optimal solution if the entire service life of the building is taken into account.

Today's building in special ages are shown in Table 5 the energy amount per m^2.

If you compare buildings with each other, this can be done with the characteristic of the amount of energy per area and time. These necessary amounts of energy for heating the building depend on different factors, as already discussed. Not only the thickness of the insulation and insulating material but also technical designs such as heating technologies, ventilation technologies and passive solar design play an important role. If the energy census of a building is determined using the same methodology, the different buildings can be compared and evaluated. Similar to a vehicle that needs so much fuel per 100 km of driving distance.

Table 5. Building rating and energy usage

Type of building /year	Class type	Energy usage
From 1950 -1960	F	160 kWh/m²a
From 1960 – 1970	E	120 kWh/m²a
From 1980 – 1990	D	100 kWh/m²a
From 1980 – 2000	C	80 kWh/m²a
From 2000 - 2010	B	70 kWh/m²a
From 2010 – 2015	B	50 kWh/m²a
From 2015 -	A	40 kWh/m²a

Source: Piccardo et al., 2019b; Meyer 2015; Krieger, 2017, own contribution

In addition, a periodic assessment of the energetic behavior of a building - despite structural changes - be made. However, energy parameters can also be used as selection criteria for further investigations, or serve as an instrument of operational management and monitoring. The value is also suitable for controlling energy-saving measures and as guideline and specification for planning new buildings, conversions and renovations.

The energy and cost-conscious planning of buildings is generally carried out with the aid of an energy balance method. Even at the design stage, the main energy flows and thus also the energy requirements of the building can be determined by calculation. Energy balance methods enable the optimization of the building under the given economic conditions. Figure 3 shows the principal method of energy balance of a building.

The energy balance of a building determines the amount of energy required. Although energy balances can be created, for example, for electrical energy.

On average, a building must receive just as much heat as it loses elsewhere.

Figure 3. Energy balance method for buildings

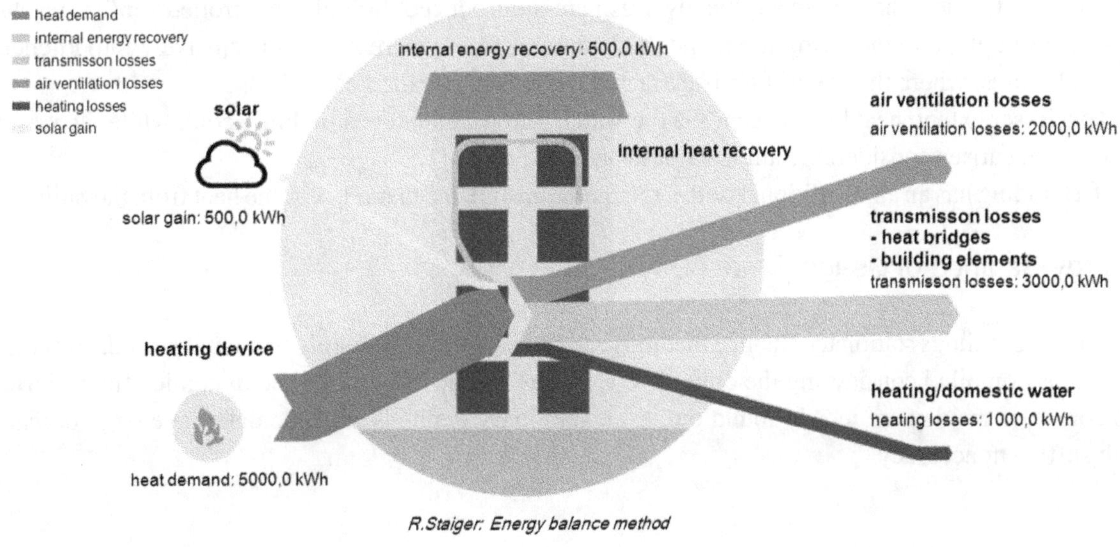

Due to the thermodynamic laws, the energy balance of a building must be balanced on average, that is to say when looking at a longer period of time (several days), just as much heat will leave the building as is being introduced into the building as a whole. Otherwise, the heat content and thus also the average temperature of the building would have to change.

Within a few hours, however, the heat balance of a building need not be compensated. Often the temperature in the building fluctuates in the rhythm of the day, for example due to strong solar radiation at lunchtime and due to the night lowering of the heating system.

It makes sense to include everything that lies within the thermal insulation perimeter. This means, for example, that unheated basements usually do not belong, unless the thermal insulation is located under these rooms. Heat can essentially enter a building through the following mechanisms:

With the help of a heating system, heating can be deliberately introduced into a building, by radiators or underfloor heating.

In addition, various internal heat sources that give internal heat gains (or internal heat gains). For example, people constantly give heat to a building; for an adult, who does not move too hard, the heat output is around 100 W, which is 2.4 kWh per day. Added to this is the heat dissipation of various devices. The electrical energy consumed by various devices (including lighting devices) often remains completely as heat in the building, as far as they cannot go directly into the open. For example, rising hot air over a cooking stove may be lost via an exhaust device, and in a washing machine or dishwasher, a substantial portion of the heat generated may escape via the waste water. Even heat that develops outside the thermal insulation perimeter (for example in unheated basements) is often largely lost.

In addition, there are so-called solar gains, which mainly arise when radiation from the sun penetrates through windows into rooms and is converted there into heat.

Heat conduction is a particularly important loss mechanism.

Every building has some transmission heat loss: Heat escapes outward through the heat conduction mechanism depending of the temperature differences between outside and inside. Such heat losses can be greatly reduced by effective thermal insulation (e.g. on facades, in the roof and on basement ceilings).

In addition, there are ventilation losses: Heat is transported outward by air, which enters as cold air and exits again in heated form. However, the ventilation losses can be greatly reduced (e.g. by a factor of 5 to 10) by using a heat recovery ventilation system; Heat is usually taken from the exhaust air here, which is used to heat the supply air. Ideally, fresh air is introduced through a controlled ventilation system, for example, into the living room and the bedroom, from where it flows exclusively into kitchens and bathrooms, where the exhaust air is extracted. Here is a so-called cascade use.

To a lesser extent, heat losses occur via the water. Tap water enters a building cold, while the wastewater often carries considerable amounts of heat.

If a building has an air conditioner with a refrigeration unit, it can remove some heat from the building.

Energy Balance of Usage Time

For simple buildings stationary balance methods can be used. Over a monthly or annual period, an energy balance is compiled considering the climatic conditions. Because in reality the dynamic effects during the course of the day are mostly found out, even stationary methods can calculate the energy demand with sufficient accuracy.

For complex buildings, dynamic simulation programs provide more accurate information about the energy balance of buildings. The used calculation models approximate the actual physical processes. They map the time course of the energetic processes within the building. Provided that the boundary conditions have been set realistically, these programs accurately represent the actual behavior of the building. However, the effort for data acquisition, input and calculation is quite high, so the simulation is costly.

Energy is the sum of working capacity and waste heat, measured in Joules, Wh, kWh or MWh. Energy is made up of exergy and anergy.

Exergy is an expression of working capacity. Exergy refers to that part of the energy that mechanical work can do.

Anergy, also called waste heat, that cannot do any mechanical work:

- Anergy + exergy = energy

Grey Energy

The energy consumption for the production of the building materials is not captured by a building simulation. The grey energy evaluates the total energy consumption for producing building elements, building process, refurbishment/maintenance and disposal/Recycling processes (Wind G., 2017). Is the material use of insulating materials, for example, worthwhile if a lot of energy is needed to produce them? Yes, it's worth it. The energy expenditure for the insulation production is compensated within a few months by the heating energy saving. Currently, the grey energy that is in a new building corresponds approximately to the necessary operating energy for 30 to 50 years. The trend is right to decrease the operating energy by more and more construction and insulation variants. However, this happens with a simultaneous increase in the grey energy that is being built, because the material costs of the structures and the technical components increase. In the future, grey energy will play an important role in the consideration of energy, energy efficiency and environmental factors. The future view must be considered in a life cycle. We shift the energy consumption from the operating phase into the production phase. From an environmental point of view, it does not matter where the CO_2 is generated. Whether through combustion processes from a fireplace or in the various process chains in the life of a building. The energy consumption must be minimized over the entire life cycle (Piccardo et al., 2019a; Hunt, 2003; Klöpffer & Grahl, 2009).

Transport Balance

The construction products must be brought from the place of their production and storage to the construction site. The energy costs incurred are summarized in the transport balance sheet.

Traffic Balance

Each building is visited by people and supplied with goods. This traffic-inducing effect is characterized by the building location and logistics of the transported persons and goods

Insulation and Technical Equipment

To reduce the heat demand in a building, the insulation of the facade, roof and floor slab is one important task. Wall and roof surfaces area are the larger part and consequently the highest transmission losses. According to different publication the proportion of individual areas in a simple single-family home shows Table 6.

The location of the insulation layer has a decisive influence on the energetic consequences and the constructive measures. The planner has the choice, the component insulated on the outside or inside, and in the case of the exterior wall, the monolithic and insulated construction.

Different Building Materials for Retrofit

Today's insulation materials are mainly used in new construction and in the renovation of buildings. Insulation materials can also be used for sound insulation, room acoustic and insulation for technical equipment's. There is a variation of natural and artificial insulation materials on the market. Table 8 shows the wide variation for different applications and usages. Insulating materials in the form of panels

Table 6. Characteristics of building elements

Building envelope family house	
Envelope component	**Total energy usage**
Floor against soil	5- 20%
Heating	30 – 35%
Roof	15 – 20%
Technical building equipment	
Ventilation	10 20%
Walls	20 – 25%
Windows	20 -25%

Source: Schmidt et al., 2017; Krieger O. 2017

Table 7. Characteristics of building elements

Building envelope Residential house 10 floors	
Envelope component	Total energy usage
Floor against soil	5-10%
Windows	30%
Walls	40%
Roof	7%
Technical building equipment	
Ventilation	10 - 20%
Heating	30 – 35%

Source: Häfele et al., 2017; Discher H., 2010; Krieger, 2017

Table 8. Insulation materials

Insulation Materials
Wood fibre
Wood wool
Flax
Cork
Cattail
Coconut
Hemp
Sheep wool
Cellulose
Natural mineral insulation
Perlite
Vermiculite
Mica
expanded Clay
Bricks
artificially manufactured insulating materials
Mineral foam 36 Mineral wool (MW) 37 Phenolic resin (PF) 39 Polystyrene, expanded (EPS) 40 Polystyrene extruded (XPS) 42 Polyurethane (PU) 44 Pyrogenic silica
Mineral wool
Phenolic resin
Polystyrene expanded EPS
Polystyrene extruded XPS
Polyurethane
Pyrogenic silica
Foam Glass
Vacuum insulation panels

Source: Kunič, 2017; Klöpffer und Grahl 2009; Krieger, 2017; Gellert et al., 2007; Dorsch et al., 2017

or mats are ideal for insulating roofs, walls and ceilings / floors. As granular fillings, insulating materials are mainly used for beamed ceilings, for certain core insulation of exterior walls and for insulation of flat roofs. Flake-shaped insulating materials are blown into larger cavities of roofs and walls (Banzhaf, 2013; Beinhauer, 2017; Canbolat, 2018; Dorsch et al., 2017; Gellert et al., 2007; Rexroth et al., 2014)

The different raw materials and additives used to produce insulating materials are characterized by certain physical properties (Gelfand und Duncan, 2012; Kunič, 2017; Wilkinson und Dixon, 2016; Yang, 2017; Gabriel und Ladener, 2018). Table 13 shows the different physical properties

Table 9. Physical properties

Physical properties	Description
Heat protection	Ambient temperature
Heat storage capacity	Capacity of energy storage in the building material over time
Diffusion resistance	How much moisture can be absorbed from the element
Water absorption capacity	
Sound proof	The element as a sound barrier
Fire behaviour	Fire protection
Elasticity	Static parameter in the building process
Temperature resistance	Resistance about higher and lower temperature levels
Primary energy usage	
Long term behaviour	Live cycle assessment figure

Source: (Yang, 2017; Gabriel & Ladener, 2018; Kunič, 2017; Wilkinson & Dixon, 2016; own contribution)

Case Study: Wall Renovation

To calculate the quality of a building element a special U Value is defined (Pfundstein et al., 2012; Filonenko, 2017). The U value defines how much power of an energy converter (for example Heat Pump or pellet boiler) is necessary to compensate for a temperature difference between inside and outside. The U value is given in the physical unit W / K m^2. The U value is determined from the reciprocal of the thermal resistance of a component (Anderson, 2014).

The following mathematical relationships are defined:

The total resistance of a component is the sum of all thermal resistances. The heat resistance depends on the material properties which is defined as the thermal conductivity and on the thickness of the material.

The following physical relationships are necessary to calculate the relation between energy usage and quality aspect of an insulation surface.

U value is defined (heat transfer coefficient) $U = \frac{1}{R_{total}}$

Thermal Resistance of a building element is: $R_{total=} R_1 + R_2 + \dots R_x = \sum_1^x R_x$

$$R_{total} = \frac{d_1}{\lambda_1} + \frac{d_2}{\lambda_2} + \dots \frac{d_x}{\lambda_x}$$

The necessary power for a constant temperature is: $P = U * A * \Delta T$
The amount of energy keeping the temperature is: $E = P * t$
The energy cost for the amount of energy is: $E_{cost} = E * Price_{kWh}$

Example: Wall Against Outside Air Limestone Wall

Layer construction (from warm to cold) (see Table 10).

- Thermal resistance $R_T = R_{si} + R_1 + R_2 + R_3 + R_{se} = 0{,}46\ m^2K/W$
- Heat transfer coefficient $U = 1/R_T = 2{,}17\ W/(m^2K)$

Temperature Distribution\

See Figure 5.

Table 10. U Value calculating EN DIN 6946

No.	Description	Thickness cm	λ W/m*K	R m^2K/W	μ_1 -	μ_2 -	ρ kg/m^3	c_p kJ/kg*K
1	Plaster	2,00	1,000	0,02	15	35	1800	1,00
2	limestone	35,00	1,400	0,25	40	50	2000	1,00
3	plaster	2,00	1,000	0,02	15	35	1800	1,00

Source: own contribution

Figure 4. Limestone wall
(picture own contribution)

Table 11. Limestone calculation

Heat transfer resistances		
Heat transfer resistance inside R_{si}	0,13	m^2K/W
Heat transfer resistance outside R_{se}	0,04	m^2K/W
Heat flow direction	horizontal	
Component is adjacent	Outside air	
Summary		
U-Wert	2,17	W/m^2K
Thermal resistance	0,29	m^2K/W
Minimum thermal resistance DIN 4108-2	1,20	m^2K/W
Effective heat storage capacity CP 3 cm	56,00	kJ/m^2K
Effective heat storage capacity CP 10 cm	196,00	kJ/m^2K
Specif. component mass	772,00	kg/m^2
thickness	39,00	cm

Source: own contribution

Figure 5. temperature curve of the wall
(source: own contribution out of a Software Simulation Tool ETU)

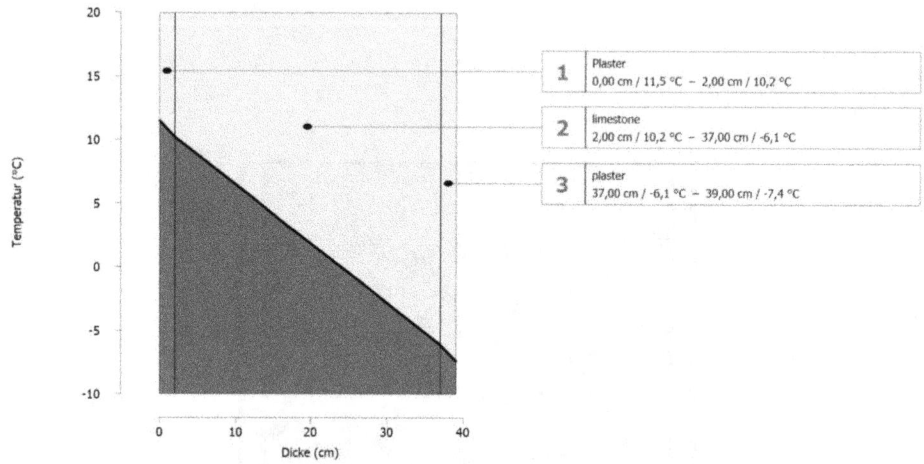

Example: Timber Frame Wall

The Table 12 shows the calculation method of the U value of a standard timber frame wall from timber frame buildings from today.

Case Study 1: U Value Comparison Building 1930 and Renovation 2019

From the individual building typologies, the following values can be defined for individual building components for the building envelope.

Table 12. Timber frame calculation

U-Wert = 0,186 W/Km²			thickness	Thermal conductivity
Component structure: Layer sequence from inside to outside			s(cm)	λ (W/mK)
A		**Inhom. layer Timber frame**		
	1	plasterboard	2,00	0,290
	2	OSB Board	1,50	0,130
	3	Timberframe	16,00	0,130
	5	wood wool	6,00	0,040
B		**Inhom. Layer Mineral wool**		
	1	plasterboard	2,00	0,290
	2	OSB Board	1,50	0,130
	3	Mineral wool (035)	16,00	0,035
	5	wood wool	6,00	0,040
		Total thickness:	25,50	

Source: own contribution

Table 13. Calculation result renovated wall

Building 1930, U Value wall 1,4 W/Km²				
Temperature difference	**formula**	**E in 1 h [kWh]**	**E in 24 h [kWh]**	**E in 1.800 h [kWh]**
E with 1 K Temp diff	$E = U * A * \Delta T * t$	0,35	8,4	630
E with 10K Temp diff	$E = U * A * \Delta T * t$	3,5	84	6.300
Building 2019, U Value wall 0,14 W/Km²				
	formula	E in 1 h [kWh]	E in 24 h [kWh]	E in 1.800 h [kWh]
E with 1 K Temp diff	$E = U * A * \Delta T * t$	0,035	0,84	63
E with 10K Temp diff	$E = U * A * \Delta T * t$	0,35	8,4	630
Energy cost over period of time thermal energy 8 cent/kWh				
Old building 10 K	Ecost = E *price	0,28 €	6,72	504 €
New building 10K	Ecost = E *price	0,03 €	0,67	50 €
Savings	New - old	0,25 €	6 €	454 €

Source: own contribution

Figure 6. Timber frame wall
(source: own contribution)

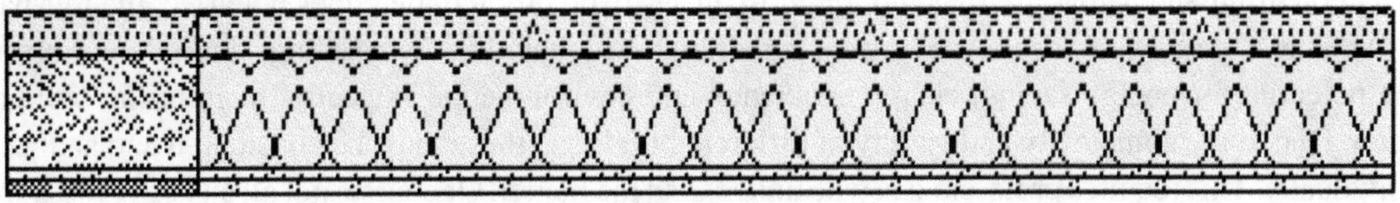

- Building from the 1930: Outside wall area 250 m^2, U Value 1,4 W/Km^2.
- Building in 2019: renovated building wall area 250 m^2 U value 0,14 W/Km^2.

This is done through an outside insulation for example mineral wool with 20 cm heat resistance of 0,032 W/mK. The case study should show how much energy you need over a period of time and a special temperature between inside and outside in compare to a renovated building with the same parameter only the U value is different.

This case study shows the method how to calculate the energy amount which is necessary for heating a building. The time period and temperature difference are depending of the location of the building. In Norway the time and temperature difference would be different in compare to Greece or Spain. This local parameter is included in special simulation software packages depending of the location. The method of calculation is the same. In our case study we did only the outside walls. The same procedure would be similar for all building's element on the outside envelope of the building like windows, roofs, floor area and flat roofs.

DISCUSSIONS AND CONCLUSION

To renovate an existing building in a sustainable way requires special attention. There are a lot of considerations that have to be addressed in the decisions.

One of the most important points in refurbishing a building is the holistic view of the different refurbishment measures and the interconnections between them.

The fundamental question arises first of all whether it makes sense to renovate an older building. As described above, personal, technical, economic, construction management and legal matters play an important role here.

If a building is to be sustainably renovated, the building must be viewed holistically. Here not only the building envelope is to be considered but also the technical building services. This view must be incorporated in the planning and execution process during the renovation. The potential consequences can be technically and economically disastrous.

Which materials are to be used for the refurbishment process depends on many other factors? Here are the investment costs and installation costs. Furthermore, the service life, the thermal quality of the materials. In order to ensure sustainable construction, the life cycle assessments of the building materials must be considered closely. Here, certain standards and eco-certificates help as a means of comparison.

Following points are often discussed about insulation:

Well planned and executed insulations last at least 40 to 50 years. However, an insulated façade, which is exposed to large temperature fluctuations and may also be overgrown with climbing plants or other mechanical influences, must be checked regularly and repaired if necessary. In a long-term study, the Fraunhofer Institute for Building Physics (IBP) has investigated that thermal insulation systems are generally no more prone to damage than uninsulated facades. And some damages, such as the algae formation discussed below, are first and foremost an aesthetic problem that does not affect the lifespan.

Around 40 percent of primary energy consumption is accounted for by buildings throughout Europe. The potential for energy saving, energy efficiency and environmental impact is huge in the building sector. Due to the complexity and variety of different buildings, these must be considered and analyzed individually. This requires professionals who understand the matter. Of course, not all austerity measures are the same: While very good savings can be achieved with an insulation of the top floor ceiling and the basement ceiling, complex façade insulation often pays off over a longer period of time.

It is generally important that only the appropriate insulation measures for a building are carried out, that the insulation is installed properly, and that the heating technology is adapted accordingly after the measures have been taken.

The assumption that the insulation materials normally do not pay for themselves over their lifecycle is simply wrong: Insulating materials can almost always recoup the energy invested through energy savings and also save a lot of energy. This applies to all commercially available insulating materials - even if, for example, significantly more energy is required for the production of polystyrene than for mineral wool. For example, in the case of a basement ceiling insulation with polystyrene installed according to the current EnEV specifications, the required production energy can be saved after about three heating periods.

And the insulation can last 40 to 50 years or even longer. How long the energy amortization takes dependence of thickness type of material and other aspects. The thicker the insulation, the longer it takes for it to pay for itself in terms of energy. This is because the effect of the insulation in the form of a hyperbola increases, so bring the first centimeter thickness of insulation the most savings. However, the energy required for production increases almost linearly and thus steeper.

Ecological insulation materials such as cellulose, wood fibers or expanded glass are no problem for the environment. Mineral wool also has no noteworthy environmental impact. And polystyrene ("Styrofoam") insulating materials consist of about 98 percent air and only 2 percent plastic - more specifically, foamed petroleum. Oil is a fossil and therefore finite resource, which produces climate-damaging CO2. However, in the case of professional material use, no environmental effects are to be expected that go beyond other common plastic-containing products or packaging.

REFERENCES

Anderson, B. (2014). Conventions for U-Value Calculations. Amazon. Retrieved from www.amazon.de/Conventions-U-Value-Calculations-Brian-Anderson/

Atkins, P. W. (2010). The laws of thermodynamics. A very short introduction. Oxford, New York: Oxford University Press.

Banzhaf, A. (2013). Impacts of 'Basic' Thermal Insulation in Residential Buildings on Fuel Wood Demand and Primary Forests (1st ed.). Aachen, Germany: Shaker.

Beinhauer, P. (2017). *Building in stock. with 213 detail design, 2.* Edition.

BP. (2018). BP Statistical Review of World Energy. With assistance of Bob Dudley.

Canbolat. (2018). Determination of Proper Insulation Thickness for Building Walls Regarding Economic Consideration 2018.

DENA. (2019). Gebäude-Energiewende: Building Energy Transformation.

Discher H., & Hinz E. (2010). dena-Sanierungsstudie_Teil_1.

Dorsch, L., Kaiser, C., Niklasch, W., Schöpgens, H., & Spritzendorfer, J. (Eds.). (2017). *Market Overview Insulation materials from renewable raw materials. Fachagentur Nachwachsende Rohstoffe (9th ed.). Gülzow-Prüzen: Fachagentur Nachwachsende Rohstoffe e.V.* Naturbaustoffe.

EBPD. (2018). New Energy Performance in Buildings Directive comes into force on 9 July 2018.

Figl, H. (2017). Ökobaudat Basis for the building life cycle assessment.

Fouad, N. A. (2015). *Bauphysik Kalender 2015. Schwerpunkt: Simulations- und Berechnungsverfahren.* Wiley. doi:10.1002/9783433605158

Gabriel, I., & Ladener, H. (Eds.). (2018). From the old building to the efficiency house. In Modernize and energetically refurbish: planning, construction practice (13th ed.). Staufen bei Freiburg: Ökobuch.

Garzon. (2018). Future requirement for sustainable building envelope.

Gelfand, L., & Duncan, C. (2012). *Sustainable Renovation. Strategies for Commercial Building Systems and Envelope.* Wiley.

Gellert, R., Pfundstein, M., Rodolphi, A., & Spitzner, M. H. (2007). Insulation materials. In Basics, Materials, Application. München: DETAIL.

Häfele, G., Oed, W., & Sabel, L. (2017). *House renovation. Repair - modernize - save energy - rebuild: ecological construction practice* (16th ed.). Staufen bei Freiburg: ökobuch.

Hermelink, S. (2013). Towards nearly zero energy buildings.

Hess, D. B., Tammaru, T., & van Ham, M. (Eds.). (2018). Housing Estates in Europe. In Poverty, Ethnic Segregation and Policy Challenges. Cham: Springer International Publishing. doi:10.1007/978-3-319-92813-5

Honig, J. M. (2013). *Thermodynamics. Principles Characterizing Physical and Chemical Processes* (4th ed.). Burlington: Elsevier Science.

Hopfe, C. J., & McLeod, R. S. (2015). *The Passivhaus Designer's Manual: A technical guide to low and zero energy buildings.* Routledge.

Hunt, R. G. (Ed.). (2003). Comparative life cycle assessment of three insulation materials: stone wool, flax and paper wool. Landsberg, Bayreuth: Eco Informa Press.

Hutter, C.-P. (2018). *Die Erde rechnet ab. Wie der Klimawandel unser tägliches Leben verändert - und was wir noch tun können.* München: Ludwig.

Klenner, W. (2009). Biodiversity in forest ecosystems and landscapes: A conference to discuss future directions in biodiversity management for sustainable forestry. Forest Ecology and Management, 258. doi:10.1016/j.foreco.2009.10.037

Klöpffer, W., & Grahl, B. (2009). *Ökobilanz (LCA). Ein Leitfaden für Ausbildung und Beruf.* Weinheim: Wiley. doi:10.1002/9783527627158

Krieger O. (2017). Gebäudestudie. Szenarien für eine marktwirtschaftliche Klima- und Ressourcenschutzpolitik 2050 im Gebäudesektor.

Kunič, R. (2017). Carbon footprint of thermal insulation materials in building envelopes.

Lüdtke, N., & Henkel, A. (Eds.). (2018). Das Wissen der Nachhaltigkeit. München: Oekom Verlag.

Meyer, C. (2015). *Energetische Sanierung von Gebäuden im Bestand. Wirtschaftlichkeit und Maßnahmen am Beispiel von drei Einfamilienhäusern*. Hamburg: Diplomica Verlag.

Oehler, S. (2018). Emissionsfreie Gebäude. Das Konzept der „Ganzheitlichen Sanierung" für die Gebäude der Zukunft.

Olander, D. (2007). *General Thermodynamics*. Hoboken, NJ: CRC Press. doi:10.1201/9781420007558

Passer, A., Kreiner, H., & Maydl, P. (2012). Assessment of the environmental performance of buildings: A critical evaluation of the influence of technical building equipment on residential buildings. *The International Journal of Life Cycle Assessment, 17*(9), 1116–1130.

Pfundstein, M., Rodolphi, A., Spitzner, M. H., & Gellert, R. (2012). Dämmstoffe. Grundlagen, Materialien, Anwendungen: DETAIL.

Piccardo, C., Dodoo, A., Gustavsson, L., & Tettey, U. Y. A. (2019a). Comparative Life-Cycle Analysis of Building Materials for the Thermal Upgrade of an Existing Building. *IOP Conf. Ser.: Earth Environ. Sci., 225*(1), 12044. 10.1088/1755-1315/225/1/012044

Piccardo, C., Dodoo, A., Gustavsson, L., & Tettey, U. Y. A. (2019b). Energy and carbon balance of materials used in a building envelope renovation. *IOP Conf. Ser.: Earth Environ. Sci., 225*(1), 12045. 10.1088/1755-1315/225/1/012045

Rexroth, S., May, F., & Zink, U. (Eds.). (2014). *Wärmedämmung von Gebäuden. Zeitgemäß und wandlungsfähig. 1. Neuerscheinung*. Berlin, Germany: VDE VERLAG GMBH.

Riley, M., & Cotgrave, A. (2018). Construction Technology 1. House Construction (4th ed.). Oxford: Red Globe Press.

Schmidt, B., Schmidt, D., & Venzmer, H. (Eds.). (2017). Energielieferant Altbau. Potenziale, Nachhaltigkeit, Objektbeispiele. Berlin: Beuth Verlag GmbH.

Schönwiese, C. (2019). *Klimawandel kompakt. Ein globales Problem wissenschaftlich erklärt*. Stuttgart: Borntraeger.

Schröpfer, T. (2016). *Dense + Green. Innovative Building Types for Sustainable Urban Architecture*. Berlin: Birkhäuser.

Schröpfer, T., & Menz, S. (2019). Dense and Green Building Typologies. In Research, Policy and Practice Perspectives. Singapore: Springer. doi:10.1007/978-981-13-0713-3

Simader, G. (2014). *Implementation of the EPBD Energy Performance of Buildings Directive*. EPBD.

Sobek, W. (2015). *Residentials by Werner Sobek*. Stuttgart: Avedition.

Steiner, T., Eßmann, F., & Dornigg, I. (Eds.). (2018). *Ökologie und Ökonomie des Dämmens. Analyse und Bewertung von Dämmmaßnahmen in der Altbausanierung. Fraunhofer IRB-Verlag*. Stuttgart: Fraunhofer IRB Verlag.

Swan, W. (Ed.). (2013). *Retrofitting the built environment*. Hoboken, NJ: Wiley Blackwell. doi:10.1002/9781118273463

Wang, J. (2012). *Modern Thermodynamics. Based on the Extended Carnot Theorem. 2011*. Berlin, Germany: Springer.

Westermann, R., Meyer, Ü., Beobachter, D. S., Zeugin, K., Bolliger, B., & Faller, K. (2013). *Mein Haus energetisch sanieren. Komfort verbessern, Kosten senken*. Klima schützen: Beobachter.

Wilkinson, S.J. & Dixon, Tim (2016). Green roof retrofit. Building urban resilience.

Wilson, J. (2013). *The greened house effect. Renovating your home with a deep energy retrofit*. White River Junction, VT: Chelsea Green Publishing.

Wind, G. (2017). Graue Energie. Retrieved from http://www.ibwind.at/download/GraueEnergie081121.pdf

Yang J. (2017). Influence of envelope insulation materials on building energy consumption.

ADDITIONAL READING

Baumann, H., & Tillman, A.-M. (2004). *The hitch hikers's guide to LCA. An orientation in life cycle assessment methodology and application*. Lund: Studentlitteratur.

Buck, M. (2019). European Energy Transition 2030: The Big Picture. Ten Priorities for the next European Commission to meet the EU's 2030 targets and accelerate towards 2050.

Day, C. (2016). *The eco-home design guide. Principles and practice for new-build and retrofit*. Cambridge, UK: Green Books.

Frey, H. (2019). *Energieautarke Gebäude. Auf dem Weg zu Smart Energy Systems*. Berlin: Springer. doi:10.1007/978-3-662-57874-2

Ibisch, P. L. (2018). Der Mensch im globalen Ökosystem. In Eine Einführung in die nachhaltige Entwicklung. München: Oekom.

International Renewable Energy Agency (IRENA). (2019). Global Energy Transformation: A Roadmap to 2050 (2019 Edition).

International Renewable Energy Agency (IRENA). (2019). Renewable Energy Capacity Statistics 2019.

Lelieveld, J. (2019). *Effects of fossil fuel and total anthropogenic emission removal on public health and climate*. National Academy of Sciences.

Pehnt, M. (2010). *Energieeffizienz*. Springer.

KEY TERMS AND DEFINITIONS

Building Envelope: The building envelope is a closed geometric structure and the physical separation between the internal and external environment of a building. It comprises all the components of a building that close it to the outside like walls, roof, windows, and floors.

Energy Balance Method: Energy balance methods enable the optimization of the building under the given economic conditions. The energy flows for the building to be calculated are determined using this method. These energy flows can be determined daily, weekly, monthly and annually. Depending on the location of the building, additional meteorological data are included in the calculations

Energy Performance Building Directive: Directive on the energy performance of buildings. The new regulation aims to make buildings "smarter" and more energy efficient. They enable saving greenhouse gases, financial savings and contribute to the creation of new jobs in the renovation and construction sector.

Insulation Materials: An insulating material is a building material that is preferably used for thermal insulation. Thermal insulation materials are materials with low thermal conductivity and reduce heat or cold losses

Live Cycle Assessment: A Life Cycle Assessment (LCA) is a systematic analysis of the environmental impact of products throughout their life cycle "from cradle to grave." Life Cycle Assessment includes all environmental impacts during production, the usage phase and the disposal of the product, as well as the related upstream and downstream processes

Nearly Energy Building: lowest energy building with a very high energy efficiency defined in Annex 1 EBPD. The main part of the energy should come from renewable energy sources.

Passive Energy building: A building standard that is truly energy efficient, comfortable, economical and environmentally friendly at the same time. The passive house is not a brand name, but a construction concept that is open to all - and has proven itself in practice. The heating demand in the passive house is less than 15 kWh / (m^2a) per heated surface.

U Value: A measure of the heat transfer is the heat transfer coefficient or U-value, its unit is given in W / (m^2K). This value takes into account the heat transfer effects of convection and radiation. A small heat transfer coefficient leads to less heat loss in the heating time - the heat remains in the building. In summer, the small U-value also has a positive effect - the heat stays outside the building.

Chapter 11
Economic Analysis for Green Residential and Non-Residential Building Envelopes

Robert Staiger
E[3]xpert, Germany, Romania

ABSTRACT

Renovating buildings is more useful than ever. Due to future rising energy prices, energy costs in poorly insulated buildings are an important component of operating costs. Another important point is the rapidly growing emissions from the combustion of fossil energy sources. Good insulation in buildings reduces the amount of primary energy and thus, less greenhouse gases are emitted. The renovation potential is high. A large part of the properties consumes more energy than would actually be necessary. Common construction without thermal insulation is responsible for this. It is advisable to invest in renovation, also, in thermal insulation. This will benefit you in two ways. The ancillary (additional) costs will be reduced massively, the living comfort increases and by today's state subsidies in many countries they will make a contribution to the investment costs.

INTRODUCTION

The aim of the chapter is to analyze the economic and ecological aspects of building renovations.

First, the energy flows that are explained in a building over a period of time. Afterwards it will be shown how the energy distributions on buildings can look like in the future. Here, Life Cycle Assessment is explained briefly and explains what is meant by grey energy.

Then, the economic basics of building renovation are presented. Based on the thermal quality of building elements, two buildings are calculated and simulated in two case studies.

In the next chapter ecological and environmental considerations for green building envelopes, the entire lifecycle of a building is considered. In doing so, a calculation is shown which the environmental authority in Switzerland uses for a life cycle observation of building elements.

DOI: 10.4018/978-1-5225-9104-7.ch011

SCOPE GENERAL CONSIDERATIONS

Renovating buildings is more useful than ever. The primary energy costs of the last 25 years worldwide are rising continuously. For the energy requirements of buildings depending on the energy requirement for heating and cooling, this means a continuous increase in energy costs. In central and northern Europe, as well as North America and Canada, the heating energy requirement is higher compared to the southern regions, where the demand for cooling energy is correspondingly higher.

The energy (heating and cooling) costs in poorly insulated buildings are an important component of operating costs. Around 30-35 percent of primary energy consumption is accounted for by buildings throughout Europe. And there is an enormous savings potential on energy usage and greenhouse gases savings through higher efficiency through better insulated building envelope in Europe (DENA, 2019).

What can the strategic decisions look like when renovating a building?

Depending on the condition of the building and its interactions, where the property is located, different strategic decisions can be made.

In the building structure analysis, the individual building and plant components are compared with the period of use, the amount of energy consumption, the structural condition of the building envelope, the room layout and the size of the apartments and the standard of the apartments. All these aspects make up a picture of the building structure.

Market Potential

Decisive influence on the choice of modernization strategy also has the general conditions and development opportunities of a property. In addition to the assessment of attractiveness in terms of location, view, development, supply and so on continue to give building regulations information about use and expansion options. Figure 1 shows a matrix between market potential and the possible building structure.

Figure 1. Modernisation strategy renovation matrix
source: Own contribution

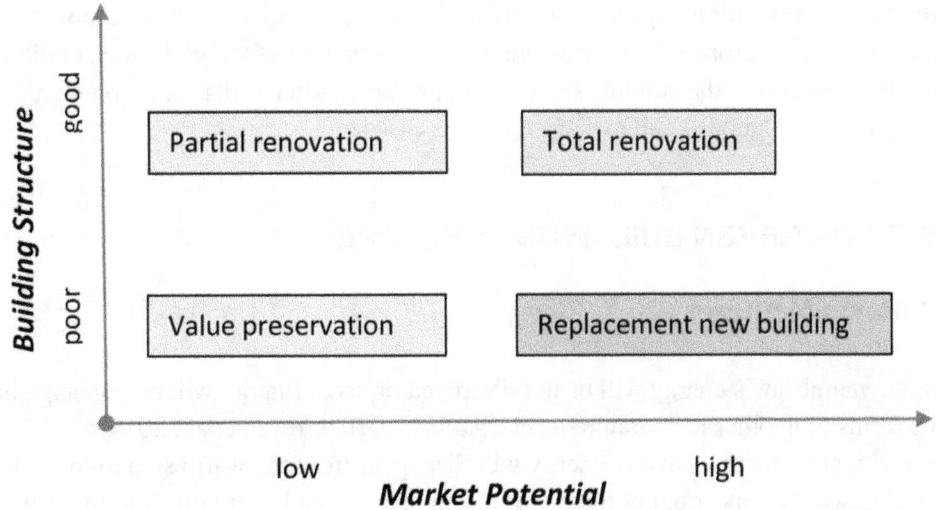

Such a matrix helps to name the building substance and market potential and thus determine the depth of intervention of the modernization strategy

Another important point is the rapidly growing emissions from the combustion of fossil energy sources. Good insulation of buildings reduces the amount of primary energy and thus less greenhouse gases are emitted (Oehler, 2018; Schmidt et al., 2017). The renovation potential is high (Wilson, 2013; Krieger, 2017). A large part of the properties consumes a multiple of energy that would actually be necessary. Responsible for this is the formerly common construction without thermal insulation.

It is advisable to invest in a renovation also in thermal insulation (Steiner, 2018). This will benefit you in two ways. The ancillary (additional) costs will be reduced massively, the living comfort increases and by today's state subsidies in many countries they will make a contribution to the investment costs. Many building owners are aware of improved thermal insulation. How exactly the punctual thermal insulation can be mounted on the building envelope and where the pitfalls are, they are less well known. Particularly important points in thermal insulation systems are the new physical conditions on the insulated building envelope. Here, due to temperature differences, moisture can form on the cover. The insulating material improves airtightness, which means less air circulation, which can lead to mold growth with moisture. These special aspects have to be professionally calculated and explained before any renovation work takes place. The following steps are important for energetic refurbishment.

The renovation work should not start indiscriminately. The individual steps have to be coordinated with each other. It takes a holistic view of all energetic renovation steps. These steps have to be optimally coordinated to achieve the desired goals.

It is important to carry out an expert consultation regarding the building envelope and its variety of insulation options. This energetic advice is recommended and should be carried out as the variety of building physics points. Comprehensive energy renovation is not very favourable and requires an accurate economic calculation with all risks and impossibilities.

It should be clarified early on which part of the investment costs you can pay from your own funds and which parts you have to finance through bank loans. Here, some banks offer special loans for renovation projects. As well the different governments give as well special subsidy for saving greenhouse gases. As part of the CO_2 savings, different energy subsidy programs for energy-efficient refurbishment are offered. In Germany, for example, refurbishments are subsidized by the KfW Bank. Here it is important to note when the funding must be submitted (before the start of construction). This funding will help the owner of the building doing more for energy savings and a higher quality standard of the renovated building.

This the chapter shows from a theoretical side how the economy of the various insulation measures for the individual elements of the building should be considered. Afterwards, an energetic refurbishment of the building will be calculated and explained on two case studies.

ECONOMIC FOR GREEN BUILDINGS ENVELOPES

Energy Balance Methods

To the thermodynamic law's energy will be not destroyed or used. Energy will only change in different energy types forms or in other temperature level. (Atkins, 2010; Greven et al., 2014)

The thermodynamic second law say energy will disappear from the warm side to the cold site. The heat in a building will disappear out of the building elements to the lower temperature level on the outside (Atkins, 2010). How fast this energy will be changed is dependent of the insulation types indirectly of the heat resistance of the different building materials (Gellert et al., 2007). To calculate these effects an energy balance method will be used for the calculation process (Piccardo et al., 2019b). However, energy can be converted to a lower level by changing the temperature of a surface. These surfaces, for example, of a building are called the building envelope. The energy balance of a building is determined by the energy that emerges from the envelope through the various materials and surfaces, and the ventilation losses caused by the leaks of a building shell. The difference that needs to be applied to keep this balance the same is provided by an energy generator (heating system). Figure 2 shows a building with building envelope and thermal envelope of a standard building.

The energy and cost-conscious planning of buildings is generally carried out with the aid of an energy balance method. Even at the design stage, the main energy flows and thus also the energy requirements of the building can be determined by calculation. Energy balance methods enable the optimization of the building under the given economic conditions.

The current calculation methodology is based on the primary energy requirements of the building. With the help of primary energy demand, greenhouse gases are calculated and evaluated. The amount of primary energy demand is usually calculated on the square meter heated surface. For the greenhouse gas emissions, it can be calculated similar in square meter.

This calculation methodology will inevitably have to change in the near future. The new guidelines for energy-efficient buildings (EPBD 2019) will mean that from 2020 only “nearly energy buildings” will be built. In order to better evaluate energy-related building structures in the future, energy components such as building elements manufacture, provision of energy in the construction phase, renovation activities during the lifecycle and finally the amount of energy needed to dispose of the building materials

Figure 2. Building envelope and energy losses through thermal and ventilation/leakages
(own contribution)

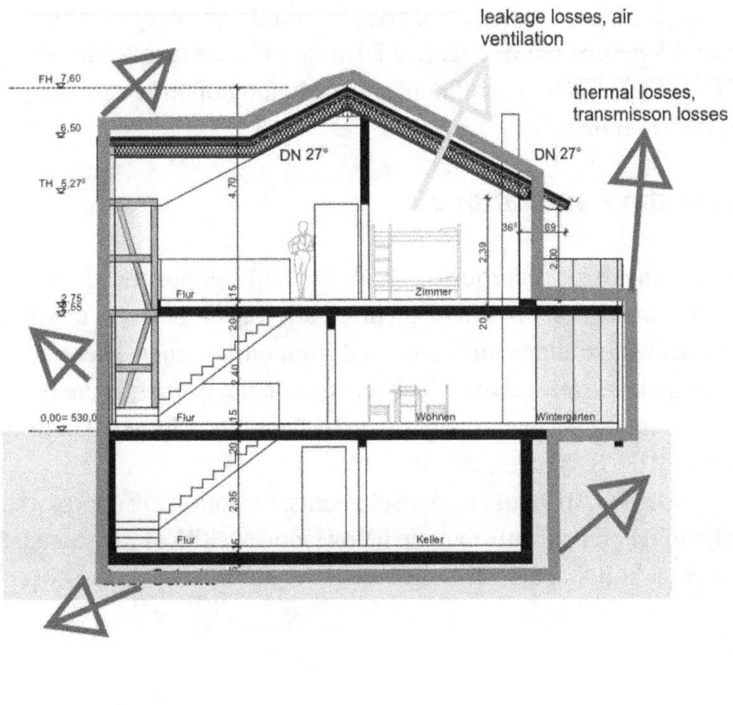

Figure 3. Energy balance method
(own contribution)

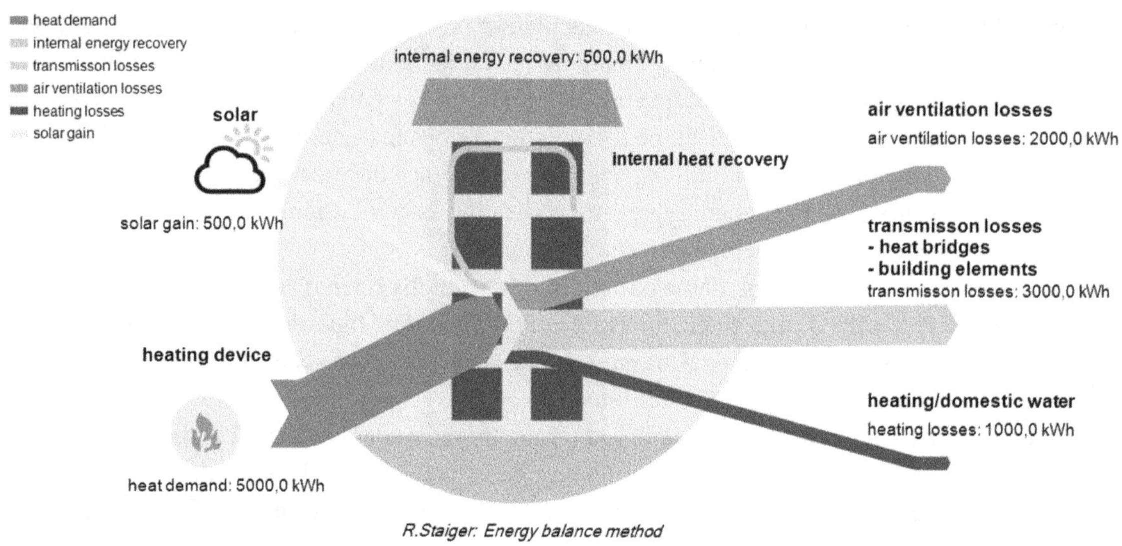

must be included in the calculation method. The primary energy requirement for heating and domestic water is playing an increasingly less role in energy considerations. Figure 4 shows the energy usage over time. This energy, not included in the calculation methods for the energy balance of buildings, is the so-called grey energy.

The complete live cycle of a building and the environmental aspect will be in the future a new method for a better balance calculation system. The grey energy will be a main energy part in the future (Wind, 2017). Figure 4 shows today's energy usage situation. This will change todays energy balance method calculation of the future valuation of buildings with grey energy.

Today's energy calculation and energy comparison is on the primary energy usage of the building. This amount will be in the future zero or negative because of more renewable energy sources in buildings. A new method must be found for calculating and evaluating of buildings and possible greenhouses taxes for the owner of the building.

Investment Cost of the Renovations

The renovation costs are considered according to the individual components of the building envelope. In the process, these refurbishment costs would be incurred anyway in the event of a refurbishment and the energy-related additional costs would be differentiated from one another. These additional costs are the so-called necessary energy investment costs which are used in the economic analysis of refurbishments. Table 1 show the possible investment costs depending of the refurbishment conditions in the event of a refurbishment (Discher, 2010; Burghardt, 2017).

From a DENA study (DENA 2019) for building elements, the following approximate investment costs were determined for building components that are below approx. 30% of the standard level in Germany. Table 2 shows the different building elements and the range of the investment cost.

Figure 4. Primary and grey energy usage today and tomorrow
(own contribution)

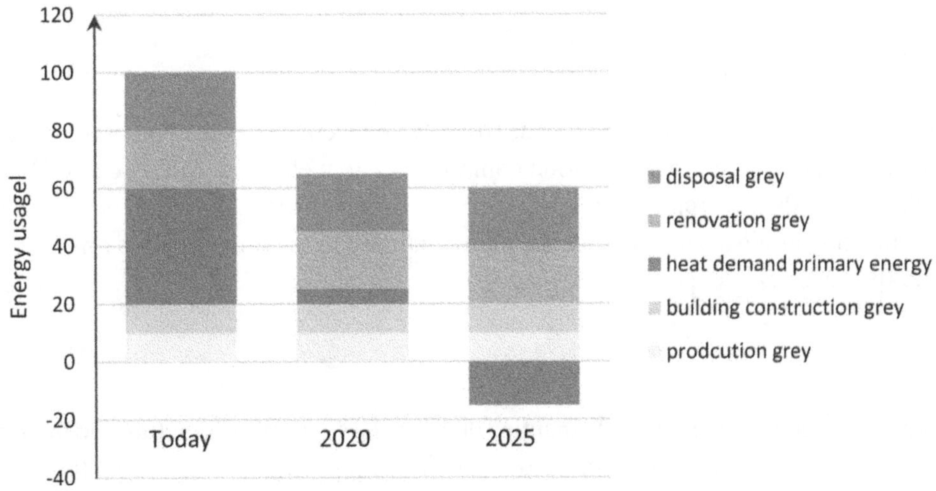

Table 1. Business as usual cost for renovations

Business as usual cost			
Building elements	refurbishment conditions		
	Not modernized	Low modernized	Modernized
	€/m²	€/m²	€/m²
facade	90	90	90
window	300	300	300
roof	145	145	145
cellar	30	30	30

Source: DENA 2018

Table 2. Business as usual cost for renovations

Building elements	Ca. investment cost
Outside wall with PS	95 -139 €/m²
Inside wall insulation	30 – 40 €/m²
Top Celling	25- 35 €/m² (soft), 35 – 50€/m² hard
Pitch roof	125 – 150 €/m²
Flat roof	70 – 100 €/m²
basement ceiling insulation	20 – 35€/m² bottom, 40 – 50€/m² top
exchange window glass	130 – 200 €/m²
Window exchange	250 – 450 €/m²

Source: Discher, 2010; DENA, 2019, 2018

Outside wall insulation prices depend on the current state of the house, including labor costs, plaster, facade color and scaffolding.

Outside Wall

The economic efficiency of the measure depends strongly on whether a renovation of the exterior plaster is necessary. By attaching a thermal insulation composite system eliminates these costs. Non-insulated walls from the 50s / 60s are prone to mold on interior corners, especially in kitchens, bedrooms and behind furniture and curtains. The costs for the pure repair are in the range of 25 € / m^2 up to 95 € / m^2 depending on the necessary scope.

Roof

An already developed attic can either be dismantled down to the supporting structure and then completely renewed. Or it can be renovated in the course of a roof covering from above. A doubling of the rafters is possible with construction wood or wooden profiles. The costs for the pure repair are in the range of 25 € / m^2 to 105 € / m^2 depending on the necessary scope.

Windows

3 glass windows cost around 350 – 400 €/m^2. For only renovation with minimum energy efficiency effort 250 €/m^2

Building Elements

There are a lot of different physical parameters which should be included in the calculation methodology for energetic considerations of building envelopes. This is the storage capacity of the building elements, moisture balances in the building components, use of passive solar energy sources and energy flows through the opaque and transparent components and many more.

To calculate the quality of a building element the simplest way to look to these is the U Value (Anderson 2002). The U value defines how much power of an energy converter (eg pellet boiler) is necessary to compensate for energy flow between higher temperature level(inside) and lower temperature level (outside). The U value is given in the physical unit W / K m^2. The U value is determined from the reciprocal of the thermal resistance of a component. The following mathematical relationships are defined. The total resistance of a component is the sum of all thermal resistances. The heat resistance depends on the material properties which is defined as the thermal conductivity and on the thickness of the material (Anderson 2002).

The Following Physical Relationships

U value is defined: $U = \frac{1}{R_{total}}$

Thermal Resistance of a building element is: $R_{total=}R_1 + R_2 + \ldots R_x = \sum_1^x R_x$

$$R_{total} = \frac{d_1}{\lambda_1} + \frac{d_2}{\lambda_2} + \ldots \frac{d_x}{\lambda_x}$$

The necessary power for a constant temperature is: $P = U * A * \Delta T$

$\Delta T = \text{temperature difference inside/outside}$

The amount of energy keeping the temperature is: $E = P * t$
The energy cost for the amount of energy is: $E_{cost} = E * Price_{kWh}$

Function of the Energy Cost

1. 1. Existing wall or building element (energy cost of the existing building element)

$$E_{existing} = U_{existing} * A * \Delta T * t * \text{Price}_{kWh}$$

$$E_{existing} = \left(\frac{1}{R_{existing}}\right) * A * \Delta T * t * \text{Price}_{kWh}$$

$$E_{existing} = \left(\frac{1}{\frac{d_{existing1}}{\lambda_{existing1}} + \frac{d_{existing2}}{\lambda_{existing}2} + \frac{d_{existing\ x}}{\lambda_{existing\ x}}}\right) * A * \Delta T * t * \text{Price}_{kWh}$$

The existing energy cost is a fix figure depending of the temperature difference the area of the wall, the heating time and the cost of energy.

New Wall or Building Element (Energy Cost of the New Building Element)

$$E_{new} = \left(\frac{1}{\frac{d_{insulation}}{\lambda_{insulation}} + R_{existing}}\right) * A * \Delta T * t * \text{Price}_{kWh}$$

Function type is:

Figure 5. Curve of energy demand and thickness normalized
(own contribution)

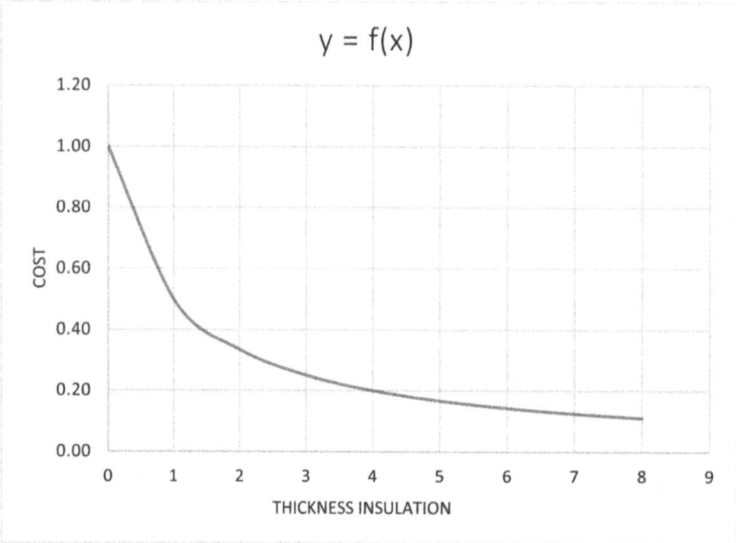

$$y\left(fx\right)=\left(\frac{1}{\frac{x}{a}+b}\right)*c$$

The general graph of the energy cost of the new wall is:

y = Energy Cost, x = thickness insulation materials, ***a*** is a variable depending of the type of insulation materials (thermal conductivity, *b* is a variable (thermal resistance) depending of the existing building element (thickness and thermal conductivity of the different layers, ***c*** is a variable depending of the building area, temperature difference, operation hour of the energy generator (boiler) and the price for a kWh

The derived curve shows the dependence between energy consumption and thickness of the insulating material. The hyperbolic course shows that the thicker the insulating material, the less energy can be saved. The economic efficiency is then no longer given.

Example:

- 1-unit insulation material needs 0,5 energy units
- 4-unit insulation materials need 0,2 energy units

The Energy Saving With the New Wall in Compare to the Old One:

$$E_{saving}=\left(\frac{1}{\frac{d_{existing}}{\lambda_{existing}}}\right)*A*\Delta T*t*\text{Price}_{kWh}-\left(\frac{1}{\frac{d_{insulation}}{\lambda_{insulation}}+\frac{d_{existing}}{\lambda_{existing}}}\right)*A*\Delta T*t*\text{Price}_{kWh}$$

Figure 6. Curve of an insulated wall with energy demand and thickness normalized
(own contribution)

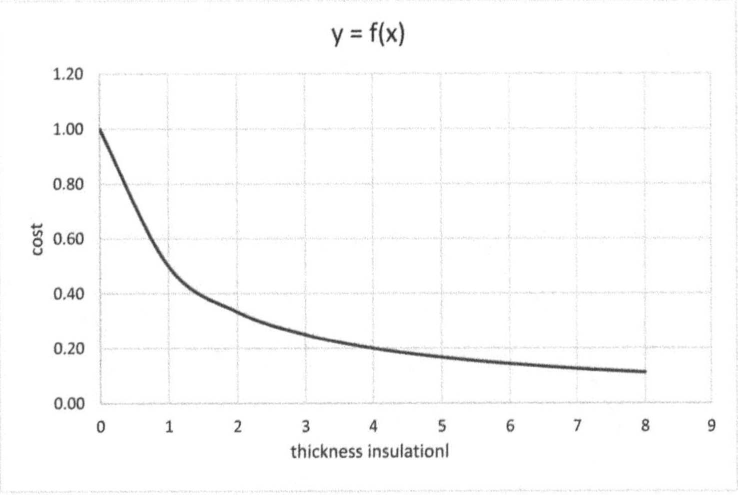

$$E_{saving} = A * \Delta T * t * \text{Price}_{kWh} \left(\left(\frac{1}{\frac{d_{existing}}{\lambda_{existing}}} \right) - \left(\frac{1}{\frac{d_{insulation}}{\lambda_{insulation}} + \frac{d_{existing}}{\lambda_{existing}}} \right) \right)$$

Function type is:

$$y\left(fx\right) = \left(\left(\frac{1}{b} \right) - \left(\frac{1}{\frac{x}{a} + b} \right) \right) * c$$

The graph of the function is:

y = Energy Cost, x = thickness of the insulation materials, *a* is a variable depending of the type of insulation materials (thermal conductivity, *b* is a variable (thermal resistance) depending of the existing building element (thickness and thermal conductivity of the different layers, ***c*** is a variable depending of the building area, temperature difference, operation hour of the energy generator (boiler) and the price for a kWh

Example 1

Outside wall with brick wall 30 cm, heat conductivity 0,25W/K m (U existing = 0,833 W/m^{2}K) and the new insulation material with a thermal conductivity of 0,035 W/ K m. The wall will be insulated with different thickness x with mineral wool 0,035 W/mK., A = 1 m^2, 15 K temperature difference and 1500 h/a.

Figure 7. Energy cost savings, example 1
(own contribution)

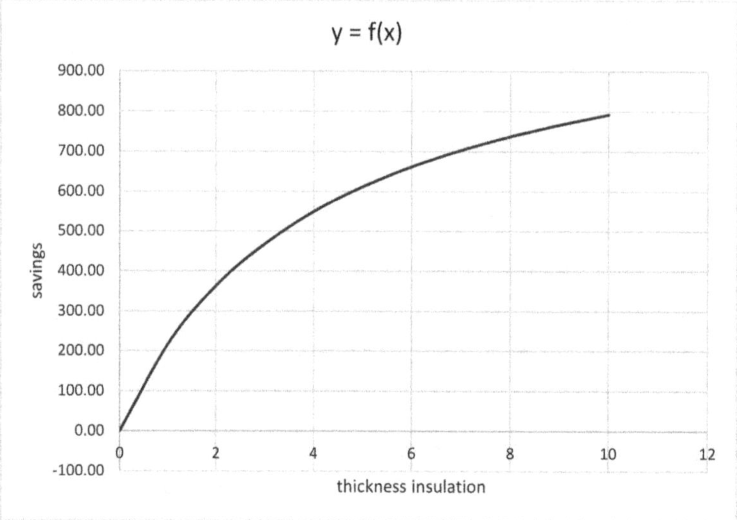

Figures in Formula

$$y\left(fx\right) = E_{savings} = \left(0,83 - \left(\frac{1}{\frac{x}{3,5} + 1,2}\right)\right) * 1.350$$

extreme point analysis:

No insulation means x = 0 → $E_{savings} = 0$
Insulation → ∞ $E_{savings} = 1.350€ / a$

Example 2

Outside wall with brick wall 30 cm, heat conductivity 0,25W/K m and the new insulation material with a thermal conductivity of 0,042 W/ K m. The wall will be insulated with x = 6 cm wood fiber insulation., A = 1 m^2, 15 K temperature difference and 1500 h/a.

Figure 8. Energy cost savings, example 2
(own contribution)

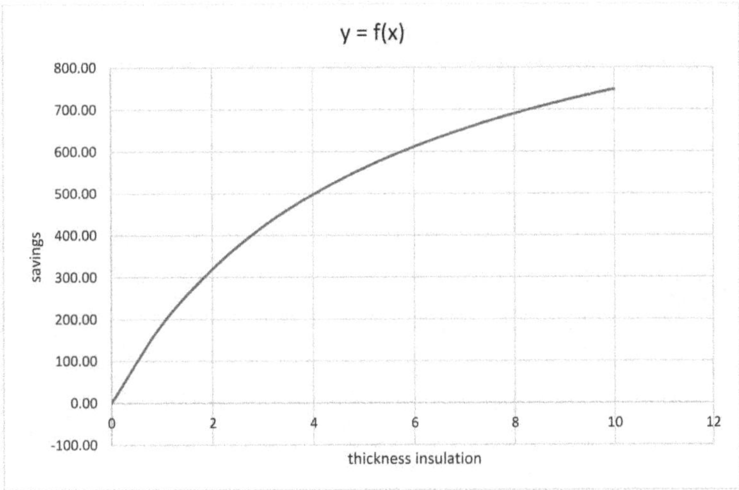

Figures in Formula

$$y\left(fx\right) = E_{savings} = \left(0,83333 - \left(\frac{1}{\frac{x}{4,2} + 1,2}\right)\right) * 1.350$$

Extreme point

No insulation means x = 0 → $E_{savings} = 0$

Insulation → ∞ $E_{savings} = 1.350€ / a$

Calculation of the Payback Time

Savings depending of the insulation materials and the dimensions. The calculation basis is the savings per m^2 for example insulation material. DENA calculation for a thermal insulation composite system is 25% for insulation, labor 50% and scaffolding 25% of the investment cost.

Calculation:

- *DENA price insulation 25% labor 50% scaffolding 25% of the cost for insulation*
- *Cost insulation complete -→ x * price/cm/0,25 through DENA*
- *Cost of different materials like Mineral wool, Wood fiber, Flax, Hemp etc.*

Figure 9. Curve payback time
Source: Own contribution

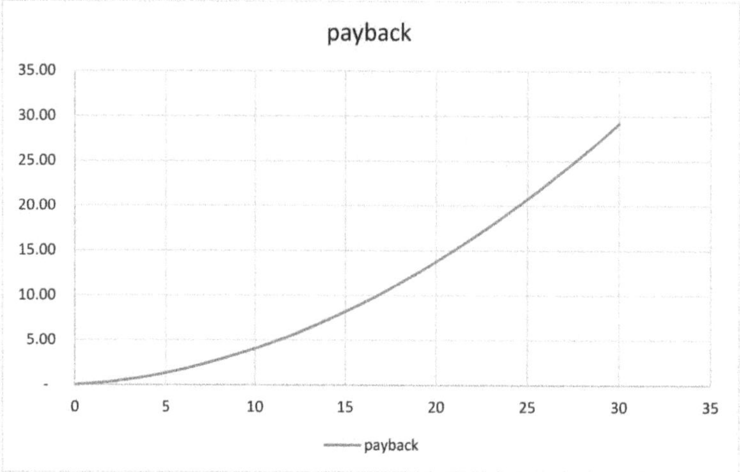

Payback time depending of the thickness and insulation Materials

$$Cost_{investment} = x * \text{price} / 0,25$$

$$Payback_{time} = \frac{\text{Cost}_{investment}}{E_{savings}}$$

First, the economics of external insulation depend on many parameters. These dependencies must be carefully considered and calculated during a refurbishment in order to disclose the pros and cons to the customer (Hassa, 2013; Holm, 2015; Canbolat, 2018).

Whether the insulation of the façade pays off, that is, whether the investment pays off with the years, largely depends on which assumptions and boundary conditions of the profitability calculation are used as a basis. The profitability calculation includes the "so-called" costs. These are costs that would have been incurred anyway for the maintenance of the facade in the near future, the projected life of the new insulation, the future development of fuel prices, the interest rate developments in the following years, the amount of government support and a realistic forecast for the saving of heating costs. All these factors are decisive, if the costs of the façade will pay off in the end.

It follows that those who refuse energetic insulation measures for the facade due to a non-existent economic efficiency, can substantiate their point easily by appropriate calculation examples. However, vice versa, those who are fundamentally positive about the topic of facade insulation also have it in their own hands to prove that they are economical. Moreover, do not forget that an energetic refurbishment increases the value of your property and that your living comfort is noticeably improved by the insulation.

Building Elements: Roofs, Floors, Windows and Outside Doors

The procedure and method for renovating and insulating a roof or a floor (to the outside boundary) is similar like on an outer wall. You calculate the new value and compare the energy savings with the investment costs over time with the existing roof. Here is also the question which parameters are included in the calculations. For example, the necessary maintenance costs, public subsidies, energy prices, interest rate, actual investment etc. The complex interrelationships, especially the actual amounts of energy that are dissipated through the individual components of the building envelope, over a year at a specific location are calculated and simulated using special software programs. Here the energy balance method is used as already mentioned. In the following text, two calculated case studies with results are shown

CASE STUDY 1: RESIDENTIAL BUILDING

Economic and technical consequences for green buildings envelopes.

Table 3 shows the existing building from 1960 with their specific building elements, technical building equipment and the dimensions of the building.

Table 4 shows the renovated building from 2019 with their specific building elements, technical building equipment and the dimensions of the building.

Economic efficiency calculation annuity method

The table shows the economic parameter for the calculation for case study 1.

The different economic results of the calculation is showing in table 6.

The next table 7 shows the different building elements (envelopes) of roof (Dach), Outside wall (AW), ground floor (BO) and windows (F). The whole area of the different building elements is calculated from the 3 D model of the building. Investment cost, live time, replacement cost and capital cost is calculated for each element. The table shows a part of the calculation.

Payback Period

See Figure 11.

Comparison Old and New

With the change of the building elements like roof, windows, and outside wall Ffigure 12 shows more than 37% can be reduced from the primary energy side. The transmission losses will be reduced more than 65%. The CO2 emission will be reduced around 37% and the energy cost for heating the building is 35% less.

Figure 13 shows the red heat demand curve. There are different energy losses and energy gains in a building. In our case study there are ventilator and transmissions heat losses. In addition, heat is released into the building through internal energy sources like humans, electrical appliances. At the same time, energy is generated through solar heat gains through transparent building elements. As a result of the energy calculations, an energy balance can be compiled in the form of a sankey diagram. This shows the primary energy requirement up to the heat losses and heat gains. Figure 14 shows the actual energy situation of a 1960 building and figure 15, the renovated building from 2019.

Figure 10. Picture of the residential building, case study 1
(own contribution)

CASE STUDY 2 NONE RESIDENTIAL BUILDING

Case study 2 is an old town hall from the 1900s. this town hall was changed in 1960 in terms of heating technology and windows. Replacement measures must now be taken for the windows and the roof. Furthermore, the facade must be renewed. The payback time and energy savings should be calculated.

Table 3. Building data: existing building 1960

Physical properties	units
Size	11 m x 9 m
Year	1960
Wall surface	207 m^2
Roof	118 m^2
Windows	31 m^2
Floor	99 m^2
Heating	gas
Building elements	U Value
Outside wall (massive)	1,4
Roof	1,4
Windows	5,0
Floor ground	1,2
Door	4,0

Source: Own contribution

Table 4. Building data renovated building 2019

Physical properties	
Size	11 m x 9 m
Year	1960
Wall surface	207 m²
Roof	118 m²
Windows	31 m²
floor	99 m²
Heating	gas
Envelope building	
Building elements	U Value
Outside wall (massive)	0,24
Roof	0,19
Windows	0,9
Floor ground	1,2
door	4,0

Source: own contribution

Existing Building 1900

Table 8 shows the existing building from 1960 with their specific building elements, technical building equipment and the dimensions of the building.

Table 5. Calculation economic values

Financial descriptions	Parameters	
Period	30,0	years
discount rate	5,50	%
Investment tax rate	32,00	%
Inflation rate investment or remedial measures	3,50	%
Inflation rate maintenance costs	4,50	%
Inflation rate for fuel in the current state	4,00	%
Inflation rate for fuel in the rehabilitated state	4,00	%
current annual fuel costs in the current state	4.795,29	€/a
current annual fuel costs in the renovated state	3.095,09	€/a

Source: own contribution

Table 6. Economic results case study 1

Taking into account the specified energy saving measures, the following annual costs are incurred for the period of observation of 30.0 years:		
Annual cost of capital Maintenance cost Reduced fuel cost	: : :	1.509 € / a 0 € / a 5.156 € / a
Total cost	:	6.665 € / a
Fuel cost without energy saving measure	:	7.989 € / a
Average savings	:	1.324 € / a
Additional cost for energy saving measures	:	21.931 €
total Investment cost	:	56.372 €

Source: own contribution

Renovation Building 2019

Table 9 shows the renovated building from 2019 with their specific building elements, technical building

Table 7. Economic data from the different building elements

Name of the measure	Invest. Cost	Live Time	Repl. Cost	Energy Cost	factors				Capital cost
	€	Year	€	€					€/a
roof 001-8 N: insulation 18 cm WLS 040	4.869	30,0	2.453	2.416	0,069	1,00	1,67	1,79	166
roof 001-9 S: insulation 18 cm WLS 040	4.869	30,0	2.453	2.416	0,069	1,00	1,67	1,79	166
wall 007 N: WDVS 14cm, auf Porenstein-Mauerwerk 1945-1972	841	30,0	512	329	0,069	1,00	1,67	1,79	23
Wall 008 W: WDVS 14cm, auf Porenstein-Mauerwerk 1945-1972	2.272	30,0	1.382	890	0,069	1,00	1,67	1,79	61
wall 005 S: WDVS 14cm, auf Porenstein-Mauerwerk 1945-1972	841	30,0	512	329	0,069	1,00	1,67	1,79	23
Window 004 O: 3-Scheiben-Wärmeschutzverglasung 3/0,8/0,8	1.160	30,0	870	290	0,069	1,00	1,67	1,79	20
Window O: 3-Scheiben-Wärmeschutzverglasung 3/0,8/0,8	1.160	30,0	870	290	0,069	1,00	1,67	1,79	20
Floor EG-1: WDVS 14cm, auf Porenstein-Mauerwerk 1945-1972	1.216	30,0	739	476	0,069	1,00	1,67	1,79	33
window 006 N: 3-Scheiben-Wärmeschutzverglasung 3/0,8/0,8	1.015	30,0	761	254	0,069	1,00	1,67	1,79	17
window 010 W: 3-Scheiben-Wärmeschutzverglasung 3/0,8/0,8	1.160	30,0	870	290	0,069	1,00	1,67	1,79	20
window 011 W: 3-Scheiben-Wärmeschutzverglasung 3/0,8/0,8	1.505	30,0	1.129	376	0,069	1,00	1,67	1,79	26

Source: Own contribution out of the simulation

Figure 11. Payback period, case study 1
(source own contribution)

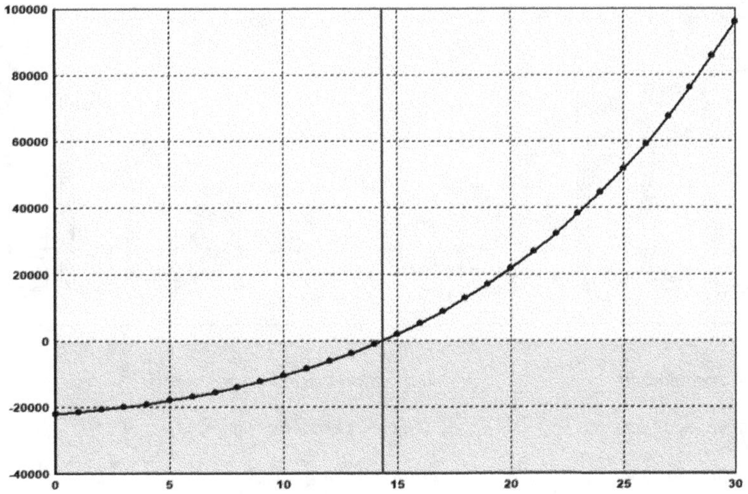

equipment and the dimensions of the building.

Economic Efficiency Calculation Annuity Method

See Tables 10 and 11 and Figure 17.

Figure 12. Comparison case study 1
(own contribution)

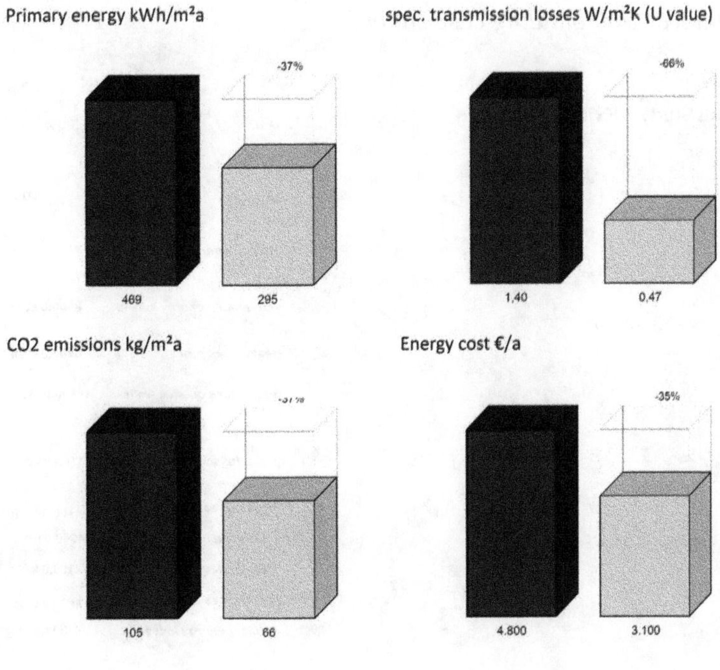

Figure 13. Energy usages and heat demand curve
(own contribution)

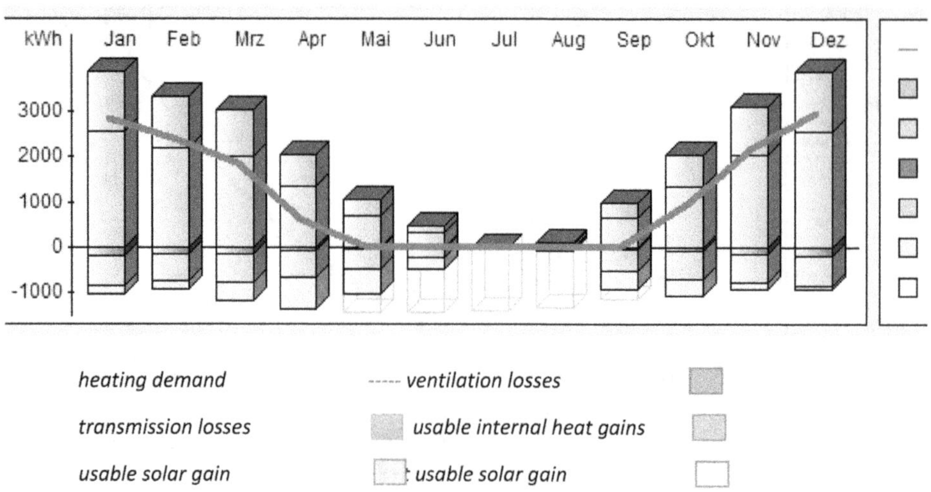

Payback Time

The next Table 11 shows all the different building elements (envelopes) of roof (Dach), Outside wall (AW) and windows (F). The whole area of the different building elements is calculated from the 3 D model of the building. Investment cost, live time, replacement cost and capital cost is calculated for each element. The table shows a small part of the calculation.

Figure 14. Energy balance 1960 building case study 1
(own contribution)

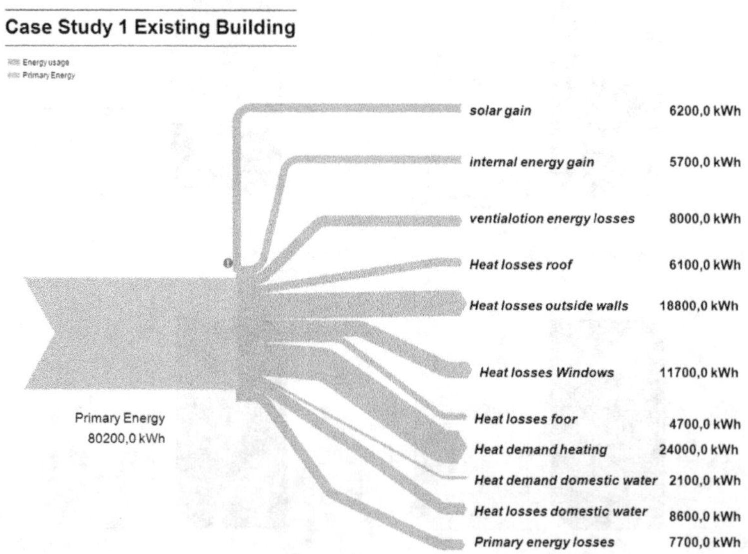

Figure 15. Energy balance 2019 building case study 1
(own contribution

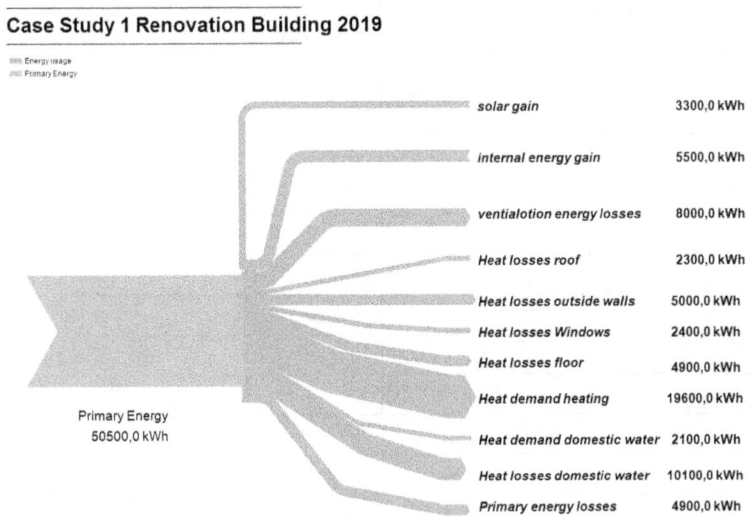

Annual Cost of Capital Case Study 2

See Tables 12 and 13.

Figure 16. Picture of the non-residential building, case study 2
Source: own contribution

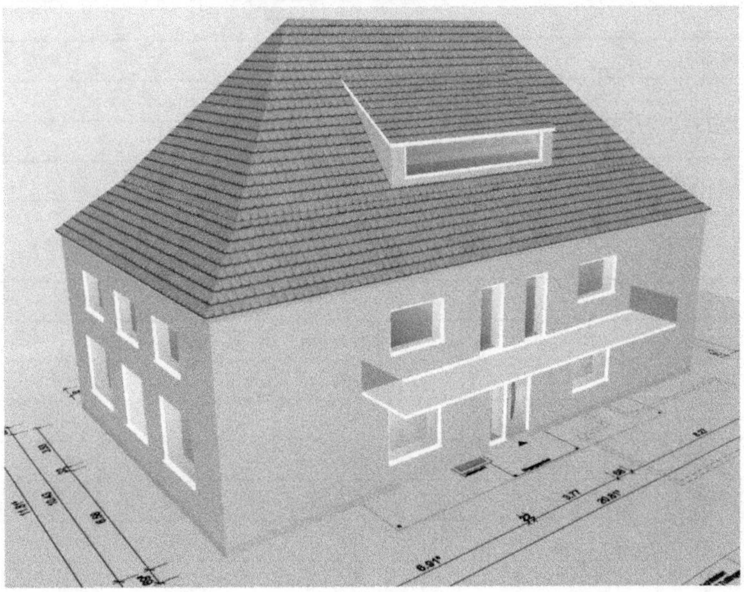

Table 8. Building data from 1960

Physical properties	units
Size	21 m x12 m
Year	1900
Wall surface	400 m^2
Roof	380 m^2
Windows	80 m^2
floor	250 m^2
Heating	gasoil
Building elements	U Value
Outside wall (massive)	1
Roof	2,8
Windows	4
Floor ground	1,2
door	4,0

Source: own contribution

ECOLOGICAL AND ENVIRONMENTAL CONSIDERATIONS FOR GREEN BUILDINGS ENVELOPES

A life cycle assessment (LCA) is a systematic analysis of the environmental impact of products throughout

Table 9. Building elements and new specification

Physical properties	units
Size	21 m x12 m
Year	1900
Wall surface	400 m^2
Roof	380 m^2
Windows	80 m^2
floor	250 m^2
Heating	gasoil
Envelope building	
Building elements	U Value
Outside wall (massive)	0,24
Roof	0,19
Windows	0,9
Floor ground	1,2
door	4,0

Source: own contribution

Table 10. Calculation economic values

Financial descriptions	Parameters	
period	30,0	year
discount rate	5,50	%
Investment tax rate	32,00	%
Inflation rate investment or remedial measures	3,50	%
Inflation rate maintenance costs	4,50	%
Inflation rate for fuel in the current state	4,00	%
Inflation rate for fuel in the rehabilitated state	4,00	%
current annual fuel costs in the current state	11.278,81	€/a
current annual fuel costs in the renovated state	8.292,57	€/a

Source: own contribution

their life cycle (Hunt, 2003; Steiner et al., 2018). The life cycle of a product includes both the production and the use phase as well as the disposal and the associated upstream and downstream processes, e.g. B. the production of raw materials and supplies (Klöpffer, 2009).

Every product, every building material, every form of energy more or less pollutes the environment through production, use and disposal. Life cycle assessments can be set up to give meaningful results on environmental impacts. The life cycle is divided into four main steps in the building sector, which include further cycle steps. These steps are production, usage, renovation and disposal.

The first step is the provision and production of materials and components for the building. Here, the basic material, locally availability, the primary energy consumption, the production process and the operation play an important role.

In the second step, the use of different materials is of importance. How are the materials used in buildings in what quantity, frequency and transported? How often do components and materials have to be overhauled and replaced? How much primary energy is necessary for this?

In the third step, the components and materials are *renovated*, disassembled and replaced. Primary energy and materials are additionally required in this regeneration process.

Table 11. Calculation results

Taking into account the specified energy saving measures, the following annual costs are incurred for the period of observation of 30.0 years:		
Annual cost of capital Maintenance cost Reduced fuel cost	: : :	3.438 € / a 0 € / a 13.816 € / a
Total cost	:	17.254 € / a
Fuel cost without energy saving measure	:	18.791 € / a
Average savings	:	1.537 € / a
Additional cost for energy saving measures	:	49.970 €
total Investment cost	:	100.005 €

Source: own contribution

Figure 17. Renovation comparison 1960 and 2019
(own contribution)

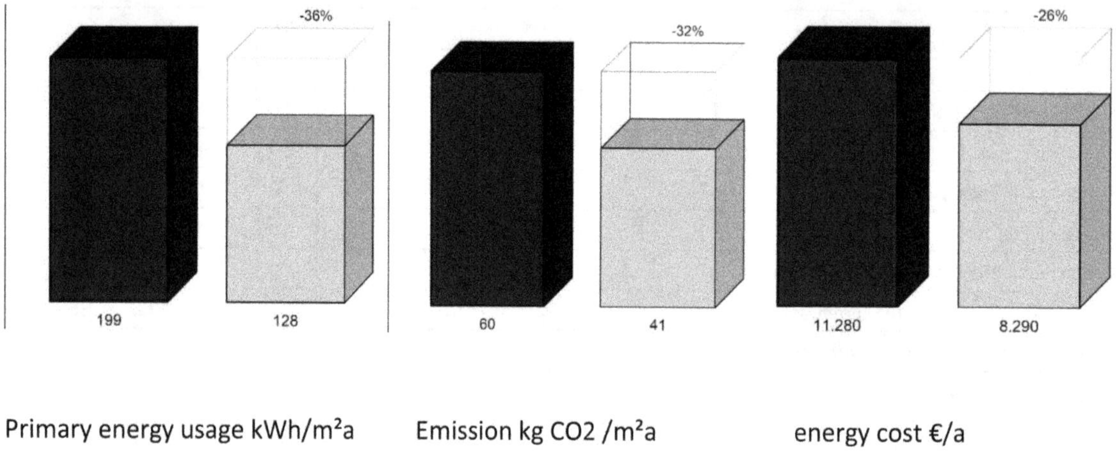

At the end of the life cycle, the various materials and components are re-used or *disposed* of. Either they are transported to waste disposal sites and stored, recycled or incinerated. All these processes require energy for the individual processes.

When a product is evaluated, it looks at the environmental impact, the resource consumption, the type of energy and the amount of energy. Depending on the LCA system, the health effects on people, flora and fauna are also examined. These calculations are very complex and difficult and mostly done by LCA specialists. Life cycle assessments can also be made for substances such as packaging, food, clothes, etc. (Piccardo et al., 2019a).

Figure 18. Payback time, case study 2
(own contribution)

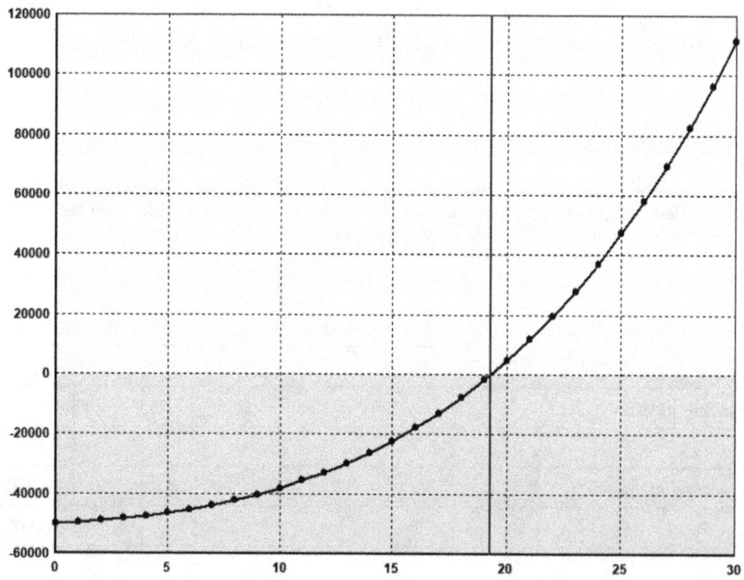

Table 12. Annual cost of capital to the different example building elements

Name of the measure	Invest. Cost	Live Time	Repl. Cost	Energy Cost	factors					Capital cost
	€	Year	€	€						€/a
roof 008-5 N: Zwischen- und Aufsparrendämmung 20cm, Sparrendach 1945-1960	7.471	30,0	4.298	3.174	0,069	1,00	1,67	1,79	218	218
roof 005-2 W: Zwischen- und Aufsparrendämmung 20cm, Sparrendach 1945-1960	2.076	30,0	1.194	882	0,069	1,00	1,67	1,79	61	61
wall 007 W: Wärmedämmverbundsystem, 16cm	685	30,0	155	530	0,069	1,00	1,67	1,79	36	36
window014 W	945	30,0	661	284	0,069	1,00	1,67	1,79	20	20
window 010 S	860	30,0	602	258	0,069	1,00	1,67	1,79	18	18
window 012 S	1.080	30,0	756	324	0,069	1,00	1,67	1,79	22	22
wall 004 N: Wärmedämmverbundsystem, 16cm	0	30,0	0	0	0,069	1,00	1,67	1,79	0	0

Source: own contribution

The life cycle assessment is specified in more detail under ISO 14040. The standard specifies different methods for creating the LCA.

Following Example Shows a Rating in CH.

The complex LCA data of various products has been presented in graphics. The simplified statements help to better understand and compare the results. In each case, the three LCA systems are presented, which are considered the most popular in Switzerland (Westermann et al., 2013; BAFU 2013; KBOB CH 2016). They are explained below:

This LCA complies with the criteria of ecological scarcity and critical loads and is based on the Swiss environmental policy. The overall assessment of the UBP 2016 shows a complete picture of the environmental impacts and calculates the amounts of environmental pollution in three partial evaluations:

- the use of energy resources, land and fresh water
- emissions into air, water and soil
- and waste disposal

The total primary energy is a subset of the UBP and calculates, in addition to the amount of non-renewable primary energy, the energy consumption of renewable energy sources. Renewable energy sources include hydropower, wood / biomass (without clear cutting of primary forests), solar, wind, geothermal and ambient energy. This parameter is used to assess the amount of energy supplied to the building (final energy) in accordance with leaflet SIA 2031 "Energy Performance Certificate for Buildings".

Non-renewable (red) The non-renewable primary energy calculates the amount of energy consumed by fossil and nuclear energy sources, as well as wood from clearcutting of primary forests. This definition differs from the FOEN Series Environmental 307 (Ecological Assessment with the help of Grey Energy). With this parameter, the reference value is evaluated according to leaflet SIA 2032 "Grey energy of buildings" *(Gibbeco, 2019)*.

Table 13. Tabular profitability calculation

Year	Energy cost present	Energy cost renovated	Savings	Interest	difference
	11.305,19	8.292,57			-49.969,75
1	11.757,40	8.624,27	3.133,12	2.748,34	-49.584,96
2	12.227,69	8.969,24	3.258,45	2.727,17	-49.053,68
3	12.716,80	9.328,01	3.388,79	2.697,95	-48.362,84
4	13.225,47	9.701,13	3.524,34	2.659,96	-47.498,46
5	13.754,49	10.089,18	3.665,31	2.612,42	-46.445,56
6	14.304,67	10.492,75	3.811,93	2.554,51	-45.188,15
7	14.876,86	10.912,46	3.964,40	2.485,35	-43.709,09
8	15.471,93	11.348,95	4.122,98	2.404,00	-41.990,11
9	16.090,81	11.802,91	4.287,90	2.309,46	-40.011,67
10	16.734,44	12.275,03	4.459,41	2.200,64	-37.752,90
11	17.403,82	12.766,03	4.637,79	2.076,41	-35.191,52
12	18.099,97	13.276,67	4.823,30	1.935,53	-32.303,75
13	18.823,97	13.807,74	5.016,23	1.776,71	-29.064,22
14	19.576,93	14.360,05	5.216,88	1.598,53	-25.445,87
15	20.360,01	14.934,45	5.425,56	1.399,52	-21.419,84
16	21.174,41	15.531,83	5.642,58	1.178,09	-16.955,35
17	22.021,39	16.153,10	5.868,28	932,54	-12.019,61
18	22.902,24	16.799,23	6.103,02	661,08	-6.577,67
19	23.818,33	17.471,19	6.347,14	361,77	-592,31
20	24.771,06	18.170,04	6.601,02	32,58	5.976,14
21	25.761,91	18.896,84	6.865,06	-328,69	13.169,89
22	26.792,38	19.652,72	7.139,66	-724,34	21.033,90
23	27.864,08	20.438,83	7.425,25	-1.156,86	29.616,01
24	28.978,64	21.256,38	7.722,26	-1.628,88	38.967,15
25	30.137,79	22.106,63	8.031,15	-2.143,19	49.141,50
26	31.343,30	22.990,90	8.352,40	-2.702,78	60.196,68
27	32.597,03	23.910,54	8.686,49	-3.310,82	72.193,99
28	33.900,91	24.866,96	9.033,95	-3.970,67	85.198,61
29	35.256,95	25.861,64	9.395,31	-4.685,92	99.279,85
30	36.667,23	26.896,10	9.771,12	-5.460,39	114.511,37

Investment: 49.969,75
End value = 114.511 €; present value = 22.976 €; annuity = 1.580 €
Source: own contribution out of the simulation

The greenhouse gas emissions are a subset of the UBP and calculate the amount of all effects of different greenhouse gases, based on the lead substance CO_2. They are a parameter for global warming

Figure 19. Environmental impact points
(Source: Gibbeco, 2019)

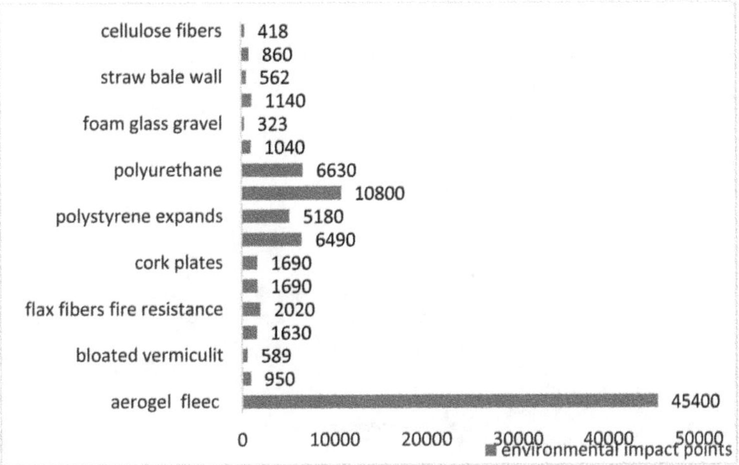

and not equivalent to the location-based CO_2 emissions, which is the subject of target agreements with the federal government in the context of the CO_2 Act.

This parameter is used to assess the amount of energy supplied to the building in accordance with leaflet SIA 2031 "Energy Performance Certificate for Buildings" as well as the reference value according to leaflet SIA 2032 "Grey Energy of Buildings".

The Eco indicator is an international system and considers the following effects: global warming potential, carcinogenicity, ozone depletion potential, winter smog, acidification potential, summer smog, eutrophication potential, pesticides, heavy metals.

Example: Outside Timber Wall With Cellulose Insulation

See Tables 14 and 15.

Result: The UBP environmental impact points are for the whole wall 559 points.

Brick Wall With EPS

See Tables 16.

For a brick wall with EPS the UBP is 2374

Limestone Wall Mineral Foam / Porous Concrete

See Table 17.

There are various methods for presenting the LCA. For example, ISO 14040 describes the product-specific life cycle assessment. Depending on the reference sizes, there are different methods (material flow balance, energy balance)

Figure 20. Primary energy in kWh/kg insulation material
(Source: Gibbeco, 2019)
PEI primary energy data (second graph, orange) renewable (green) and non-renewable (red)

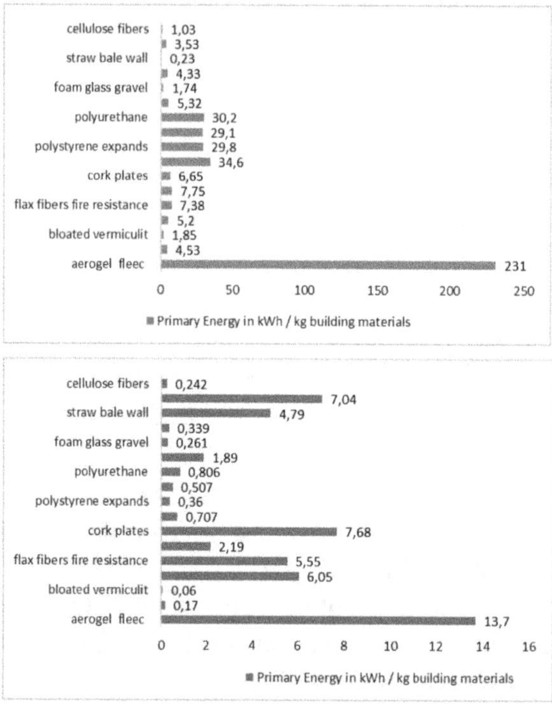

In our example, we have presented a possible method for determining the eco-balance. most of these principles are used in a similar way in all procedures. The question of these methods are the evaluation

Figure 21. Greenhouse gas in kg CO_2 per kg insulation material
source: (Gibbeco, 2019)
Greenhouse gas potential CO2, soil acidification SO2 etc. (third graph, blue)

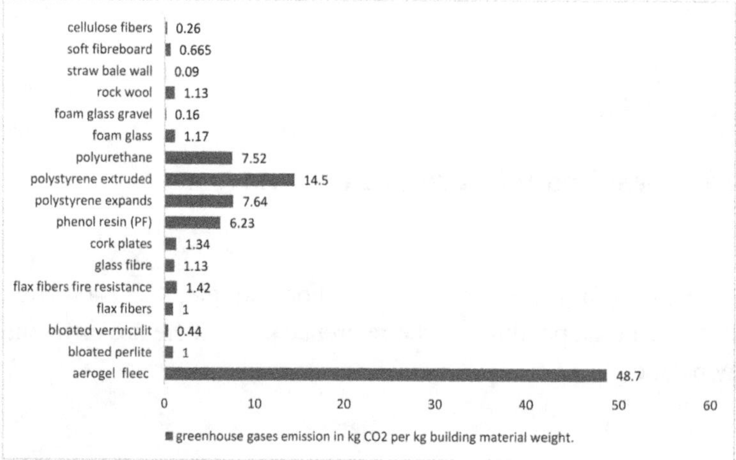

Table 14. Timber wall

Construction:		**Outside wall timber, cellulose fibres wood surface**						
Component ECG:		E4.214.200						
Mat. Nr	Function	Materials	Unit Mass	Density Kg/m^3	Unit kg	Thickness m	**Qty per element kg**	**Live time /a**
7,010	Wooden facade Spruce	Solid wood air-dried, planed	Mass	465		0,020	9,30	40
7,009	Spaulders fir raw	Solid wood air-dried, raw	Mass	485		0,040	2,00	40
7,013	Reinforcement	OSB	Mass	605		0,012	7,26	80
7,009	wood stand	Solid wood air-dried, raw	Mass	485	parts	0,200	12,00	80
10,010	thermal insulation	Cellulose, injected	Mass	60	20cm		11,00	80
9,006	wind seal	Kraft paper	Mass	650		0,000	0,13	80
7,009	Spaulders fir raw	Solid wood air-dried, raw	Mass	485	parts	0,030	2,03	80
3,007	plaster base	Plasterboard	Mass	1200		0,015	18,00	40
4,001	interior plaster	Gypsum / white plaster	Mass	1100		0,005	5,50	40
			Per m2			**0,322**	**67,22**	

Source: (KBOB CH 2016)

criteria.

DISCUSSIONS AND CONCLUSION

The main points in the energetic renovation of building envelopes is the question of meaningfulness and answering the question makes economic sense.

In order to answer the question of economy, there is always the question of which criteria will be used to compare and calculate the economic reality.

Here are two questions to answer. a) if the investment is only an actual energy measure or b) it is inevitable, for reasons of age, for example, to repaint the walls or cover the roof or replace the old windows anyway. In most cases, an energy-efficient investment involves an additional replacement measure. This means that there are one area replacement costs and one area energy-related renovation costs for the energy-efficient refurbishment.

The examples show that a refurbishment can be advantageous. Here, in particular, the individual investment costs for the building envelope are of importance.

If replacement measures are carried out anyway, the additional energy-related costs are usually only marginal. An amortization calculation using dynamic economic calculations such as capital value or annuity method shows a picture over a lifetime of more than 30 years. As a rule, the energy measures together with the replacement measures are economically positive.

Table 15. Timber wall with cellulose insulation

Environmental stress points UBP		Primary energy content PEI in kWh				greenhouse gas emissions CO_2	
		renewable		Not renewable			
UBP/unit	**UBP/a**	**kWh**	**kWh/a**	**kWh**	**kWh/a**	**kg CO2**	**kg CO2/a**
424	98,58	**5,780**	1,34	**0,690**	0,16	**0,1250**	0,0291
347	17,35	**5,510**	0,28	**0,504**	0,03	**0,1010**	0,0051
883	80,13	**8,240**	0,75	**2,760**	0,25	**0,6140**	0,0557
347	52,05	**5,510**	0,83	**0,504**	0,08	**0,1010**	0,0152
418	57,48	**0,242**	0,03	**1,030**	0,14	**0,2570**	0,0353
3040	4,94	**14,500**	0,02	**8,050**	0,01	**1,6900**	0,0027
347	8,81	**5,510**	0,14	**0,504**	0,01	**0,1010**	0,0026
488	219,60	**0,086**	0,04	**2,460**	1,11	**0,5370**	0,2417
146	20,08	**0,062**	0,01	**0,692**	0,10	**0,1470**	0,0202
	559		**3,44**		**1,88**		**0,4075**
	UBP		**kWh**		**kWh**		**kg CO2**

Source: (KBOB CH 2016)

Of course, as shown in the theoretical calculations previously, an increase in the insulating strength makes only limited economic sense. In this case, technical and economic conditions for the customer must first be calculated and presented when the concept is drawn up. With this information, the owner of the building who wants to carry out renovations can make clear and unambiguous decisions.

Around 30-35% percent of primary energy consumption is accounted for by buildings throughout Europe. There is a huge energy saving potential in today's existing building stock. Of course, not all aus-

Table 16. Brick wall with EPS Insulation

Environmental stress points UBP		Primary energy content PEI in kWh				greenhouse gas emissions CO2	
		renewable		Not renewable			
UBP/unit	**UBP/a**	**kWh**	**kWh/a**	**kWh**	**kWh/a**	**kg CO2**	**kg CO2/a**
1230	369,00	**0,227**	0,07	**5,250**	1,58	**0,9510**	0,2853
3430	20,58	**0,433**	0,00	**22,700**	0,14	**5,4300**	0,0326
1000	334,00	**0,146**	0,05	**4,200**	1,40	**0,7580**	0,2532
5180	828,80	**0,360**	0,06	**29,800**	4,77	**7,6400**	1,2224
1000	334,00	**1,460**	0,49	**4,200**	1,40	**0,7580**	0,2532
217	427,22	**0,078**	0,15	**0,791**	1,56	**0,2580**	0,5079
146	60,23	**0,062**	0,03	**0,692**	0,29	**0,1470**	0,0606
	2374		**0,84**		**11,13**		**2,6152**
	UBP		**kWh**		**kWh**		**kg CO_2**

Source: (KBOB CH 2016)

Table 17. Limestone wall with mineral insulation

Environmental stress points UBP		Primary energy content PEI in kWh				greenhouse gas emissions CO2	
		renewable		Not renewable			
UBP/unit	**UBP/a**	**kWh**	**kWh/a**	**kWh**	**kWh/a**	**kg CO2**	**kg CO2/a**
255	63,75	**0,122**	0,03	**0,728**	0,18	**0,2690**	0,0673
7730	38,65	**1,010**	0,01	**39,800**	0,20	**9,6000**	0,0480
348	133,40	**0,065**	0,02	**0,925**	0,35	**1,1700**	0,4485
427	49,82	**0,166**	0,02	**1,600**	0,19	**0,4060**	0,0474
157	480,81	**0,037**	0,11	**0,398**	1,22	**0,4170**	1,2771
146	60,23	**0,062**	0,03	**0,692**	0,29	**0,1470**	0,0606
	827		**0,22**		**2,43**		**1,9488**
	UBP		**kWh**		**kWh**		**kg CO2**

Source: (KBOB CH 2016)

terity measures are the same: While very good savings can be achieved with an insulation of the top floor ceiling and the basement ceiling, complex façade insulation often pays off over a longer period of time.

It is generally important that only the appropriate insulation measures for a building are carried out, that the insulation is installed properly and that the heating technology is adapted accordingly after the measures have been taken.

Ecological and environmental considerations will be a major part of renovating buildings in the future. Due to the ever-lower energy costs for heating and domestic hot water, other energy components in a building are moved to the foreground. This energy component, which is defined as grey energy, will play an important role in the energy balance of a building in the future. Depending on the materials used for the building envelope, life cycle considerations are made. Depending on the life cycle model, all considerations are made from the production to the recycling of building components. The ideal solution would be a 100% recycling of the building materials used. This cradle and cradle principle should already be taken into consideration in the construction of the building and the manufacture of building materials. Conclusions can be drawn on ecological and economic issues.

REFERENCES

Anderson, B. (2002). Conventions for U-value calculations. Watford: CRC Construction Research Communications Ltd.

Atkins, P. W. (2010). The laws of thermodynamics. A very short introduction. Oxford, New York: Oxford University Press.

BAFU. (2013). Ökofaktoren Schweiz 2013 gemäss der Methode der ökologischen Knappheit.

Burghardt, J., & Fox, A. (2017). *Wirtschaftlichkeitsberechnungen im Bereich der Gebäudedämmung.* Ilmenau.

Canbolat. (2018). Determination of Proper Insulation Thickness for Building Walls Regarding Economic Consideration 2018.

DENA. (2018). dena CONCISE 2018 BUILDING REPORT. Energy efficiency in the building stock – statistics and analyses.

DENA. (2019). Gebäude-Energiewende: Building Energy Transformation. Retrieved from https://www.geea.info/fileadmin/Downloads/Pressemitteilungen/2019/geea_PB_2019_final.pdf

Discher, H., & Hinz, E. (2010). dena-Sanierungsstudie_Teil_1. Retrieved from https://shop.dena.de/fileadmin/denashop/media/Downloads_Dateien/bau/9122_dena-Sanierungsstudie_Teil_1.pdf

Gellert, R., Pfundstein, M., Rodolphi, A., & Spitzner, M. H. (2007). *Dämmstoffe. Grundlagen, Materialien, Anwendungen*. München: DETAIL.

Gibbeco. (2019). Ökobilanz - Wärmedämmstoffe - gesundes Haus. Retrieved from https://www.gesundes-haus.ch/dachdeckungen-flachdach/oekobilanz-waermedaemmstoffe.516.html

Greven, A., Keller, G., & Warnecke, G. (2014). Entropy. Princeton: Princeton University Press.

Hassa, S. (2013). Wirtschaftlichkeitsbetrachtung von nachträglicher Wärmedämmung am Wohnhausbestand in Deutschland. Aachen: Shaker.

Holm, A. (2015). Wirtschaftlichkeit_daemmung_gdi_studie_2015.

Hunt, R. G. (Ed.). (2003). Comparative life cycle assessment of three insulation materials: stone wool, flax and paper wool. Unter Mitarbeit von Anders Schmidt. Landsberg, Bayreuth: ECO INFORMA Press.

KBOB CH. (2016). Ökobilanzdaten im Baubereich 2009/1:2016. Retrieved from https://www.kbob.admin.ch/kbob/de/home/publikationen/nachhaltiges-bauen/oekobilanzdaten_baubereich.html

Klöpffer, W., & Grahl, B. (2009). *Ökobilanz (LCA). Ein Leitfaden für Ausbildung und Beruf*. Weinheim: WILEY. doi:10.1002/9783527627158

Krieger O. (2017). Gebäudestudie. Szenarien für eine marktwirtschaftliche Klima- und Ressourcenschutzpolitik 2050 im Gebäudesektor.

Oehler, Stefan (2018). Emissionsfreie Gebäude. Das Konzept der „Ganzheitlichen Sanierung" für die Gebäude der Zukunft.

Piccardo, C., Dodoo, A., Gustavsson, L., & Tettey, U. Y. A. (2019a). Comparative Life-Cycle Analysis of Building Materials for the Thermal Upgrade of an Existing Building. *IOP Conf. Ser.: Earth Environ. Sci., 225*(1), 12044. 10.1088/1755-1315/225/1/012044

Piccardo, C., Dodoo, A., Gustavsson, L., & Tettey, U. Y. A. (2019b). Energy and carbon balance of materials used in a building envelope renovation. *IOP Conf. Ser.: Earth Environ. Sci., 225*(1), 12045. 10.1088/1755-1315/225/1/012045

Schmidt, B., Schmidt, D., & Venzmer, H. (Eds.). (2017). Energielieferant Altbau. Potenziale, Nachhaltigkeit, Objektbeispiele. Beuth Verlag. Auflage. Berlin: Beuth Verlag GmbH.

Steiner, T. (Ed.). (2018). *Ökologie und Ökonomie des Dämmens. Analyse und Bewertung von Dämmmaßnahmen in der Altbausanierung. Fraunhofer IRB-Verlag*. Stuttgart: Fraunhofer IRB Verlag.

Steiner, T., Eßmann, F., & Dornigg, I. (Eds.). (2018). *Ökologie und Ökonomie des Dämmens. Analyse und Bewertung von Dämmmaßnahmen in der Altbausanierung. Fraunhofer IRB-Verlag*. Stuttgart: Fraunhofer IRB Verlag.

Westermann, R.; Meyer, Ü.; Beobachter, D. S.; Zeugin, K.; Bolliger, B.; Faller, K. (2010). Mein Haus energetisch sanieren. Komfort verbessern, Kosten senken, Klima schützen: Beobachter-Edition.

Wilson, J. (2013). *The greened house effect. Renovating your home with a deep energy retrofit*. White River Junction, VT: Chelsea Green Publishing.

Wind G. (2017). Graue Energie.

ADDITIONAL READING

Day, C. (2016). *The eco-home design guide. Principles and practice for new-build and retrofit*. Cambridge: Green Books.

Frey, H. (2019). *Energieautarke Gebäude. Auf dem Weg zu Smart Energy Systems*. Berlin: Springer Vieweg.

International Renewable Energy Agency (IRENA). (2019). Global Energy Transformation: A Roadmap to 2050 (2019 Edition).

International Renewable Energy Agency (IRENA). Renewable Energy Capacity Statistics 2019. (2019).

KEY TERMS AND DEFINITIONS

Cradle and Cradle: At the heart of the Cradle to Cradle principle is the idea of thinking in complete product cycles right from the start, and in the first place, not creating waste in the traditional sense. Terms such as ecological, environmentally friendly or sustainable become obsolete. Products should be made in such a way that their end is considered from the beginning. All used material can be reused after use or composted without harmful residues.

Cradle and Crave: Often, construction products are disposed of as waste after use. The materials are used only once and are either lost to landfills or incinerated and are therefore no longer available for further use. The term "from cradle to grave" stands for this resource-consuming principle

Energy Balance Method: Energy balance methods enable the optimization of the building under the given economic conditions. The energy flows for the building to be calculated are determined using this method. These energy flows can be determined daily, weekly, monthly and annually. Depending on the location of the building, additional meteorological data are included in the calculations

Grey Energy: Grey energy is the primary energy necessary to construct a building. Grey energy includes energy used to extract materials, to manufacture and process components, to transport people, machines, components and materials to the construction site, to install building components and to dispose of them

Insulation Materials: An insulating material is a building material that is preferably used for thermal insulation. Thermal insulation materials are materials with low thermal conductivity and reduce heat or cold losses

Live Cycle Assessment: A Life Cycle Assessment (LCA) is a systematic analysis of the environmental impact of products throughout their life cycle “from cradle to grave”. Life Cycle Assessment includes all environmental impacts during production, the usage phase and the disposal of the product, as well as the related upstream and downstream processes

Primary Energy: Is the energy that is available with the originally occurring forms of energy or energy sources, such as fuel (e.g. coal or natural gas), but also energy sources such as sun, wind and tides. Primary energy can be converted to secondary energy through a conversion process. Primary or secondary energy, after transmission losses, becomes consumer-usable final energy.

Chapter 12
Heating Systems:
A Comparative Assessment of Alternative Solutions

Teodora Melania Şoimoşan
https://orcid.org/0000-0002-5929-1071
Technical University of Cluj-Napoca, Romania

Raluca Andreea Felseghi
https://orcid.org/0000-0001-6310-5932
Technical University of Cluj-Napoca, Romania

Maria Simona Răboacă
National R&D Institute for Cryogenic and Isotopic Technologies Râmnicu Vâlcea, Romania

Constantin Filote
https://orcid.org/0000-0001-6640-5308
Ştefan cel Mare University of Suceava, Romania

ABSTRACT

Within the current context of energy, there are several ways to meet the challenges of durable development. Efficiency in energy use, considered to be the fifth energy source, as well as the use of sustainable energy sources represent critical objectives. Nowadays, almost 50% of the total energy consumption in Europe is consumed by building heating and cooling. The current heat demand is mainly covered by conventional energy—fossil fuels. Consequently, there is a significant growth potential for the use of renewable energy sources (RES) in order to produce heat. One can expect in the near future that the energy systems would include a larger percentage of renewable sources, so the increase of the RES share is one of the main objectives of the thermoenergetic field. This chapter approaches heating system typology, the performance indicators used to asses the hybrid heating systems, and at the same time synthetising the assumptions of ensuring the optimum operating conditions.

DOI: 10.4018/978-1-5225-9104-7.ch012

INTRODUCTION

The emergence and accelerated incurrence of centralized heating systems related to urban localities represent the superior stage of heating technique. This stage implied a ratiocination based on the prediction of the link between social and economic growth, the incurrence of urban congestion and the need for technological optimization and the adjustment of existing heating systems to the new demands of the population. Worldwide, 50% of the population lives in urban areas. This percentage is forecasted to increase to 60,8% by 2030 (Cohen, 2006). In Europe, 72% of the population lives in urban areas, with an increase of up to 84% by 2050 (Eicker, 2012).

Due to the significant impact of urban heating systems on the environment, the result of their span and the nature of traditional used fuels, the optimization of existing and future systems has become a priority with the awareness of the need for sustainable energy development. The technological development and the increase of the energetic consumption over the last two centuries have generated a number of unprecedented negative effects over the environment, such as air pollution in localities, soil and water pollution, greenhouse gas accumulation in the atmosphere, with an important role in climate change and the incurring of ecosystem imbalances, minimization of fossil energy resources, etc. Throughout history, the mankind has developed an energy intensive society, with negative impact on the environment, being necessary to focus all the efforts in trying to limit, reduce and minimize the negative effects and optimize actions in the field of energy. In the context of the exhaustion of traditional energy resources, it is imperative to bring major changes in terms of production, transport and last but not least, energy consumption. The competitiveness and security of energy supply, the moral and legal goals to mitigate the impact on the environment, and the fight against climate change require technological modernization and optimization of installations and increase efficiency throughout the entire energy - producing - transport - consumption chain. Reducing fossil energy consumption and eliminating energy wastage without adversely affecting indoor comfort in heated spaces and without compromising the users' ability to set their own comfort levels (indoor temperatures, durations, consumption levels, etc.) is one of the main objectives in the field. Thus, the following objectives are outlined: optimization of heating systems, development and diversification of alternative, clean and renewable energy harnessing technologies.

The followings are addressed by this chapter:

- Configurations associated with conventional heating systems.
- Identification of the optimization opportunities and trends for conventional heating systems, in order to increase variation of the energy mix and operational performances.
- Assessment of correspondence between energy efficiency improvement and harnessing energy from renewable sources.
- The identification and analysis of factors that influence the RES share in the production of thermal energy within heating systems.
- Assessment of thermo-energy performances of heating systems, by securing the specific performance indicators.

The conventional heating systems are presented and analysed within this chapter, together with the modern concepts as energetic hub that allows the association of one single energy output - energy demand, to several energy inputs - related to different energy sources (conventional, renewable or alternative). At the level of hybrid energy hub, the thermo-energy indicators used for assessing the performances of the heating systems are emphasized, together with a comparative assessment of the conventional heating systems, and of the hybrid heating systems, respectively.

Several measures are proposed for optimizing the heating system and increasing the annual thermo-energy efficiency:

- Increase of the buildings' energy performance class within urban areas.
- Reduction of operating temperatures within the existing district heating system.
- Configuration of hybrid heating systems by integrating renewable energy sources (RES).
- Increase of RES quota in the heat production mix.
- Integration of RES systems in the existing district heating systems, in a centralized manner - at the thermic source or in a non-centralized manner - within the heating systems.

BACKGROUND

Centralized heating involves the centralized production of the heat at the thermic source - the districtual or central power plant, the cogeneration / trigeneration plant and the tariff distribution of thermal energy to consumers, through transport and distribution networks and intermediary stations for the transformation of thermal agent parameters - state of phase, pressure and temperature, respectively. Urban heat demand represents the heat demand delivered to ensure the living conditions, respectively the work conditions imposed inside the heated indoors, within urban localities. This category includes the heat demands for space heating related to urban and industrial spaces, heat demands for ventilation and air conditioning of urban and industrial buildings and heat demands for the preparation of hot water for the urban and industrial buildings.

Nowadays, in order to produce thermal energy, the following forms of primary energy are used:

- Fossil energy: deposit energy: coal, oil, natural gases.
- Solar energy.
- Biomass and biogas related energy.
- Geothermal energy.
- Nuclear energy.
- Mechanical energy / electromagnetic energy / electrical energy.

There are several variants of classification of heating systems related to urban localities, depending on the chosen criteria, as follows:

According to the type of heat source, the heating systems related to the urban localities are classified into:

- District heating / cogeneration systems: are centralized thermal energy production systems based on combined heat and power generation in district heating plants, respectively.
- Trigeneration systems: centralized systems for the production of thermal energy in power plants with trigeneration, based on the combined heat, electric and cold generation; centralized heat supply systems: with district heating or quartz.
- Decentralized heat supply systems: with low-capacity, building or individual thermal power plants.

- Mixed heat supply systems: with individual sources combined with centralized system.

According to the type of heating agent produced, the urban heating systems are classified into:

- Hot water heating systems with a maximum temperature of 100 °C.
- Hot water heating systems with temperatures in the range of 100 - 160 °C.
- Steam heating systems with average steam pressure of 6 - 15 bar.
- Steam and hot water heating systems.

Depending on the thermal power, it can take into consideration the next classification of the urban thermal power plants and of urban centralized systems of thermal supply:

- Of small capacity: with thermal capacities installed under 0,1 MW.
- Of medium capacity: with thermal capacities installed between 0,1 MW and 2 MW.
- Of high capacity: with capacities installed over 2 MW (Source: I 13-02).

Based on the number of embedded technologies, thermal systems are classified into:

- Monoagent systems: they include a single thermic source.
- Hybrid systems: they include at least two thermic sources and at least one of them is using renewable fuels.

Once prefigured the future energetic concepts, such as the approach regarding the target objectives, the transition will be carried out gradually, possibly materialized in intermediate stages. The importance of this optimization action derives from the obligation of having these intermediate steps, mostly, because the vast majority of urban localities are based on existing energy systems. The optimization of existing energetic systems represents one of the methods at hand in order to achieve these intermediate steps.

In Figure 1, an urban heating system is represented in a general manner, dismantled in the following subsystems: the heat source (1), the primary thermal transport network of the thermal agent (2), the thermal stations (3), the secondary distribution thermal network of the thermal agent (4), respectively the heat consumers (5).

Depending on the degree of centralization, decentralization respectively of an urban heating system, the subsystems presented above can exist independently or can be integrated in a subsystem.

The urban heating systems can be equiped with the following heating sources, synthesized in the Table 1.

There are currently the following technologies, for producing thermal energy, related to the urban systems listed in Table 1:

1. The applied technology is district heating or cogeneration. Whilst cogeneration can be organized on the basis of heat and electric energy production gaps, the heat representing the secondary product, residual, in the case of district heating, heat itself is the main product. Cogeneration is the system of simultaneous production of thermal and electrical energy using multiple sources of combustibles to generate energy, yielding up to 20% of production from a single source. Coupling the production of two types of energy leads to increased thermo-energy efficiency and also lowering the cost of

Figure 1. A schematic diagram of a conventional centralized system for producing thermal energy, along with an electrical thermal power plant
(Source: Şoimoşan T-M., 2015)

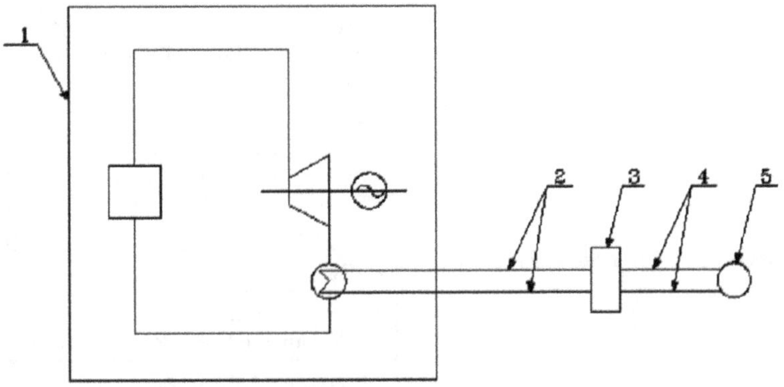

the heat supplied. Common fuel used in cogeneration is fossil fuel: lignite, fuel oil, natural gases. There are other situations in which alternative fuels are used, such as: biogas, biomass, LPG, diesel. In the case of thermic sources equiped with cogeneration systems, no matter what the cogeneration technology used, there are two energetic equipment categories: cogeneration equipments and peak thermic equipment. If, in addition to producing heat and electricity, a thermal cooling agent is also produced, the process is called trigeneration. The centralised cooling within trigeneration additionaly requires industrial absorption of refrigerating machines, which use heat as an energy source to produce cooled water. This technology offers the possibility to use renewable cooling sources, eg., seawaters, lakes or underground waters, and the use of the trigeneration system during the summer when excess heat is available. In the field of low-power cogeneration there is also the concept of cogeneration group. This represents a set of pre-assembled installations, mounted in a monoblock casing, usually equiped with internal combustion thermal engines. In this case, the used fuels are conventional or gaseous waste: diesel fuels, biodiesel, vegetable oil, methan and diesel combinations, etc. Motor cycles can use the folowings heat sources:

a. Conventional:
    i. Fossil fuels;
    ii. Nuclear (fission reactors).
b. Unconventional:
c. Heat recovered from energotechnical processes;
d. Solar energy, geothermal;
e. Biomass.

2. Nuclear power plants provide heat to urban systems in the same way as district heating plants. The place of the boiler is taken by the nuclear reactor in this case, and the fuel used is uranium, thorium and plutonium ore. The applied technology is chain-induced nuclear fission.
3. The technology applied in this case is the thermal conversion of biomass into a secondary fuel by thermal gasification (pyrolysis) and anaerobic fermentation of organic raw materials. Gasification represents the pyrolysis process of biomass at high temperatures, 1.100 – 1.300 °C, in order to obtain a gas having the chemical composition of variable methane (biogas). The anaerobic fermentation of

Table 1. Heating Sources

Crt. no.	Source of thermal energy	Heat generator
1.	Heating Power Plant Cogeneration Plant	• Gas turbine • Recovery boiler • Peak boiler: classic boiler, or recovery boiler with additional combustion
		• Steam generator • Backpressure steam turbine • Backpressure steam turbine and adjustable socket • Steam turbine with condensation and adjustable sockets • Peak boiler
		• Mixed steam-gas cycle
		• Vapor generator • ORC turbine (Organic Rankine Cycle)
		• Internal combustion heat engine (Otto Cycle) • Recovery boiler • Peak boiler
		• Stirling engine
		• Fuel cell
2.	Nuclear Power Plant	• Nuclear reactor • Steam turbine
3.	Biomass or biogas thermal plant	• Boiler with gasification
		• Biogas plant • Boiler on biogas
4.	Central heating	• Boiler
5.	Waste incineration plant	• Incineration plant • Boiler air recovery - water/steam
6.	Heat pump	• Geothermal heat pump
		• Industrial heat pumps (extracting the heat from seawater, river or water lake, district sewage, waste heat from industrial processes)
		• Solar heat pump
7.	The direct use of geothermal water heat	• Heat exchanger
8.	Solar thermal power plant	• Solar collectors field • Heat pump • Energy storage technology

Source: Şoimoşan, 2015

organic raw materials, called metanogenesis, is a complex microbiological process through which raw materials are converted into biogas.

4. The thermal power plants (see Figure 2 and Figure 3) are equipped with heating boilers and provied thermal energy. Heating boilers are exclusively designed based on the heat demand of the serviced customers. The applied technology is the direct combustion of primary fuels in the outbreaks and the recovery of the heat from combustion gases.
5. The technology applied in this case is the recovery of heat from municipal waste by one of the following methods: waste incineration - any type of waste, pyrolysis and gasification of waste -

Figure 2. Medium capacity thermal power plant (400 kW) for the production of the heating agent, working on natural gases
(Source: Şoimoşan, 2015)

organic waste, co-incineration - incineration of certain types of industrial waste, drying processes - municipal waste, municipal sludge, production waste. The thermal energy recovery is made from the combustion of gas burning and takes place in the heat recovery installations.

6. The technology applied is the transfer of the heat from a lower temperature - cold environment, to a higher temperature - warm environment, by consuming mechanical work. The amount of heat

Figure 3. High-capacity thermal power plant (10,5 MW) for the production of the heating agent, working on wood waste
(Source: Şoimoşan, 2015)

transmitted is greater than the mechanical power consumed. There is a wide variety of heat pumps adapted to the heat environments, respectively adapted to the nature of the thermal agents used.

7. The technology applied in this case is the direct use of hydro-thermal energy for heating. Geothermal reservoirs are located a few kilometers deep in the earth's crust and can be used for heating in applications with the direct use of geothermal energy. In order to increase the efficiency in using of geothermal energy, the applied technology in district heating systems is the use of exploited geothermal water, in cascade. This type of use implies the existence of several consumers, with different thermal potential of the energy demands, that are connected in series: electric energy production → space heating → domestic hot water production → technological consumption in industry → balneology → recovery of waste heat through heat pumps. In the usual way, the consumer distribution network is hydraulically separated from the geothermal water network through heat exchangers.
8. The applied technology is the capture and conversion of solar energy and the storage technology of solar energy thus obtained. Thermosolar collectors capture and convert into thermal energy, the energy cotained in incident solar rays on the capture surface. Since almost the entire spectrum of solar radiation is used in the production of thermal energy, this type of solar collectors has high conversion efficiency, $60 \div 90\%$. SDH (solar district heating or smart district heating applications involving more renewable energy technologies) have been developed lately, with the use of large fields of solar collectors, usually assembled on the ground or, in some smaller applications, on the roofs of serviced building, parking lots, etc., with the storage of thermal energy in underground reservoirs or drilling wells. This storage mode assumed seasonal thermal energy storage (STES) and involves the storage of heat up to a couple of months or more. The energy thus obtained does not usually entirely cover consumers' consumption peaks, so they are complementary covered by obtaining thermal energy in alternative technologies, usually combustion in cogeneration, simple combustion in boilers, geothermalism, etc.

THERMAL NETWORKS

The thermal networks of the district heating system represent the transport and distribution of the thermal agent produced by the thermic sources to the heat consumers networks. Nowadays the following district heating systems are known, which are listed in Table 2.

Following the way of domestic hot water supply, urban heating systems are classified in closed systems and open systems. Closed systems are equipped with heat exchangers for the preparation of hot water, whilst in open systems, domestic hot water is taken directly from the heating network. Since open systems are directly connected to the atmosphere and the flow of water is significantly higher, very high water treatment costs are generated. In addition to this disadvantage, open systems involve complicated systems of control losses and the increase of the thermal power of the source due to the circulation of an increased flow of heat. Because of these disadvantages, nowadays, these systems are rarely applied. The advantages of this type of system are strictly economical: the lack of heat exchangers and the reduced size of the expansion system.

Single duct systems are most efficient, but can only be used if the average flow rate for the heating system coincides with the average domestic hot water flow rate. This system is a bit more flexible and

Table 2. District heating systems classification

<table>
<tr><th>Crt. no.</th><th colspan="2">District heating system</th><th>Applied technology</th></tr>
<tr><td rowspan="2">1.</td><td rowspan="2">Arborescent networks</td><td>Closed</td><td>Bitubular piping system
Three-pipe system
Four-pipe system</td></tr>
<tr><td>Open</td><td>Monotubular piping system
Bitubular piping system</td></tr>
<tr><td>2.</td><td>Ring networks</td><td>Closed</td><td>Bitubular piping system</td></tr>
<tr><td>3.</td><td>Interconnected networks (network loop)</td><td>Closed</td><td>Bitubular piping system</td></tr>
</table>

Source: Şoimoşan, 2015

does not usually support later extensions. The largest spread system is the two-pipe system, flow and return closed system, respectively, being economically efficient and with the greatest degree of flexibility.

Three-pipe systems involve two flow pipes and a return duct that are used in the case of settlements with constant technological consumption (with independent pipe for seasonal consumers).

Four-pipe systems (two ducts, two return pipes) are found in small towns with significant industrial consumers. They are not economical for large localities. There are also mixed heating systems in which two or more types of systems are located in the case of centralized heat supply systems for large urban localities.

In order to increase the safety of the heat supply of consumers as well as due to the decrease of the degree of centralization in certain localities, which also led to an increase in the number of thermic sources with the decrease of their installed thermal power, interconnected power network loops type - the thermal agent is delivered to the consumers from at least two directions, respectively from multiple sources.

The pipes for modern district heating systems are pre-insulated steel pipes type and they are mechanically protected on the outside, normally with high-density polyethylene PE-HD, or alike. They are relatively easy to install and easily allow for further expansion, giving the system flexibility. In case of secondary distribution networks, pre-installed pipelines can also be used: PE-Xa cross-linked type a polyethylene, polybutylene PM, etc. The modern transport and distribution networks are equipped with monitoring systems. Through continuous monitoring, heat and thermal agent losses are detected, prompting rapid intervention and limiting the damages.

Thermal Stations

Depending on the existence or absence of thermal stations and their degree of equipping, existing urban heating systems are classified according to Table 3.

Due to the hydraulic separation of the heat transfer agents from the transport, distribution and indoor installations of the heat consumers, through the heat exchangers embedded in the thermal stations, the heating system becomes much more flexible, the operation regime being possible to be independently optimized in the primary and secondary circuits. Thermal stations, modules or thermal substations are intermediate - where the hydraulic networks separate the transport networks from the distribution ones, or the final, mounted to the consumer or group of consumers.

Thermal stations are equipped with:

Table 3. Thermal stations. Classification

Crt. no.	Urban heating system	Applied technology	
1.	Through direct junction	Not blended	Temperature controller
		Blended	Elevator Flow controller
			With mixing pump Flow controller
2.	Through indirect junction	Thermal station	

Source: Şoimoşan, 2015

- **Heat Exchangers:** Designed to hydraulically separate heating circuits and modify thermal agent parameters: pressure, temperature, chemical composition.
- **Expansion and Safety System:** To take over the variations in the volume of the thermal agent, for instance to remove any accidental overpressures from the plant resulting from the pressure and temperature increases.
- **Circulation System:** With the role of ensuring the forced circulation of the thermal agent in the system.
- **Thermal Agent Filtration System:** With the role of maintaining the water quality in the plant.
- **Adjustment And Control System:** To regulate and limit the parameters of the thermal agent in the secondary circuit.
- **Measuring And Metering System:** Designed to measure heat meter parameters and record energy consumption.

Operating Modes

The operating mode is mainly characterized by three parameters related to the thermal agent: temperature [°C], differential pressure [bar], volumic flow [mc/h] and massic flow [kg/h], respectively.

In the case of existing centralized systems, when the thermal agent is hot water, the flow temperature in the transport pipes is limited up to 120 °C. Low-temperature operation modes allow the efficient use of energy resources with low temperature potential. Because of this, the consumer-related installations require sizing for the lowest thermal agent temperature values. The temperature of the supplied thermal agent varies, within certain limits, for each region (see Table 4). Higher return values often limit the possibility of using renewable energy resources with low temperature potential. On the other hand, low return water temperatures have a positive influence on the thermal energy efficiency in cogeneration, where the thermal agent is used as a cooling agent. The values of return temperatures in centralized heating systems are between 75 - 40 °C. The lowest return temperatures are found in countries where the use of renewable resources is highest.

In terms of pressures in the system, they are limited in the case of distribution systems where the connection of the consuming installations is carried out directly by the minimum nominal pressure in the system, typically the nominal pressure of the heating elements, 6 bar. If the connection of the users is done indirectly through the thermal stations, in the transport pipes, the pressure can reach up to 25 bar but is usually technologically limited to 16 bar. From the point of view of the flow rates in the system, the operating mode can be of constant flow or variable flow, depending on the system possibilities.

Table 4. System temperatures in the distribution network. Existing centralized heating systems

Country	Thermal agent temperature supplied to consumers	
	Flow [°C]	Return [°C]
Russia	95	75
Romania	95	75
Poland	85	71
Great Britain	82	70
Germany	80	60
Coreea	70	50
Finland	70	40
Denmark	70	40

Source: Skagestad, & Mildenstein, In press

The main heating systems used in urban areas are district heating systems using cogeneration technology due to the many advantages they present compared to individual heating systems. Typically, they exhibit higher efficiency at source due to the simultaneous generation of electrical and thermal energy. This is also an advantage if the carbon footprint (IEE, Ecoheatcool and Euroheat & Power 2005-2006) is to be reduced. On the other hand, the flue gas exhaust systems from thermic sources are more technologically advanced than conventional boilers, resulting in lower emissions of pollutants into the atmosphere.

Large-scale urban energy power plants are often based on biomass for economically advantageous reasons in countries with "green feed-in" (Madlenera, & Vögtli, 2008).

The use of certain resources, e.g. industrial waste heat, waste heat from sewage system, municipal waste, etc. is only centralized and not accessible to individual heating systems. From the point of view of final energy costs, centralized systems are economically advantageous due to higher energy efficiency compared to decentralized heating systems. As long-term investments (50 years), the initial investment costs are justified. Internationally, there are many studies on the replacement of classical heating systems with systems that incorporate technologies based on unconventional resources. The most common are heat pumps and solar collectors. Unlike heating systems that integrate cogeneration technology, integration of heat pumps, solar installations, and other technologies that harness primary sources of clean and renewable energy, good results have been achieved, in addition to reducing fossil fuel consumption, emissions reductions greenhouse gas emissions, GHG and reduction of pollutant emissions, NOx, in inhabited areas (Genon, Torchio, & Poggio, 2009; Gaofeng, 2012).

ENERGY PERFORMANCE

In the present context, energy efficiency can be defined as the efficient use of energy resources to create and maintain vital conditions, respectively the level of comfort, in line with current technology and technology developments.

Energy efficiency is already considered to be the "fifth energy source" after air, water, earth and fire, identified in ancient Greece by Empedocle philosopher (490 BC - 430 BC) (Source: ONU Information Center for Romania).

The notion of energy efficiency has two aspects. The initial aspect, which was the basis for the emergence of the notion of heating systems, was purely economic, being related to the level of the price of energy supplied to the heat consumer and the specific heat energy expenditure. The second aspect relates to the efficient use of primary energy from the point of view of reducing environmental impact, reducing GHG emissions, reducing fossil fuel consumption and increasing fuel supply security and having the meaning of energy performance. Increasing the energy performance of a heating system positively influences the initial energy efficiency, while in reverse the effect obtained is not always a positive one.

Optimizing Heating Systems

Since urban heating systems are usually large-scale thermal systems, their impact on the social and environmental environments is significant. In this regard, one of the most important objectives in the field of engineering is the optimization of heating systems in order to reduce the negative impact. Central-

Table 5. Optimization goals and possibilities for heating systems

Type of optimization	Objectives of optimization	Opportunities for optimization
Energetics	Increasing the energy efficiency of the heating system.	Increasing the energy performance of each subsystem.
		Increasing the economic efficiency of each subsystem and the heating system as a whole.
		Optimizing the distribution of thermal loads on equipment related to the thermic sources in a preset regime.
		Optimal dimensioning of the associated networks and thermoinsulations.
		Technological optimization of installations.
		Reducing heat loss.
		Recycle all heat loss.
		Increasing the sealing capacity of the plant.
		Use of renewable resources insufficiently capitalized by increasing their participation quota.
Technology	Accelerating the transition to green energy use and reducing environmental impact.	Technological optimization of RES-based installations.
		Use of renewable resources insufficiently capitalized by increasing the participation quota.
		Optimization of RES storage technologies.
		Replacement of high temperature heating systems with low temperature systems.
Functional	Increased flexibility of operation of heating systems.	Optimizing the distribution of thermal loads on thermal equipment in case of a given regime.
		Ensuring optimal heat storage at source level.
		Optimization of thermal networks.
		Optimization of the information system attached to the thermal system.
Operational	Increasing productivity, quality and safety throughout the energy chain: transport -consumption.	Technological optimization of installations.
		Optimization of the information system attached to the thermal system.

Source: Şoimoşan T-M., 2015

ized heating systems are mostly existing systems, therefore their optimization is done under off-design conditions. In general, the possibilities of optimizing a heating system include the following directions: energy optimization, technological optimization, functional optimization and operational optimization. Table 5 summarizes the objectives and possibilities for optimization of heating systems identified for the directions listed above.

The negative effects of the global warming phenomenon, as well as the significant reduction in the future of the fossil fuel resources, have increased the role, with a focus on the use of renewable energy RES resources, in the configuration of future energy systems. The multi-energy district heating system, energy hub type, efficient from the point of view of RES and energy storage systems, is a desideratum and has a high potential for optimization with all the aspects involved: energetic, technological, functional and operational (Geidl, & Andersson, 2005; Corrado, Fabrizio, & Filippi, 2007; Sheikhi, Ranjbar, & Safe, 2011).

Regarding the thermosolar resource in particular, it is an important component in the configuration of future district heating systems. Medium and high capacity thermal systems (> 100 m^2 of collector surface), effectively integrated into existing or future district systems, are one of the cleanest and most innovative methods of energy optimization. In Europe, there are a number of applications, such as the modernization of the seasonal thermal heating system of the Hamburg-Bramfeld locality by integrating it into the district heating system of Hamburg, Germany (Schmidt, & Mangold, 2010) or pilot applications in different monitoring stages (Schmidt, Mangold, & Muller-Steinhagen, 2003). Their monitoring provides the possibility to obtain significant information about the influence of operating conditions and strategies on the efficiency of hybrid systems. In some applications, the annual operating efficiency of thermal insulation systems, energy storage systems or hybrid heating systems is well below the initial designed efficiencies in the project phases. This is due to various causes, the most common being parametric regimes and inadequate operating strategies for hybrid heating systems.

Energy Performance Indicators

The energy performance of a heating system can be evaluated relatively quickly using indicator-based methods. This assessment, although somewhat approximate, is useful in the assessment and monitoring process, by reference to benchmark performance indicators of a heating system, at various stages of design, exploitation and retrofitting - upgrading and optimization. Energy performance indicators can be a simple parameter, a simple ratio or a complex pattern and are set for each significant energy consumer or thermal system if their performance is quantifiable. Energy performance indicators can be expressed as an absolute or relative values, by comparison with a reference value. Currently, a numeric value is attached to a performance indicator as follows:

- **Base Value:** Defined by a calculated field from which a value is derived, e.g. the results of energy balances;
- **Target Value:** Defined by a calculated field from which a value or an absolute value is derived, e.g. the result of an optimized balance sheet, reference value, etc.
- **Status Threshold:** Is defined by the range between a minimum threshold and a maximum threshold. The status threshold is defined to allow for the difference between the state of the baseline value and the target value.

Table 6. Indicators of energy performance

Performance indicators	Symbol	U.m.
Seasonal (or annual) global yield of a subsystem	η_c	%
Annual RES fraction	α_a	%
Collectors field' annual efficiency, related to the aperture area	Ef_a^c	kWh/(m² year)
Annual efficiency of the hybrid system	Ef_a	-
Maximum annual economy of fossil fuels	B_{ec}	kg/year
Maximum quantity of CO_2 emissions, annualy avoided (reference fuel - natural gases)	Em_{CO2}	kg_{CO2}/year

Source: Şoimoşan, 2015

Table 7. Working hypotheses

Scen.	Fuel used for the thermic source	Integrated thermal energy class of consumers		Performance indicators		
		Hypothesis H1	Hypothesis H2			
S1	Fossil fuel	before R*	after R	Maximum annual economy of fossil fuel, natural gases	The maximum amount of CO_2 emissions annually avoided	The annual energy efficiency of the thermal system
S2	Fossil fuel & Low RES share	Before R	after R			
S3	Fossil fuel & Average RES share	Before R	after R			
S4	Fossil fuel & High share of RES	Before R	after R			

*R - Thermoenergetic retrofitting of the urban buildings
Source: Şoimoşan, 2015

Reference indicators are values obtained in systems equipped with the best state-of-the-art world-known technologies previously obtained reference values.

Performance indicators can refer to the global thermal system (global) or subsystems, machinery, thermal (individual) equipment. The followings are the global performance indicators applicable to heating systems. In this case, as a rule, the core values attached to energy performance indicators are annual or seasonal values. In order to evaluate and comparatively analyze the performances of the hybrid heating system achieved in different scenarios of retrofitting, the following indicators of energy performance, synthetized in Table 6 were determined in the case study, within the thermo-energy balances.

Case Study

During this subchapter, some indicators that quantify the performance of hybrid heating systems were highlighted and analysed. Table 7 summarizes the working framework for the scenarios addressed in the case study. The thermo-energetic simulation of the district heating system was made for the following proposed and configured study scenarios.

The criteria for a global evaluation of the performances achieved by the thermal energy systems in accordance with the particularities of the working hypotheses and objectives tracked were established. By following the specific indicators defined above, in order to validate them in analyzing and comparing the heating systems, case studies were carried out for urban areas characterized by different thermal energy densities, in two scenarios of energy efficiency of the associated buildings (existing, thermoenergetic optimized respectively) and for different hypotheses of configuration of the thermic sources. The computational modeling and setting the operational frames of heating systems' simulation for the suggested working scenarios was achieved. The simulations were carried on by using Polysun computing simulation software (Velasolaris 2016), during an entire calendaristic year.

The case study was carried out on urban areas noted $A_1 - A_5$ from the reference locality (city of Oradea, Romania), where the identified thermal consumers (residencial and administrative buildings, commercial, office and industrial buildings, respectively) were analyzed from a thermoenergetic point of view, in two hypotheses of energy efficiency:

- H1: the initial hypothesis presumed the envelopes of the existing buildings that were not energetically rehabilitated or modernized.
- H2: the final hypothesis presumed the envelopes of the existing buildings that were energetically rehabilitated and modernized.

The characteristics of the thermal urban areas were centralized in the followings tables, Table 8 and Table 9, for the two working hypotheses H1 and H2, that were previously set.

In order to compare the thermal energy performances of the hybrid heating systems of the identified areas, the same hypothetical territorial surface was taken into account, for all thermal areas, respectively the territorial surface of the reference area A_2. The evaluations of the heating demands for urban areas A_1, A_3 and A_4 were achieved by using the method of thermal heating characteristics of the thermal representative consumers of area A_2, respectively based on thermal-energy density of every urban area (Şoimoşan, 2015). The values centralized in Table 9 were secured by respecting the methodologies in force in Romania (Mc 001/rev. 2016; SR EN 1907/ rev. 2016; C107-3/2005; AllEnergy® V7.2 2014).

Table 8. Thermal areas characteristics

Thermal area	Description of the thermal area	Territorial surface of reference	Thermal density		Built volume	No. of pers. (DHW consumers)
-	-	S_t	ρ_c		V_c	N
			H1	H2		
-	-	m^2	W/m^2		m^3	pers.
A_1	Urban area - very high thermal density	32.000	236,84	138,94	185.390	2.755
A_2^{ref}	Urban area - high thermal density	32.000 ref	171,56	116,91	176.950	2.600
A_3	Urban area - average thermal density	32.000	43,75	29,69	52.521	1.048
A_4	Urban area - low thermal density	32.000	20,94	12,81	23.530	277
A_5	Industrial area	32.000	23,81	12,84	63.234	134

Source: Şoimoşan, 2015

Table 9. Thermal-energy characteristics of the urban areas

Thermal area	Hourly heating demand for space heating		Hourly heating demand for DHW		Hourly heating demand for space heating and DHW		Annual heating demand for space heating and DHW		Specific annual heating demand for space heating and DHW	
-	H1	H2	H1	H2	H1	H2	H1	H2	H1	H2
-	MW		MW		MW		MWh/year		kWh/ (m^2 year)	
A_1	6,711	3,578	0,868	0,868	7,579	4,446	15.441,60	9.686,00	304,28	191,80
A_2	4,787	3,038	0,703	0,703	5,490	3,741	12.602,17	8.781,98	241,76	168,47
A_3	1,250	0,800	0,150	0,150	1,400	0,950	3.575,38	2.819,75	297,95	235,98
A_4	0,560	0,360	0,110	0,110	0,670	0,470	1.363,00	1.027,78	214,31	161,60
A_5	0,458	0,229	0,304	0,182	0,762	0,411	1297,25	774,516	207,47	123,87

Source: Şoimoşan T-M., 2015

The hybrid heating system is the result of integrating thermal-solar fields to the level of district heating source. The optimized configuration of the hybrid heating systems and the attached control algorithm were developed on Polysun platform (Velasolaris) and are schematically represented in Figure 4.

The nominal temperature values of the thermal agent in the district heating network are, in the first hypothesis H1, 90/60 °C, respectively 70/40 °C in the second hypotheses, H2.

The optimized configurations of hybrid heating systems for the study areas A_1 - A_5 and the assumed four study scenarios S1 - S4 are presented synthetically in Figure 4.

The followings were considered:

- Average annual global solar radiation used in the simulation is 1.249,27 kWh/m^2 (Source: NASA, Atmospheric Science Data Center 2015).
- The operating regime in the heating system is with variable flow. For an uniform and safe circulation of the solar-thermal agent in the entire collector field, without any stagnation areas of the solar agent with risk of increasing the system's temperature, the volumetric flow must be high enough. In this instance, the volumetric flow is automatically set, function of the temperature in the storage tank and of the temperature in the collector's field, respectively (Şoimoşan, T.-M, Felseghi R.-A. 2015). The thermal-solar collector field presents the following characteristics, according to the Table 10.

In the following Tables 11 to 14, the performance indicators for the hybrid systems configured in Figure 4 are centralized, for the study scenarios proposed in Table 7.

In the 1st Scenario, the classical heating source is represented by fossil-fired heating boilers, the first one, with fuel oil burning, designed for the production of thermal energy for hot water preparation and the second one, with charcoal burning, destined for the production of thermal energy for space heating. The thermo-energy indicators achieved within the case study in Scenario 1 are centralized in Table 11.

In the 2nd and 3rd Scenario, the classical heating source is represented by fossil-fired natural gas (NG) fired boilers, the first designed for the production of thermal energy for hot water preparation, the second one for the production of thermal energy for space heating. The thermo-energy indicators achieved within the case study, in the 2nd and 3rd Scenario are centralized in Table 12, respectively Table 13.

Figure 4. Thermoenergetic hub. Hydraulic schematic (Velasolaris, PolySun Software Solarthermal)

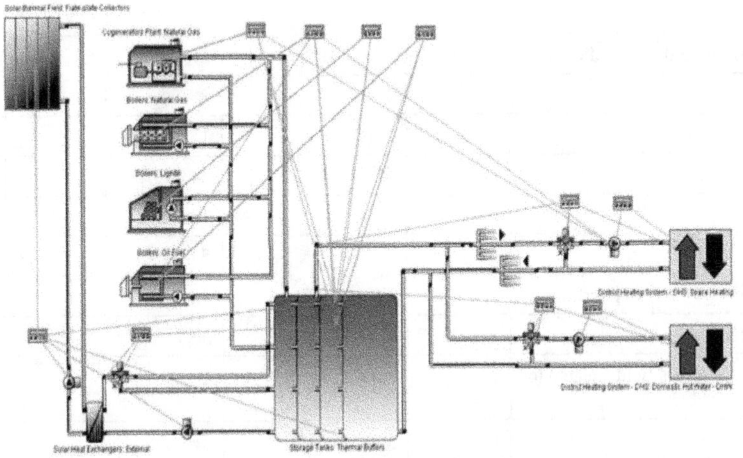

Table 10. Characteristics of the solar thermal collectors

Flate-plate solar collectors / No tracking			
Optical yield	$\eta_{o\ (lam.\text{-}turb.)}$	0,75 - 0,80	%
1st order heat loss coefficient	$k_{1\ (lam.\text{-}turb.)}$	3,5 – 4,0	$W/(m^2K)$
2nd order heat loss coefficient	k_2	0,02	$W/(m^2K^2)$
Collectors tilt / Azimuth angle	α / β	45 / 0	°
Flow rate (variable)	m_{cc}	15 - 35	$kg/(hm^2)$

Source: Velasolaris, PolySun Software Solarthermal

Table 11. Classic thermo-energy hub. Balance sheet and performance indicators secured within the 1st Scenario

Thermo-energy balance indicators			u.m.	value	
-			-	H1	H2
Hourly thermal energy demand		Q_h	MWh	15,9	10,0
Annual thermal energy demand		Q_a	MWh/year	34.279,4	23.090,0
Balance sheet component		Q_{aux}	MWh/year	42.819,8	33.413.9
Annual energy efficiency		Ef_a	-	0,74	0,76
Seasonal yield of boiler	Black oil	η_a	%	84,8 (85)	
	Charcoal			75,0 (75)	
Annual consumption of fossil fuels	Black oil	B_{aux}	kg/year	734.423,9	738.069,6
	Charcoal		to/year	28.698,9	21.296,8
Maximum annual economy of fossil fuels	Black oil	B_{ec}	kg/year	-	-
	Charcoal		to/year		7.402,1
Maximum amount of CO_2 emissions annually avoided		Em_{CO2}	kg_{CO2}/year	-	4.580.445,9

Source: Şoimoşan T-M., 2015

Table 12. Hybrid thermo-energy hub. Balance sheet and performance indicators obtained in the 2nd Scenario

Thermo-energy balance indicators			u.m.	value	
-			-	H1	H2
Hourly thermal energy demand		Q_h	MWh	15,9	10,0
Annual thermal energy demand		Q_a	MWh/year	34.279,4	23.090,0
Annual solar fraction		α_a	%	**10,0**	**17,4**
Collector field area	gross	S	m^2	5.000	
	of aperture	A		4.500	
Annual yield of the collector		η_c	%	37,7	50,9
Annual efficiency of the collectors field (related to the aperture area)		Ef_a^c	kWh/(m^{2}year)	765,4	1.033,5
Exergy components of thermo-energy balance		Q_{sol}	MWh/year	3.444,4	4.650,8
		Q_{aux}	MWh/year	30.826,8	22.068,1
Seasonal boiler yield		η_{aux}	%	89,3/89,9 (90)	89,7/89,9 (90)
Annual energy efficiency		Ef_a	-	0,97	1,07
Annual consumption of fossil fuels consumption		B_{aux}	m$^3_{GN}$/year	3.264.505,6	2.338.662,2
Maximum annual economy of fossil fuels		B_{ec}	m$^3_{GN}$/year	364.689,0	492.178,5
Maximum amount of CO2 emissions annually avoided		Em_{CO2}	kg$_{CO2}$/year	886.804,7	1.196.817,7

Source: Şoimoşan T-M., 2015

Table 13. Hybrid-thermo energy & natural gases (NG) hub. Thermo-energy balance indicators achieved in Scenario 3

Thermo-energy balance indicators			u.m.	value	
-			-	H1	H2
Hourly thermal energy demand		Q_h	MWh	15,9	10,0
Annual thermal energy demand		Q_a	MWh/year	34.279,4	23.090,0
Annual solar fraction		α_a	%	29,0	46,6
Collector field area	total	S	m^2	20.000	
	of aperture	A		18.000	
Annual yield of the collector		η_c	%	31,7	39,3
Annual efficiency of the collector field (related to the aperture surface)		Ef_a^c	kWh/m^{2}year	643,0	798,1
Exergy components of thermo-energy balance		Q_{sol}	MWh/year	11.574,7	14.365,0
		Q_{aux}	MWh/year	28.325,2	16.493,8
Seasonal boiler yield		η_{aux}	%	89,1/89,9 (90)	89,7/89,9 (90)
Annual energy efficiency		Ef_a	-	1,23	1,65
Fossil fuel consumption		B_{aux}	m$^3_{GN}$/year	2.999.288,0	1.747.996,5
Maximum annual economy of fossil fuels		B_{ec}	m$^3_{GN}$/year	1.230.503,7	1.520.415,5
Maximum amount of CO_2 emissions annually avoided		Em_{CO2}	kg$_{CO2}$/year	2.992.183,9	3.697,154,6

Source: Şoimoşan T-M., 2015

Table 14. Balance sheet and performance indicators obtained in the 4th Scenario

Thermo-energy balance indicators			u.m.	value
-			-	H2
Hourly thermal energy demand		Q_h	MWh	10,0
Annual thermal energy demand		Q_a	MWh/year	23.090,0
Annual solar fraction		α_a	%	~ 50
Collector field area	gross	S	m^2	20.000
	of aperture	A		18.000
Annual yield of the collector		η_c	%	36,6
Annual efficiency of the collectors field (related to the aperture area)		Ef_a^c	kWh/(m^2year)	744,0
Exergy components of thermo-energy balance		Q_{sol}	MWh/year	13.392,3
		Q_{aux}^{boiler}	MWh/year	3.509,8
		Q_{aux}^{cogen}	MWh_t/year	8,586,5
Electricity produced in cogeneration		E_e	MWh_e/year	6.804,4 6.396,0
Seasonal yield of energy equipment (thermal, electrical and global)		η_t^{boiler}	%	89,8 (90)
		η_t^{cogen}		49,4
		η_e^{cogen}		39,1
		η_{gl}^{cogen}		88,5
Annual energy efficiency		Ef_a	-	1,50
Coefficient of cogeneration		α_{cogen}	-	0,50
Number of cogeneration plant functioning hours		n	h/year	4.431
Fossil fuels consumption		B_{aux}	m^3_{GN}/year	2.027.760,7
Maximum annual economy of fossil fuels		B_{ec}	m^3_{GN}/year	2,296,604.2
Maximum amount of CO_2 emissions annually avoided		Em_{CO2}	kg_{CO2}/year	5,584,593.5

Source: Şoimoşan T-M., 2015

In the 4th Scenario, the classic heating source is represented by cogeneration plants running on fossil fuel, natural gases (NG). As a premise, it was intended to obtain the solar fraction from the 3rd Scenario - Hypotheses 2 (previous case). The production of thermal energy in cogeneration involves complex equipment with high investment values. In order to be profitable, it is necessary that the operating time for thermal loads as close to the maximum value does not fall below the value of 4.000 hours/year (α_{cogen} = 0,3 - 0,5). In order to achieve the profitability requirement, it is necessary to divide the thermal load on several thermal energy production equipments (~ 4,4 MWt in cogeneration system), the peak load to be taken over by heating boilers. The thermo-energy indicators achieved in the case study in Scenario 4 are centralized in Table 14.

Figure 5 illustrates graphically and comparatively, the annual energy efficiency values of the heating system in the four S1 - S4 thermic source configuration scenarios, respectively the two hypotheses H1 and H2.

When reffering to the buildings' green retrofitting, in particular to the heating systems related to buildings, the relevant performance indicators are:

- Fossil fuels consumption (B_{aux}) used in heating process.
- Maximum annual economy of fossil fuels (B_{ec}) or quota of RES' use in energy mix (α_a).
- Maximum amount of CO_2 emissions annually avoided (Em_{CO2}).
- Annual efficiency of the hybrid heating system (Ef_a).
- And annual efficiency of the RES system, in this particular case the annual solar yield (η_c).

Generally, the target indicators are imposed by the chosen criteria (imposed targets) within the retrofitting process (e.g. limiting energy consumption or carbon footprint). If the imposed criterion have more than one valence, the most relevant performance indicator remains annual efficiency of the hybrid heating system, indicator that reflects the degree of the achievement of the other indicators.

Within the case study, the highest values of the annual efficiency of the heating system were secured in Scenario 3 - Hypothesis 2 (S3 - H2) and Senario 4 - Hypothesis 2 (S4 - H2), respectively (Figure 5), both scenarios based on RES and naturale gases heating systems, with natural gases fuels used in boilers, and combined boilers and cogeneration, respectively.

Another common feature of the scenarios with high-value of this indicator is the reducing of the heat demands (heat consumptions) related to the buildings after retrofitting process of the buildings' enevelopes. Increasing the enveloping degree of the buildings' built fond has led, on one side to the reducing of the heat demands and heat consumptions, Q_h and Q_a respectively, and on the other side has created the possibility of turning to the low temperatures regimes in operation of the existing heating systems, regimes that are perfectly compatible with the harnessing of renewable energy sources with low temperature potential in urban areas, as an example.

Figure 5. Annual energy efficiency of the heating system. Study scenarios S1 - S4

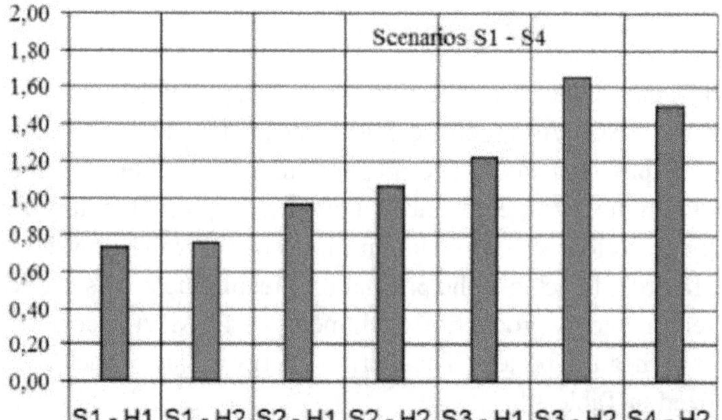

	S1 - H1	S1 - H2	S2 - H1	S2 - H2	S3 - H1	S3 - H2	S4 - H2
■ Annual efficiency of the heating system, Efa	0,74	0,76	0,97	1,07	1,23	1,65	1,50

SOLUTIONS AND RECOMMENDATIONS

Several measures are proposed so as to optimize the heating systems, in order to increase the annual thermo-energy efficiency:

- Increase of the energy performance class of the heat consumers from urban areas.
- Reduction of operating temperatures within the district heating systems.
- Configuration of hybrid heating systems by integrating renewable energy sources (RES).
- Increase of RES quota in the production of heat.
- Integration of RES systems in the existing district heating systems, in a centralized manner - at the thermic sources or in a non-centralized manner - within the heating systems, based on feed-in principle.

In the latter case, the final user's involvement and co-participation in the green retrofitting process in energy management is additionally obtained.

FUTURE RESEARCH DIRECTIONS

The current state of the field of heating systems confirms that future energy systems will include high levels of renewable energy. The scenarios carried out in the context of the study, through the results obtained, but also with the aspects related to the issues raised, have allowed identifying and establishing future research directions in the field of the subject addressed:

- Expanding fundamental and applied research on clean primary energy sources as inputs into the urban thermo-energy hub and synergistic optimization of processes within it.
- The optimization of heat storage technologies from discontinuous renewable energy sources, equipment, thermal storage agents and their operating modes.
- Optimizing the information system attached to the heating system to reduce energy losses, increase system flexibility and operational safety.
- The elaboration of exhaustive studies for estimating the potential for utilization of renewable energies with low temperature potential in urban areas characterized by high thermal densities.
- Conducting studies to optimize socio-economic perception, viability and public acceptance by end-users regarding the use of renewable energies as an energetic alternative in urban areas and accelerate the transition to the sustainable heating systems.

CONCLUSION

This chapter deals synergistically with the thermo-energy efficiency of urban thermal consumers, the efficiency of exploitation of renewable energy sources with low temperature potential, identified in urban areas, solarthermal energy in particular, in buildings which are connected to the district heating system.

Concluding remarks can be synthetised as follows:

In the case of existing heating systems, passing to an operation at low temperatures of its thermal agent is possible after the development and completion of energy efficiency actions for its consumers, by rehabilitating and retrofitting the buildings' anvelopes and of their interior thermal installations.

The aim is to ensure an optimal balance between energy efficiency measures and use of energy produced by renewable sources. Meeting such aim is mandatory for integrating RES under optimal energy efficiency conditions and passing from "using" renewable energy sources to "harnessing" thereof.

The integration of technologies aimed at harnessing renewable or alternative energy sources within urban areas, as characterized by elevated thermal densities for heat consumption, within a decentralized manner for the existing thermal systems or at the consumers, allows on one hand to cover some of the share of their own consumption, and on the other hand to redirect such excess of thermal energy into the district system, based on the feed-in principle, securing in this manner elevated values of the performance thermo-energy indicators.

The incurrence and accelerated development of high energy density urban agglomerations and associated increasing energy consumption in the context of the fact that the quantitative and polluting fossil energy is no longer a sustainable solution, the intelligent approach of energy systems increasingly presumes more and more different synergy forms of energy in transition and the synergy of renewable and clean forms of energy in the future energy.

ACKNOWLEDGMENT

This research received no specific grant from any funding agency in the public, commercial, or not-for-profit sectors.

REFERENCES

C107-3. (2005). *Romanian Standard regarding Thermotechnical calculation of construction elements of the building*. FAST PRINT.

Algorithm+ SRL & IPCT Installations. AllEnergy® V7.2. (2014). Software for determining the energy performance of buidings. Romania.

Cohen, B. (2006). Urbanization in developing countries: Current trends, future projections, and key challenges for sustainability. *Technology in Society*, *28*(1-2), 63–80. doi:10.1016/j.techsoc.2005.10.005

Corrado, V., Fabrizio, E., & Filippi, M. (2007, june). Modelling and optimization of multi-energy source building systems in the design concept phase. *Proceedings of Clima 2007 WellBeing Indoors.*

Eicker, U. (2012). Renewable Energy Sources within Urban Areas: Results From European Case. *Academic Journal of ASHRAE Transactions*, (118), 73.

Gaofeng, Ch. (2012). *Energy efficiency and carbon dioxide emissions reduction*. Elsevier Journal, Resources. *Conservation and Recycling*, *60*, 147–158.

Geidl, M., & Andersson, G. (2005). A modeling and optimization approach for multiple energy carrier power flow. *Proc. IEE PES Power Tech*, 1-7. 10.1109/PTC.2005.4524640

Geidl, M., Koeppel, G., & Favre-Perrod, P. (2007, March). The energy hub - A powerfull concept for future energy systems. *Third Annual Carnegie Mellon Conference on the Electricity Industry.*

Genon, G., Torchio, M. F., & Poggio, A. (2009). Energy and environmetal assessment of small district heating systems: Global and local effects in two case-studies. *Energy Conversion and Management, 50*(3), 522–529. doi:10.1016/j.enconman.2008.11.010

IEE Ecoheatcool and Euroheat & Power. (2005-2006). *Ecoheatcool. Work package 3. Guidelines for assessing the efficiency of district heating and district cooling.* Retrieved from http:// www.ecoheatcool.org

John, A., & Duffie Beckman, W. A. (2013). Solar Engineering of Thermal Processes. John Wiley & Sons, Inc.

Madlenera, R., & Vögtli, S. (2008). Diffusion of bioenergy in urban areas: A socio-economic analysis of the Swiss wood-fired cogeneration plant in Basel. *Biomass and Bioenergy, 32*(9), 815–828. doi:10.1016/j.biombioe.2008.01.006

NASA, Atmospheric Science Data Center. (n.d.). *Surface meteorology and Solar Energy, RETScreen Report August 2015.* Retrieved from https://eosweb.larc.nasa.gov

ONU Information Center for Romania. (in press). *Climate changes - A global problem.* Retrieved from http://www.onuinfo.ro/resurse/schimbari_climatice

Romanian Ministry of Transport, Construction, and Tourism. (2006). Calcul methodology of the energy performance of buildings. Indicative Mc 001/2006. Published in Official Monitor, Bucharest, Romania.

Schmidt, T., & Mangold, D. (2010, November). *The multi-functional heat storage in Hamburg-Bramfeld - innovative extension of the oldest German solar energy housing estate.* Solites, IRES 2010, Berlin, Germany.

Schmidt, T., Mangold, D., & Muller-Steinhagen, H. (2003, June). *Seasonal thermal energy storage in Germany.* Solites, ISES Solar World Congress 2003, Goteborg, Schweden.

SDH. (n.d.). *Solar District Heating - The SDH Projects.* Retrieved from http://www.solar-district-heating.eu

Sheikhi, A., Ranjbar, A. M., & Safe, F. (2011). *Optimal dispatch of a multiple energy carrier system equipped with a CCHP.* Tehran: Department of Electrical Engineering Sharif University of Technology. doi:10.24084/repqj09.675

Skagestad, B., & Mildenstein, P. (in press). District Heating and Cooling Connection Handbook. International Energy Agency IEA, District Heating and Cooling DHC. *Combined Heat & Power Association.*

Şoimoşan, T.-M. (2015). *Contributions regarding the optimization of the heating systems in urban localities* (Unpublished doctoral dissertation). Technical University of Cluj-Napoca, Romania.

Şoimoşan, T.-M., Felseghi, R.-A., & Corsiuc, G. D. (2015). Influence of the Operating Regime on the Performances of Thermal- Solar Systems Integrated in Heating Networks. *Applied Mechanics and Materials, 772*, 531–535. doi:10.4028/www.scientific.net/AMM.772.531

SR EN 1907-1. (2014). *Romanian Standard regarding Heating installations. Heating demand calculation.* Matrix Rom Publisher.

Sřrensen, P. A., Nielsen, J. E., & Battisti, R. (2012). Solar district heating. Collection of fact sheets WP3 - D3.1 & D3.2. Inteligent Energy Europe, IEE.

United Nations. (2004). *World urbanization prospects: the 2003 revision Data tables and highlights.* New York: United Nations.

ADDITIONAL READING

Agrella, P. J. (2005). Economic and environmental efficiency of district heating plants. *Energy Policy, 33*(10), 1351–1362. doi:10.1016/j.enpol.2003.12.011

Allegrini, J., Orehounig, K., Mavromatidis, G., Ruesch, F., Dorer, V., & Evins, R. (2015). A review of modelling approaches and tools for the simulation of district-scale energy systems. *Renewable & Sustainable Energy Reviews, 52*, 1391–1404. doi:10.1016/j.rser.2015.07.123

Bagheri, A., Feldheim, V., & Ioakimidis, C. S. (2018). Review On the Evolution and Application of the Thermal Network Method for Energy Assessments in Buildings. *Energies, 11*(890), 1–20.

Delia, M. F. (Matrix Rom). (2004). Use of thermo-hygro-energetic analyzes in the design of residential buildings. Bucharest, Romania, Matrix Rom Publisher.

Evins, R. (2013). A review of computational optimisation methods applied to sustainable building design. *Renewable & Sustainable Energy Reviews, 22*, 230–245. doi:10.1016/j.rser.2013.02.004

Lake, A., Rezaie, B., & Beyerlein, S. (2017). Review of district heating and cooling systems for a sustainable future. *Renewable & Sustainable Energy Reviews, 67*, 417–425. doi:10.1016/j.rser.2016.09.061

Morille, B., Musy, M., & Malys, L. (2016). Preliminary study of the impact of urban greenery types on energy consumption of building at a district scale: Academic study on a canyon street in Nantes (France) weather conditions. *Energy and Building, 114*, 275–282. doi:10.1016/j.enbuild.2015.06.030

Olsthoorn, D., Haghighat, F., & Mirzaei, P. A. (2016). Integration of storage and renewable energy into district heating systems: A review of modelling and optimization. *Solar Energy, 136*, 49–64. doi:10.1016/j.solener.2016.06.054

Swan, L. G., & Ugursal, V. I. (2009). Modeling of end-use energy consumption in the residencial sector: A review of modeling techniques. *Renewable & Sustainable Energy Reviews, 13*(8), 1819–1835. doi:10.1016/j.rser.2008.09.033

Verda, V., Guelpa, E., & Cona, A. (2012). Reduction of primary energy needs in urban areas trough optimal planning of district heating and heat pump installations. *Energy, 48*(1), 40–46. doi:10.1016/j.energy.2012.07.001

Yamaguchi, Y., & Shimoda, Y. (2010). District-scale simulation for multi-purpose evaluation of urban energy systems. *Journal of Building Performance Simulation, 3*(4), 289–305. doi:10.1080/19401491003746621

KEY TERMS AND DEFINITIONS

Annual Efficiency of the Hybrid Heating System: Ratio between annual heat demand, or heat consumption in the case of an existing system and total amount of primary energy used at the auxiliary thermic source, based on conventional, fossil fuels.

Annual Solar Fraction: Percentage of annual demand/consumption of heat, covered through solar energy.

Conventional Thermic Source: The assembly of installations, technological equipments that produce heat in a centrally manner, by burning fossil fuels within outbreaks.

Hybrid Heating System: A heating system that uses, to produce heat, a mix of conventional, regenerable or alternative energies and includes technologies for harness renewable and alternative energy sources, respectively.

Performance Indicator: Numeric expression that quantifies the performance of a thermal system, at the present case.

Thermal Density: In the case of an urban area, it represents the ratio between thermal energy demand/consumption for space heating, domestic hot water preparation and industrial processes and considered territorial surface.

Thermal-Solar Field: An assembly of technical systems with the role of collecting the solar energy, and converting it in the thermal energy, respectively.

Thermoenergetic Hub: Technological way of organising, that allows the association of one single energy output-energy demand, to several energy inputs related to different energy sources (conventional, renewable, clean or alternative).

Compilation of References

Abdul-Wahab, S. A. (2011). *Sick Building Syndrome*. Berlin: Springer-Verlag. doi:10.1007/978-3-642-17919-8

Abrudan, A., Rus, T., & Mare, R. (2017). Thermal rehabilitation influence upon the comfort in hospitals. In International Conference on Advancements of Medicine and Health Care through Technology, Cluj-Napoca, Romania (pp. 155-158). Springer. 10.1007/978-3-319-52875-5_35

Adelantado, F., Vilajosana, X., Tuset-Peiro, P., Martinez, B., Melia-Segui, J., & Watteyne, T. (2017). Understanding the limits of LoRaWAN. *IEEE Communications Magazine*, *55*(9), 34–40. doi:10.1109/MCOM.2017.1600613

Aldrich, F. K. (2003). Smart Homes: Past, Present and Future. In R. Harper (Ed.), *Inside the Smart Home* (p. 17). London: Springer. doi:10.1007/1-85233-854-7_2

Algorithm+ SRL & IPCT Installations. AllEnergy® V7.2. (2014). Software for determining the energy performance of buidings. Romania.

Al-Kodmany, K. (2014). Green Retrofitting Skyscrapers: A Review. *Buildings*, *4*(4), 683–710. doi:10.3390/buildings4040683

Allen, J. G., Bernstein, A., Xiadong, C., Eitland, E. S., Flanigan, S., & Gokhale, M. (2017). *The 9 Foundations of a Healthy Building*. Harvard T.H. Chan School of Public Health. Retrieved from http://forhealth.org/Harvard.Building_Evidence_for_Health_the_9_Foundations.pdf

Almeida, C. P., Ramos, A. F., & Silva, M. (2018). Sustainability assessment of building rehabilitation actions in old urban centres. *Sustainable Cities and Society*, *36*, 378-385.

Amin, N. D. M., Akasah, Z. A., & Razzaly, W. (2015). Architectural evaluation of thermal comfort: Sick building syndrome symptoms in engineering education laboratories. *Procedia: Social and Behavioral Sciences*, *204*, 19–28.

Anderson, B. (2002). Conventions for U-value calculations. Watford: CRC Construction Research Communications Ltd.

Anderson, B. (2014). Conventions for U-Value Calculations. Amazon. Retrieved from www.amazon.de/Conventions-U-Value-Calculations-Brian-Anderson/

Araki, A., Kawai, T., Eitaki, Y., Kanazawa, A., Morimoto, K., Nakayama, K., ... Kishi, R. (2010). Relationship between selected indoor volatile organic compounds, so-called microbial VOC, and the prevalence of mucous membrane symptoms in single family homes. *The Science of the Total Environment*, *408*(10), 2208–2215. doi:10.1016/j.scitotenv.2010.02.012 PMID:20188399

Ardente, F., Beccali, M., Cellura, M., & Mistretta, M. (2011). Energy and environmental benefits in public buildings as a result of retrofit actions. *Renewable & Sustainable Energy Reviews*, *15*(1), 460–470. doi:10.1016/j.rser.2010.09.022

Aşchilean, I. (2009). Revamping and upgrade methods for wells. *Acta Technica Napocensis, Section: Civil Engineering – Architecture*, *52*, 271-282.

Aşchilean, I. (2014). *Reabilitarea si modernizarea sistemelor de alimentare cu apa a localitatilor urbane*. Cluj-Napoca, Romania: Ed. RISOPRINT.

Aslani, A., & Mohaghar, A. (2013). Business structure in renewable energy industry: Key areas. *Renewable & Sustainable Energy Reviews*, *27*, 569–575. doi:10.1016/j.rser.2013.07.021

Atanasie, B. (n.d.). Net-Zero Energy Building (nZEB) Implementation In Romania. *Buildings Performance Institute Europe (BPIE)*. Retrieved from www.bpie.eu

Atkins, P. W. (2010). The laws of thermodynamics. A very short introduction. Oxford, New York: Oxford University Press.

Augustin, A., Yi, J., Clausen, T., & Townsley, W. (2016). A Study of LoRa: Long Range & Low Power Networks for the Internet of Things. *Sensors (Basel)*, *16*(9), 1466. doi:10.339016091466 PMID:27618064

Autodesk. (n.d.). Retrieved from https://www.autodesk.com

Ayyad, K., & Fekry, A. (2016). Green Retrofitability Index (GRI) as an Indicator for Decision Makers in Green Retrofitting Projects. *Procedia Environmental Sciences*, *34*, 233–244. doi:10.1016/j.proenv.2016.04.022

Badea, G., Felseghi, R. A., Aşchilean, I., Răboacă, S. M., & Şoimoşan, T. M. (2017). The role of hydrogen as a future solution to energetic and environmental problems for residential building. In AIP Conference Proceedings (Vol. 1917, No. 1, p. 030004). AIP Publishing. doi:10.1063/1.5018277

BAFU. (2013). Ökofaktoren Schweiz 2013 gemäss der Methode der ökologischen Knappheit.

Baheti, R., & Gill, H. (2011). Cyber-physical systems. *The impact of control technology*, *12*(1), 161–166.

Baker, H. K., & English, P. (Eds.). (2011). *Capital Budgeting Valuation. Financial Analyses for Today's Investment Projects*. New Jersey: Wiley. doi:10.1002/9781118258422

Baldwin, A. N., Loveday, D. L., Li, B., Murray, C., & Yu, W. (2018). A research agenda for the retrofitting of residential buildings in China – A case study. *Energy Policy*, *113*, 41–51. doi:10.1016/j.enpol.2017.10.056

Banzhaf, A. (2013). Impacts of 'Basic' Thermal Insulation in Residential Buildings on Fuel Wood Demand and Primary Forests (1st ed.). Aachen, Germany: Shaker.

Bardyn, J.-P., Melly, T., Seller, O., & Sornin, N. (2016). IoT: The era of LPWAN is starting now. In *Proceedings of the 42nd European Solid-State Circuits Conference, ESSCIRC Conference 2016* (pp. 25–30). IEEE. 10.1109/ESSCIRC.2016.7598235

Beinhauer, P. (2017). *Building in stock. with 213 detail design, 2*. Edition.

Bell, C., Nadel, S., & Hayes, S. (2011, December). *On-bill financing for energy efficiency improvements: A review of current program challenges, opportunities and bets practices*. American Council for an Energy-Efficient Economy (ACEEE). Retrieved from http://www.aceee.org/sites/default/files/publications/researchreports/e118.pdf

Belli, P., Anderson, J.R., Barnum, H.N., Dixon, J.A., Tan, J-P, 2001. Economic analysis of investment operations: analytical tools and practical applications. The World Bank.

Bendel, O. (2018). *Smart Home Definition*. Gabler Wirtschaftslexikon. Retrieved from https://wirtschaftslexikon.gabler.de/definition/smart-home-54137/version-277191

Berk, J., & DeMarzo, P. (2017). *Corporate Finance*. Harlow, UK: Pearson Education.

Bertoldi, P., Hirl, B., & Labanca, N. (2012). *Energy Efficiency Status Report 2012. Electricity Consumption and Efficiency Trends in the EU-27.* Retrieved from https://setis.ec.europa.eu/sites/default/files/reports/energy-efficiency-status-report-2012.pdf

Beu, D., & Pop, F. (2001). *Tehnica iluminatului în spaţii industriale, birouri şi locuinţe.* Cluj-Napoca, Romania: Mediamira.

Beunder, A., & Groot, L. (2015). Energy consumption, cultural background and payment structure. *Journal of Cleaner Production, 94*, 137–143. doi:10.1016/j.jclepro.2015.01.083

Blenn, N., & Kuipers, F. (2017). LoRaWAN in the wild: Measurements from the things network. *arXiv preprint arXiv:1706.03086.* Retrieved from https://arxiv.org/abs/1706.03086v1

Boardman, N. E., Greenberg, D. H., Vining, A. R., & Weimer, D. L. (2018). *Cost-benefit Analysis: Concepts and Practice* (4th ed.). Cambridge University Press. doi:10.1017/9781108235594

Boons, F., & Lüdeke-Freund, F. (2013). Business models for sustainable innovation: State-of-the-art and steps towards a research agenda. *Journal of Cleaner Production, 45*, 9–19. doi:10.1016/j.jclepro.2012.07.007

Bor, M., Vidler, J. E., & Roedig, U. (2016). LoRa for the Internet of Things. In *Proceedings of the 2016 International Conference on Embedded Wireless Systems and Networks.*

Bouhal, T., Fertahi, S.-D., Agrouaz, Y., El Rhafiki, T., Kousksou, T., Zeraouli, Y., & Jamil, A. (2018). Technical Assessment, Economic Viability and Investment Risk Analysis of Solar Heating/Cooling Systems in Residential Buildings in Morocco. *Solar Energy, 170*, 1043–1062. doi:10.1016/j.solener.2018.06.032

Bourbeau, J., Brisson, C., & Allaire, S. (1997). Prevalence of the sick building syndrome symptoms in office workers before and six months and three years after being exposed to a building with an improved ventilation system. *Occupational and Environmental Medicine, 54*(1), 49–53. doi:10.1136/oem.54.1.49 PMID:9072034

BP. (2018). BP Statistical Review of World Energy. With assistance of Bob Dudley.

BPIE. (2013) A Guide to Developing Strategies for Building Energy Renovation - Delivering Article 4 of the Energy Efficiency Directive. Retrieved from www.bpie.eu/.../BPIE/Developing_Building_Renovation_Strategies.pdf

Bradski, G., & Kaehler, A. (2008). *Learning OpenCV: Computer vision with the OpenCV library.* O'Reilly Media, Inc.

Broman, G., & Robert, K. (2016). A framework for strategic sustainable development. *Journal of Cleaner Production, 140*(Part 1), 17–31.

Buildings Performance Institute Europe - BPIE. (2011, October). *Europe's buildings under the microscope. A country-by-country review of the energy performance of buildings.* Retrieved from http://bpie.eu/publication/europes-buildings-under-the-microscope/

Burge, S., Hedge, A., Wilson, S., Bass, J. H., & Robertson, A. (1987). Sick Building Syndrome: A Study Of 4373 Office Workers. *The Annals of Occupational Hygiene, 31*(4a), 493–504. PMID:3439759

Burghardt, J., & Fox, A. (2017). *Wirtschaftlichkeitsberechnungen im Bereich der Gebäudedämmung.* Ilmenau.

C107-3. (2005). *Romanian Standard regarding Thermotechnical calculation of construction elements of the building.* FAST PRINT.

Cambridge Dictionary. (n.d.). Refurbishment. Retrieved from https://dictionary.cambridge.org/dictionary/english/refurbishment?q=+refurbishment

Cambridge Dictionary. (n.d.). Renovation. Retrieved from https://dictionary.cambridge.org/dictionary/english/renovation

Cambridge Dictionary. (n.d.). Retrofitting. Retrieved from https://dictionary.cambridge.org/dictionary/english/retrofitting

Canbolat. (2018). Determination of Proper Insulation Thickness for Building Walls Regarding Economic Consideration 2018.

Cascone, S., Catania, F., Gagliano, A., & Sciuto, G. (2018). A comprehensive study on green roof performance for retrofitting existing buildings. *Building and Environment*, *136*, 227–239. doi:10.1016/j.buildenv.2018.03.052

Castells-Rufas, D., & Carrabina, J. (2006). Camera-based digit recognition system. *Networks*, *5*, 6.

Cayla, J. M., Maizi, N., & Marchand, C. (2011). The role of income in energy consumption behaviour: Evidence from French households' data. *Energy Policy*, *39*, 7874–7883. doi:10.1016/j.enpol.2011.09.036

Centenaro, M., Vangelista, L., Zanella, A., & Zorzi, M. (2016). Long-range communications in unlicensed bands: The rising stars in the IoT and smart city scenarios. *IEEE Wireless Communications*, *23*(5), 60–67. doi:10.1109/MWC.2016.7721743

Christersson, M., Vimpari, J., & Junnila, S. (2015). Assessment of financial potential of real estate energy efficiency investments – A discounted cash flow approach. *Sustainable Cities and Society*, *18*, 66–73. doi:10.1016/j.scs.2015.06.002

Claudy, M. C., Peterson, M., & O'Driscoll, A. (2013). Understanding the Attitude-Behavior Gap for Renewable Energy Systems Using Behavioral Reasoning Theory. *Journal of Macromarketing*, *33*(4), 273–287. doi:10.1177/0276146713481605

Clements-Croome, D. (2018). *Creating the Productive Workplace: Places to Work Creatively* (D. Clements-Croome, Ed.). London: Routledge.

Cohen, B. (2006). Urbanization in developing countries: Current trends, future projections, and key challenges for sustainability. *Technology in Society*, *28*(1-2), 63–80. doi:10.1016/j.techsoc.2005.10.005

COM. (2011a). 112 final. *A Roadmap for moving to a competitive low carbon economy in 2050.* Retrieved from http://eur-lex.europa.eu/LexUriServ/LexUriServ.do?uri=CELEX:52011DC0112:EN:NOT

COM. (2011b). 885 final. *Energy Roadmap 2050.* Retrieved from http://eur-lex.europa.eu/LexUriServ/LexUriServ.do?uri=CELEX:52011DC0885:EN:NOT

COM. (2011c). 571 final. *Roadmap to a Resource Efficient Europe.* Retrieved from: http://ec.europa.eu/environment/resource_efficiency/pdf/com2011_571.pdf

Communication from the commission to the European Parliament. (2011). *A Roadmap for moving to a competitive low carbon economy in 2050.* Retrieved from https://eur-lex.europa.eu/legal-content/RO/TXT/?uri=celex%3A52011DC0112

Copleand, B. J. (2019). *Artificial Intelligence.* Encyclopedia Britannica. Retrieved from https://www.britannica.com/technology/artificial-intelligence

Corrado, V., Fabrizio, E., & Filippi, M. (2007, june). Modelling and optimization of multi-energy source building systems in the design concept phase. *Proceedings of Clima 2007 WellBeing Indoors.*

Cortez, P., Felix, J., Girão, A., Frota, J., & Bessa, J. (2014). An OCR system for numerals applied to energy meters. *IEEE Latin America Transactions*, *12*(6), 957–964. doi:10.1109/TLA.2014.6893986

Costanza, R., Fisher, B., Ali, S., Beer, C., Bond, L., Boumans, R., ... Snapp, R. (2007). Quality of life: An approach integrating opportunities, human needs, and subjective well-being. *Ecological Economics*, *61*(2-3), 267–276. doi:10.1016/j.ecolecon.2006.02.023

Cravioto, J., Yamasue, E., Okumura, H., & Ishihara, K. (2011). *Performance analysis between well-being, energy and environmental indicators using Data Envelopment Analysis.* Japan: Kyoto University; doi:10.1007/978-4-431-53910-0_6

Crockford, D. (2006). *The application/json media type for javascript object notation (json)* (No. No. RFC 4627).

Crook, B., & Burton, N. C. (2010). Indoor Moulds, Sick Building Syndrome and Building Related Illness. *Fungal Biology Reviews*, *24*(3), 106–113. doi:10.1016/j.fbr.2010.05.001

Csoknyai, T., Hrabovszky-Horvátha, S., Georgiev, Z., Jovanovic-Popovic, M., Stankovic, B., Villatoro, O., & Szendro, G. (2015). Building stock characteristics and energy performance of residential buildings in Eastern-European countries. *Energy and Buildings,* 132(15), *39-52.* Retrieved from www.sciencedirect.com

Dabija, A.-M. (2018). Sustainability - A Principle of Education in Architecture (and not only). *Paper presented at Conference on Sustainable Energy*. Retrieved from www.researchgate.net

Danna, K., & Griffin, R. (1999). Health and Well-Being in the Workplace:A Review and Synthesis of the Literature. *Journal of Management*, *25*(3), 357–384. doi:10.1177/014920639902500305

Daugherty, P. R., & Wilson, H. J. (2018). *Human + machine: Reimaging work in the age of AI*. Harvard Business Press.

de Dear, R., & Brager, G. S. (1998). Developing an adaptive model of thermal comfort and preference. UC Berkeley. Retrieved from www.escholarship.org

Deac, A., & Iacob, C. A. (2017). Reverse osmosis, the optimal solution for water treatment with membranes technology. *Bulletin of the Transylvania - Series I: Engineering Sciences*, *10*(59).

DENA. (2018). dena CONCISE 2018 BUILDING REPORT. Energy efficiency in the building stock – statistics and analyses.

DENA. (2019). Gebäude-Energiewende: Building Energy Transformation.

DENA. (2019). Gebäude-Energiewende: Building Energy Transformation. Retrieved from https://www.geea.info/fileadmin/Downloads/Pressemitteilungen/2019/geea_PB_2019_final.pdf

Designing Buildings Wiki. (n.d.). Retrieved from https://www.designingbuildings.co.uk/wiki/

Discher H., & Hinz E. (2010). dena-Sanierungsstudie_Teil_1.

Discher, H., & Hinz, E. (2010). dena-Sanierungsstudie_Teil_1. Retrieved from https://shop.dena.de/fileadmin/denashop/media/Downloads_Dateien/bau/9122_dena-Sanierungsstudie_Teil_1.pdf

Dixon, T., Lannon, S., & Eames, M. (2018). Reflections on disruptive energy innovation in urban retrofitting: Methodology, practice and policy. *Energy Research & Social Science*, *37*, 255–259. doi:10.1016/j.erss.2017.10.009

Domniţa, F., Popovici, T., & Hoţupan, A. (2011). *Instalaţii de ventilare şi condiţionare* (Vol. 2). Cluj-Napoca, Romania: U.T. Press.

Dorsch, L., Kaiser, C., Niklasch, W., Schöpgens, H., & Spritzendorfer, J. (Eds.). (2017). *Market Overview Insulation materials from renewable raw materials. Fachagentur Nachwachsende Rohstoffe (9th ed.). Gülzow-Prüzen: Fachagentur Nachwachsende Rohstoffe e.V.* Naturbaustoffe.

Dougall, S. (2018). *Smart city projects to watch in 2018*. Cbronline. Retrieved from https://www.cbronline.com/internet-of-things/smart-cities/smart-city-projects/

Dragotă, V., Obreja-Braşoveanu, L., & Dragotă, I. M. (2016). *Management financiar. Management financiar strategic*. Bucharest: Editura Economică.

Duan, Y., Edwards, J. S., & Dwivedi, Y. K. (2019). Artificial intelligence for decision making in the era of Big Data – evolution, challenges and research data. *International Journal of Information Management*, *48*, 63–71. doi:10.1016/j.ijinfomgt.2019.01.021

Eames, M., Dixon, T., Lannon, S., Hunt, M., De Laurentis, C., Marvin, S., . . . Georgiadou, M. C. (2014). Retrofit 2050: Critical Challenges for Urban Transitions. Retrieved from http://centaur.reading.ac.uk/36187/1/critical%20challenges%20briefing-March%202014.pdf

Ebel, M., Drath, R., & Sauer, O. (2008). Automatische Projektierung eines Produktionsleitsystems der Fertigungstechnik mit Hilfe des Datenaustauschformats CAEX. *Automatisierungstechnische Praxis*, *50*(5), 40.

EBPD. (2018). New Energy Performance in Buildings Directive comes into force on 9 July 2018.

Eicker, U. (2012). Renewable Energy Sources within Urban Areas: Results From European Case. *Academic Journal of ASHRAE Transactions*, (118), 73.

El-Darwish, I., & Gomaa, M. (2017). Retrofitting strategy for building envelopes to achieve energy efficiency. *Alexandria Engineering Journal*, *56*(4), 579-589. Retrieved from www.sciencedirect.com

El-Darwish, I., & Gomaa, M. (2017). Retrofitting strategy for building envelopes to achieve energy efficiency. *Alexandria Engineering Journal*, *56*, 579–589. doi:10.1016/j.aej.2017.05.011

ElMaraghy, H. A. (2005). Flexible and reconfigurable manufacturing systems paradigms. *International Journal of Flexible Manufacturing Systems*, *17*(4), 261–276. doi:10.100710696-006-9028-7

Elrefaei, L. A., Bajaber, A., Natheir, S., AbuSanab, N., & Bazi, M. (2015). Automatic electricity meter reading based on image processing. In *2015 IEEE Jordan Conference on Applied Electrical Engineering and Computing Technologies (AEECT)* (pp. 1–5). IEEE. doi:10.1109/AEECT.2015.7360571

Energy Audit Tool Manual (EATM), Energy Efficiency Management Office, Republic of Mauritius. Retrieved from http://eemo.govmu.org/English/Documents/Manual_Audit_Tool.pdf

Engineering Toolbox. (n.d.). Combustion of Fuels - Carbon Dioxide Emission. Retrieved from https://www.engineeringtoolbox.com/co2-emission-fuels-d_1085.html

Engvall, K., Norrby, C., & Norback, D. (2001). Sick building syndrome in relation to building dampness in multi-family residential buildings in Stockholm. *International Archives of Occupational and Environmental Health*, *74*(4), 270–278. doi:10.1007004200000218 PMID:11401019

Environmental Protection Agency (EPA). (1991). *Indoor Air Facts No. 4 (Revised) Sick Building Syndrome*. Retrieved from https://www.epa.gov/sites/production/files/201408/documents/sick_building_factsheet.pdf

Epstein, Y., & Moran, D. S. (2006). Thermal comfort and the heat stress indices. *Industrial Health*, *44*(3), 388–398. doi:10.2486/indhealth.44.388 PMID:16922182

Eremia, M., Toma, L., & Sanduleac, M. (2016). The Smart City Concept in the 21st Century. *Procedia Engineering*, *181*, 14. Retrieved from https://www.sciencedirect.com/science/article/pii/S1877705817309402

Eriksson, E., Nenonen, S., Junghans, A., & et al, . (2015). Nordic campus retrofitting concepts - Scalable practices. *Procedia Economics and Finance*, *21*, 329–336. doi:10.1016/S2212-5671(15)00184-7

Erol-Kantarci, M., & Mouftah, H. T. (2011). Wireless sensor networks for cost-efficient residential energy management in the smart grid. *IEEE Transactions on Smart Grid*, *2*(2), 314–325. doi:10.1109/TSG.2011.2114678

European Comission. (2015). Europe 2020 Strategy. Retrieved from: https://ec.europa.eu/info/business-economy-euro/economic-and-fiscal-policy-coordination/eu-economic-governance-monitoring-prevention-correction/european-semester/framework/europe-2020-strategy_en

European Commission DG Environment News Alert Service. (2015). Energy efficiency policies for home renovations and retrofitting should consider the social factors. *Science for Environment Policy*. Retrieved from http://ec.europa.eu/environment/integration/research/newsalert/pdf/energy_efficiency_policies_home%20renovations%20_retrofitting_should_consider_social_factors_49si5_en.pdf

European Commission. (2015). *EU Buildings Database*. Retrieved from https://ec.europa.eu/energy/en/eu-buildings-database#eu_buildings_countries_tab

European Commission. (2015). *Guide to Cost-Benefit Analysis of Investment Projects - Economic appraisal tool for Cohesion Policy 2014-2020*. European Union.

European Parliament. (2012). Directive 2012/27/UE of The European Parliament and of the Council of 25 October 2012 on energy efficiency, amending Directives 2009/125/EC and 2010/30/EU and repealing Directives 2004/8/EC and 2006/32/EC. *Official Journal of the European Union.*

European Union. (n.d.). European Commission Directives. Retrieved from www.ec.europa.eu

Eurostat. (n.d.). Retrieved from https://ec.europa.eu/eurostat

Evans, D. (2011). *The Internet of Things*. How the Next Evolution of the Internet Is Changing Everything. CISCO.

Evans, G. W., & McCoy, J. M. (1998). When buildings don't work: The role of architecture in human health. *Env Pschy*, *18*(1), 85–94. doi:10.1006/jevp.1998.0089

Facebrands. (2018). Retrieved from www.facebrands.ro

Fanger, P. O. (1967). Calculations of thermal comfort: Introduction of a basic comfort equation. *ASHRAE Trans*, *73*, 1–4.

Fantana, N. L., Riedel, T., Schlick, J., Ferber, S., Hupp, J., Miles, S., ... & Svensson, S. (2013). IoT applications—value creation for industry. Internet of things: Converging technologies for smart environments and integrated ecosystems, 153.

Fay, M., & Opal, C. (1999). Urbanization without Growth: A not so uncommon Phenomenon. World Bank. doi:10.1596/1813-9450-2412

Feitosa, R. C., & Wilkinson, S. J. (2018). Attenuating heat stress through green roof and green wall retrofit. *Building and Environment*, *140*, 11–22. doi:10.1016/j.buildenv.2018.05.034

Felseghi, R. A. (2015). *Contributions regarding the application of fuel cells in the passive house domain.* Unpublished doctoral dissertation, Technical University of Cluj-Napoca, Romania.

Figl, H. (2017). Ökobaudat Basis for the building life cycle assessment.

Filippi, M. (2015). Remarks on the green retrofitting of historic buildings in Italy. *Energy and Building*, *95*, 15–22. doi:10.1016/j.enbuild.2014.11.001

Fill, C. (2009). *Marketing Communications: Interactivity, Communities and Content*. Financial Times/ Prentice Hall.

Fouad, N. A. (2015). *Bauphysik Kalender 2015. Schwerpunkt: Simulations- und Berechnungsverfahren.* Wiley. doi:10.1002/9783433605158

Fouquet, R., & Pearson, P. (2011). The Long Run Demand for Lighting: Elasticities and Rebound Effects in Different Phases of Economic Development. *Economics of Energy & Environmental Policy*, *1*(1), 83–100.

Franca, C., Broman, G., Robert, K., & et al, . (2016). An approach to business model innovation and design for strategic sustainable development. *Journal of Cleaner Production, 140*, 155–166. doi:10.1016/j.jclepro.2016.06.124

Frej, A., & Browning, W. (2005). *Green office buildings: a practical guide to development.* Washington, D.C.: Urban Land Institute.

Friedman, H., & Piette, M. A. (2001). Comparison of Emerging Diagnostic Tools for Large Commercial HVAC Systems. In 7th National Conference on Building Commissioning, Cherry Hill, NJ, May 9-11. Retrieved from https://eta.lbl.gov/sites/all/files/publications/lbnl-47698.pdf

Friedman, C., Becker, N., & Erell, E. (2014). Energy retrofit of residential building envelopes in Israel: A cost-benefit analysis. *Energy, 77*, 183–193. doi:10.1016/j.energy.2014.06.019

Fritz, R. (2018). Is X-10 an obsolete technology? *Lifewire*. Retrieved from https://www.lifewire.com/is-x-10-obsolete-818408

Fritz, R. (2018). What Is the Best Home Automation Technology? *Lifewire*. Retrieved from https://www.lifewire.com/best-home-automation-technology-817601

Gabriel, I., & Ladener, H. (Eds.). (2018). From the old building to the efficiency house. In Modernize and energetically refurbish: planning, construction practice (13th ed.). Staufen bei Freiburg: Ökobuch.

Gann, D., Barlow, J., & Venables, T. (1999). *Digital futures – making homes smarter.* Coventry: Chartered Institute of Housing.

Gaofeng, Ch. (2012). *Energy efficiency and carbon dioxide emissions reduction.* Elsevier Journal, Resources. *Conservation and Recycling, 60*, 147–158.

Garbuzova, M., & Madlener, R. (2012). Towards an efficient and low carbon economy post-2012: Opportunities and barriers for foreign companies in the Russian energy market. *Mitigation and Adaptation Strategies for Global Change, 17*(4), 387–413. doi:10.100711027-011-9332-8

Garzon. (2018). Future requirement for sustainable building envelope.

Geidl, M., & Andersson, G. (2005). A modeling and optimization approach for multiple energy carrier power flow. *Proc. IEE PES Power Tech*, 1-7. 10.1109/PTC.2005.4524640

Geidl, M., Koeppel, G., & Favre-Perrod, P. (2007, March). The energy hub - A powerfull concept for future energy systems. *Third Annual Carnegie Mellon Conference on the Electricity Industry.*

Geissdoerfer, M., Savaget, P., & Evans, S. (2017). The Cambridge Business Model Innovation Process. *Procedia Manufacturing, 8*, 262–269. doi:10.1016/j.promfg.2017.02.033

Gelfand, L., & Duncan, C. (2012). *Sustainable Renovation. Strategies for Commercial Building Systems and Envelope.* Wiley.

Gellert, R., Pfundstein, M., Rodolphi, A., & Spitzner, M. H. (2007). Insulation materials. In Basics, Materials, Application. München: DETAIL.

Gellert, R., Pfundstein, M., Rodolphi, A., & Spitzner, M. H. (2007). *Dämmstoffe. Grundlagen, Materialien, Anwendungen.* München: DETAIL.

Genon, G., Torchio, M. F., & Poggio, A. (2009). Energy and environmetal assessment of small district heating systems: Global and local effects in two case-studies. *Energy Conversion and Management, 50*(3), 522–529. doi:10.1016/j.enconman.2008.11.010

Ghaffarianhoseini, A., AlWaer, H., Omrany, H., Ghaffarianhoseini, A., Alalouch, C., Clements-Croome, D., & Tookey, J. (2018). Sick building syndrome: Are we doing enough? *Architectural Science Review*, *61*(3), 99–121. doi:10.1080/00038628.2018.1461060

Ghiţă A.-M. (2017). Training platform "Sustainability Consultant". *Solar Decathlon Bucharest.* Retrieved from www.efden.org

Gibbeco. (2019). Ökobilanz - Wärmedämmstoffe - gesundes Haus. Retrieved from https://www.gesundes-haus.ch/dachdeckungen-flachdach/oekobilanz-waermedaemmstoffe.516.html

Goldman, L. (1996). How Sick is Your Building. *Utilities Law Review*, *12*(3), 14–15.

Gomzi, M., & Bobić, J. (2009). Sick Building Syndrome. Do We Live and Work in Unhealthy Environment? *Periodicum Biologorum*, *111*(1), 79–84.

Gomzi, M., Bobić, J., Radosevic-Vidacek, B., Macan, J., Varnai, V. M., Milkovic-Kraus, S., & Kanceljak-Macan, B. (2007). Sick Building Syndrome: Psychological, Somatic, and Environmental Determinants. *Archives of Environmental & Occupational Health*, *62*(3), 147–155. doi:10.3200/AEOH.62.3.147-155 PMID:18400655

Gooch, D., Wolff, A., Kortuem, G., & Brown, R. 2015. Reimagining the Role of Citizens in Smart City Projects. In *Proceedings of the 2015 ACM International Joint Conference on Pervasive and Ubiquitous Computing and Proceedings of the 2015 ACM International Symposium on Wearable Computers UbiComp '15.* New York, NY: ACM, pp.1587–1594 10.1145/2800835.2801622

Gooding, L., & Gul, M. S. (2016). Energy efficiency retrofitting services supply chains: A review of evolving demands from housing policy. *Energy Strategy Reviews*, *11-12*, 29–40. doi:10.1016/j.esr.2016.06.003

Gouldson A.,Kerr N., Millward-Hopkins J., Freeman M., Topi C., & Sullivan R. (2018). Innovative financing models for low carbon transitions: Exploring the case for revolving funds for domestic energy efficiency programmes. *Energy Policy*, *86*, 739-748.

Grade, F., & Donn, M. (2014). Solution sets and net zero energy buildings: A review of 30 Net ZEBs case studies worldwide. IEA Joint SHC Task 40, International Energy Agency.

Greer, C. (2007). Something in the Air: A Critical Review of Literature on the Topic of Sick Building Syndrome. *World Safety Journal*, *16*(1), 23–26.

Greven, A., Keller, G., & Warnecke, G. (2014). Entropy. Princeton: Princeton University Press.

GRI. (2014). ISO GRI G4 Guidelines and ISO 26000:2010, How to use the GRI G4 and ISO 26000 in conjunction, Global reporting initiative and ISO.

Guerreiro, B. V., Lins, R. G., Sun, J., & Schmitt, R. (2018). Definition of Smart Retrofitting: First steps for a company to deploy aspects of Industry 4.0. In *Advances in Manufacturing* (pp. 161–170). Cham: Springer.

Gustav, T. (1938). Reading machine. US Patent 2,115,563.

Häfele, G., Oed, W., & Sabel, L. (2017). *House renovation. Repair - modernize - save energy - rebuild: ecological construction practice* (16th ed.). Staufen bei Freiburg: ökobuch.

Hamblen, M. (2018). *Economic Benefits Of A Smart City*, Retrieved from https://www.smartresilient.com/economic-benefits-smart-city

Hamilton, L. (2010). *Sick Building Syndrome. When good air goes bad.* Lynchburg, VA: Strategic Sustainability Consulting.

Hannelius, T., Salmenpera, M., & Kuikka, S. (2008). Roadmap to adopting OPC UA. In *6th IEEE International Conference on Industrial Informatics INDIN 2008* (pp. 756–761). IEEE.

Harper, R. (2003). Inside the Smart Home: Ideas, Possibilities and Methods. In R. Harper (Ed.), *Inside the Smart Home*. London: Springer. doi:10.1007/1-85233-854-7_1

Hassa, S. (2013). Wirtschaftlichkeitsbetrachtung von nachträglicher Wärmedämmung am Wohnhausbestand in Deutschland. Aachen: Shaker.

Hast, A. Alimohammadisagvand, & B, Syri, S. (2015). Consumer attitudes towards renewable energy in China - The case of Shanghai. *Sustainable Cities and Society*, *17*, 69–79. doi:10.1016/j.scs.2015.04.003

Haves, P. (1999). Overview of Diagnostic Methods. In *Proceedings of the Workshop on Diagnostics for Commercial Buildings: From Research to Practice*, Pacific Energy Center, San Francisco, CA, June 16-17.

Hermelink, S. (2013). Towards nearly zero energy buildings.

Hess, D. B., Tammaru, T., & van Ham, M. (Eds.). (2018). Housing Estates in Europe. In Poverty, Ethnic Segregation and Policy Challenges. Cham: Springer International Publishing. doi:10.1007/978-3-319-92813-5

He, Y., Liao, N., Bi, J., & Guo, L. (2019). Investment decision-making optimization of energy efficiency retrofit measures in multiple buildings under financing budgetary restraint. *Journal of Cleaner Production*, *215*, 1078–1094. doi:10.1016/j.jclepro.2019.01.119

Hillary, R. (1993). *The eco-management and audit scheme: a practical guide*. Hertfordshire, UK: Technical Communications Letchworth.

Hollands, R. G. (2008). Will the real smart city please stand up? *City*, *12*(3), 303–320. doi:10.1080/13604810802479126

Holm, A. (2015). Wirtschaftlichkeit_daemmung_gdi_studie_2015.

Honig, J. M. (2013). *Thermodynamics. Principles Characterizing Physical and Chemical Processes* (4th ed.). Burlington: Elsevier Science.

Hopfe, C. J., & McLeod, R. S. (2015). *The Passivhaus Designer's Manual: A technical guide to low and zero energy buildings*. Routledge.

Huijben, J., & Verbong, G. (2013). Breakthrough without subsidies? PV business model experiments in the Netherlands. *Energy Policy*, *56*, 362–370. doi:10.1016/j.enpol.2012.12.073

Hunkeler, U., Truong, H. L., & Stanford-Clark, A. (2008). MQTT-S—A publish/subscribe protocol for Wireless Sensor Networks. In *Proceedings of the 3rd international conference on Communication systems software and middleware and workshops COMSWARE 2008* (pp. 791–798). IEEE.

Hunt, R. G. (Ed.). (2003). Comparative life cycle assessment of three insulation materials: stone wool, flax and paper wool. Landsberg, Bayreuth: Eco Informa Press.

Hunt, R. G. (Ed.). (2003). Comparative life cycle assessment of three insulation materials: stone wool, flax and paper wool. Unter Mitarbeit von Anders Schmidt. Landsberg, Bayreuth: ECO INFORMA Press.

Hutter, C.-P. (2018). *Die Erde rechnet ab. Wie der Klimawandel unser tägliches Leben verändert - und was wir noch tun können*. München: Ludwig.

Iacob, C., Deac, A., Chira, T., & Bolboaca, A. (2018). *Study of performance indicators at a waste water treatment plant in Romania, 18th International Multidisciplinary Scientific GeoConference SGEM 2018*. Bulgaria: Abena.

IEA. (2013). *Transition to Sustainable Buildings. Strategies and Opportunities to 2050.* Retrieved from https://www.iea.org/publications/freepublications/publication/Building2013_free.pdf

IEE Ecoheatcool and Euroheat & Power. (2005-2006). *Ecoheatcool. Work package 3. Guidelines for assessing the efficiency of district heating and district cooling.* Retrieved from http:// www.ecoheatcool.org

IEEE. (2014). *The Institute Special Report: The Internet of Things.* Retrieved from http://theinstitute.ieee.org/static/special-report-the-internet-of-things

IETF. (2010). *The Internet of Things - Concept and Problem Statement.* Retrieved from http://tools.ietf.org/id/draft-lee-iot-problem-statement-00.txt

Imai, N., & Imai, Y. (2011). Psychosocial Factors That Aggravate the Symptoms of Sick Building Syndrome and a Cure for Them. In S. A. Abdul-Wahab (Ed.), *Sick Building Syndrome* (pp. 105–111). Berlin: Springer. doi:10.1007/978-3-642-17919-8_6

Imtiaz, J., & Jasperneite, J. (2013). Scalability of OPC-UA down to the chip level enables "Internet of Things. In *2013 11th IEEE International Conference on Industrial Informatics (INDIN)* (pp. 500–505). IEEE.

International Well Building Institute. (n.d.). Well-Being Building Standard, v1 with Q3 2018 addenda.

Internet World Stats. (2018). *Internet Usage Statistics.* Retrieved from https://www.internetworldstats.com/stats.htm

IPCC. (2018). Special Report on Global Warming of 1.5 C (SR15), Intergovernmental Panel on Climate Change. Retrieved from http://www.ipcc.ch/report/sr15/

ISO. (2014). ISO 50002 Energy audits – Requirements with guidance for use. Retrieved from https://www.sis.se/api/document/preview/917504/

ITU. (2005). *The Internet of Things Executive Summary, ITU Internet Reports 2005.* Retrieved from http://www.itu.int/osg/spu/publications/internetofthings/InternetofThings_summary.pdf

Iwata, K., Katayama, H., & Arimura, T. H. (2015). Do households misperceive the benefits of energy-saving actions? Evidence from a Japanese household survey. *Energy for Sustainable Development, 25,* 27–33. doi:10.1016/j.esd.2014.12.005

Jagarajana, R., Asmonia, M., Mohammeda, A., & et al, . (2017). Green retrofitting – A review of current status, implementations and challenges. *Renewable & Sustainable Energy Reviews, 67,* 1360–1368. doi:10.1016/j.rser.2016.09.091

Jain, A. K., Mao, J., & Mohiuddin, K. M. (1996). Artificial neural networks: A tutorial. *Computer, 29*(3), 31–44. doi:10.1109/2.485891

Jansz, J. (2011a). Introduction to Sick Building Syndrome. In S. A. Abdul-Wahab (Ed.), *Sick Building Syndrome* (pp. 1–24). Berlin, Germany: Springer.

Jansz, J. (2011b). Theories and Knowledge About Sick Building Syndrome. In S. A. Abdul-Wahab (Ed.), *Sick Building Syndrome* (pp. 25–58). Berlin, Germany: Springer. doi:10.1007/978-3-642-17919-8_2

Jeschke, S., Brecher, C., Song, H., & Rawat, D. B. (Eds.). (2017). *Industrial Internet of Things.* Springer International Publishing. doi:10.1007/978-3-319-42559-7

Joachim Schenk, G. R. (2010). *Mensch-Maschine-Kommunikation.* Springer-Verlag GmbH. Retrieved from https://www.ebook.de/de/product/9280122/joachim_schenk_gerhard_rigoll_mensch_maschine_kommunikation.html

John, A., & Duffie Beckman, W. A. (2013). Solar Engineering of Thermal Processes. John Wiley & Sons, Inc.

Joyce, A., & Paquin, R. (2016). The triple layered business model Canvas: A tool to design more sustainable business models. *Journal of Cleaner Production, 135*, 1474–1486. doi:10.1016/j.jclepro.2016.06.067

Judson, E. P., & Maller, C. (2014). Housing renovations and energy efficiency: Insights from homeowners' practices. *Building Research and Information, 42*(4), 501–511. doi:10.1080/09613218.2014.894808

Juniper Research. (2018). Smart cities – what's in it for citizens? Retrieved from https://newsroom.intel.com/wp-content/uploads/sites/11/2018/03/smart-cities-whats-in-it-for-citizens.pdf

Kamaruzzaman, S., Sabrani, N.A. (2011). The Effect of Indoor Air Quality (IAQ) Towards Occupants' Psychological Performance in Office Buildings. *Jurnal Rekabentuk dan Binaan, 4*, 49-61.

Kapalo, P., Domnița, F., Bacoțiu, C., & Spodyniuk, N. (2018). The Impact of Carbon Dioxide Concentration on the Human Health-Case Study. *Journal of Applied Engineering Sciences, 8*(1), 61–66. doi:10.2478/jaes-2018-0008

Kastner, W., & Szucsich, S. (2011). Accessing KNX networks via BACnet/WS. In *Proceedings of the 2011 IEEE International Symposium on Industrial Electronics (ISIE)* (pp. 1315–1320).

KBOB CH. (2016). Ökobilanzdaten im Baubereich 2009/1:2016. Retrieved from https://www.kbob.admin.ch/kbob/de/home/publikationen/nachhaltiges-bauen/oekobilanzdaten_baubereich.html

Khorshidi, M. (2018). Research for developing sustainable business models for a national oil company in the crude oil supply surplus situation [PhD. Thesis]. The Bucharest University of Economic Studies, ASE Bucharest.

Klenner, W. (2009). Biodiversity in forest ecosystems and landscapes: A conference to discuss future directions in biodiversity management for sustainable forestry. Forest Ecology and Management, 258. doi:10.1016/j.foreco.2009.10.037

Klöpffer, W., & Grahl, B. (2009). *Ökobilanz (LCA). Ein Leitfaden für Ausbildung und Beruf.* Weinheim: Wiley. doi:10.1002/9783527627158

Koezjakov, A., Urge-Vorsatz, D., Crijns-Graus, W., & van den Broek, M. (2018). The relationship between operational energy demand and embodied energy in Dutch residential buildings. *Energy and Building, 165*, 233–245. doi:10.1016/j.enbuild.2018.01.036

Komninos, N. (2013). What makes cities intelligent? In *M. Deakin (Ed.), Smart Cities: Governing, Modelling and Analysing the Transition* (p. 77). Taylor and Francis.

Koo, J., & Kim, Y.-G. (2017). Interoperability of device identification in heterogeneous IoT platforms. In *Proceedings of the 2017 13th International Computer Engineering Conference (ICENCO)* (pp. 26–29). IEEE.

Krieger O. (2017). Gebäudestudie. Szenarien für eine marktwirtschaftliche Klima- und Ressourcenschutzpolitik 2050 im Gebäudesektor.

Krutwig, M., & Starosta, K. (2017). Characterization, Classification and Assessment of Non-Energy Benefits of Energy Efficiency Measures. In *Proceedings of 30th IBIMA Conference, Madrid 2017.* IBIMA.

Kukec, A., & Dovjak, M. (2014a). *Prevention and control of Sick Building Syndrome (SBS). Part 1: Identification of risk factors, Slovenia n Public and Environmental Health Professionals.* Ljubljana, Slovenia: University of Ljubljana, Faculty of Medicine. Retrieved from https://journal.institut-isi.si/prevention-and-control-of-sick-building-syndrome-sbs-part-1-identification-of-risk-factors/

Kukec, A., & Dovjak, M. (2014b). *Prevention and control of Sick Building Syndrome (SBS). Part 2: Design of a preventive and control strategy to lower the occurrence of SBS, Slovenia n Public and Environmental Health Professionals.* Ljubljana, Slovenia: University of Ljubljana, Faculty of Medicine. Retrieved from https://journal.institut-isi.si/wp-content/uploads/2014/10/Prevention-and-control-of-sick-building-syndrom-SBS.-Part-2.pdf

Kunič, R. (2017). Carbon footprint of thermal insulation materials in building envelopes.

Kurnitskia, J., Saarib, A., Kalameesc, T., Vuolled, M., Niemeläd, J., & Tark, T. (2011). Optimal and nearly zero (nZEB) energy performance calculations for residential buildings with REHVA definition for nZEB national implementation. *Energy and Buildings, 43*(11). Retrieved from www.sciencedirect.com

Lauren, N., Fielding, K. S., Smith, L., & Louis, W. R. (2016). You did, so you can and you will: Self-efficacy as a mediator of spillover from easy to more difficult pro-environmental behavior. *Journal of Environmental Psychology, 48*, 191–199. doi:10.1016/j.jenvp.2016.10.004

Lauridsen, M., Kovacs, I. Z., Mogensen, P., Sorensen, M., & Holst, S. (2016). Coverage and Capacity Analysis of LTE-M and NB-IoT in a Rural Area. In *Proceedings of the 2016 IEEE 84th Vehicular Technology Conference (VTC-Fall).* IEEE. doi:10.1109/vtcfall.2016.7880946

Lee, J., Je, H., & Byun, J. (2011). Well-Being index of super tall residential buildings in Korea. *Building and Environment, 46*(5), 1184–1194. doi:10.1016/j.buildenv.2010.12.010

Leitner, S.-H., & Mahnke, W. (2006). *OPC UA–service-oriented architecture for industrial applications.* ABB Corporate Research Center.

Leonardo energy. (2015). How many types of energy audit are there? Copper Alliance Initiative and partners. Retrieved from https://help.leonardo-energy.org/hc/en-us/articles/203598661-How-many-types-of-energy-audit-are-there-

Litti, G., Audenaert, A., & Lavagna, M. (2018). Life cycle operating energy saving from windows retrofitting in heritage buildings accounting for technical performance decay. *Journal of Building Engineering, 17*, 135–153. doi:10.1016/j.jobe.2018.02.006

Liu, Y., Liu, T., Ye, S., & Liu, Y. (2018). Cost-benefit analysis for Energy Efficiency Retrofit of existing buildings: A case study in China. *Journal of Cleaner Production, 177*, 493–506. doi:10.1016/j.jclepro.2017.12.225

Lizana, J., Barrios-Padura, A., Molina-Huelva, M., & Chacartegui, R. (2016). Multi-criteria assessment for the effective decision management in residential energy retrofitting. *Energy and Building, 129*, 284–307. doi:10.1016/j.enbuild.2016.07.043

Long, N., Hirsch, A., Lobato, C., & Daniel Macumber, D. (2010). Commercial building design pathways using optimization analysis, National Renewable Energy Laboratory. In *2010 ACEEE Summer Study on Energy Efficiency in Buildings.* Retrieved from https://aceee.org/files/proceedings/2010/data/papers/2007.pdf

Lüdtke, N., & Henkel, A. (Eds.). (2018). Das Wissen der Nachhaltigkeit. München: Oekom Verlag.

Ma, Z., Cooper, P., Daly, D., & Ledo, L. (2012). Existing building retrofits: Methodology and state-of-the-art. *Energy and Buildings, 55.* doi:10.1016/j.enbuild.2012.08.018

Madlenera, R., & Vögtli, S. (2008). Diffusion of bioenergy in urban areas: A socio-economic analysis of the Swiss wood-fired cogeneration plant in Basel. *Biomass and Bioenergy, 32*(9), 815–828. doi:10.1016/j.biombioe.2008.01.006

Mahapatra, K., Gustavsson, L., Haavik, T., & Aabrekk, S. (2011). *Report on possible financing schemes for one-stop-shop service for sustainable renovation of single-family house.* Nordic Innovation Publications.

Mahnke, W., Leitner, S.-H., & Damm, M. (2009). *OPC Unified Architecture*. Springer Berlin Heidelberg; doi:10.1007/978-3-540-68899-0

Marino, A., Bertoldi, P., Rezessy, S., & Boza-Kiss, B. (2011). A snapshot of the European energy service market in 2010 and policy recommendations to foster a further market development. *Energy Policy*, *39*(10), 6190–6198. doi:10.1016/j.enpol.2011.07.019

Markets Insider. (n.d.). CO2 European emission allowances. Retrieved form https://markets.businessinsider.com/commodities/co2-emissionsrechte

Marszal, A. J., Heiselberga, P., Bourrelle, J. S., Musall, E., Voss, K., Sartori, I., & Napolitanoe, A. (2010). Zero Energy Building – A review of definitions and calculation methodologies. *Energy and Buildings, 43*(4). Retrieved from www.sciencedirect.com

Mârza, C., Hoţupan, A., Moldovan, R., & Corsiuc, G. (2013). *Surse neconvenţionale de energie*. Cluj-Napoca, Romania: U. T. Press.

Matzler, S., Wollschlaeger, M., Fernbach, A., Kastner, W., & Huschke, M. (2013). An OPC UA cross-domain information model for energy management in automation systems. In *Industrial Electronics Society, IECON 2013-39th Annual Conference of the IEEE* (pp. 7513–7518). IEEE.

Medina, B. E., & Manera, L. T. (2017). Retrofit of air conditioning systems through an wireless sensor and actuator network: An IoT-based application for smart buildings. In *Proceedings of the 2017 IEEE 14th International Conference on Networking, Sensing and Control (ICNSC)* (pp. 49–53). IEEE.

Meyer, C. (2015). *Energetische Sanierung von Gebäuden im Bestand. Wirtschaftlichkeit und Maßnahmen am Beispiel von drei Einfamilienhäusern*. Hamburg: Diplomica Verlag.

Milani, B. (2005). *Building materials in a green economy: Community-based strategies for dematerialization* [PhD Dissertation]. University of Toronto, Canada

Miller, S. (2018). AI: Augmentation, more so then automation. Asian Management Insights, 5(1), 1-20.

Ministerul Dezvoltării Regionale, Administraţiei Publice şi Fondurilor Europene. (n.d.). Normative for the Design, Execution an Operation of Ventilation and Air-conditioning Systems Indicative I5.

Ministerul Dezvoltării Regionale. (2016). Administraţiei Publice şi Fondurilor Europene - MDRAP Strategia Energetică a României 2016-2030, cu perspectiva anului 2050. Retrieved from http://www.mmediu.gov.ro/app/webroot/uploads/files/2017-03-02_Strategia-Energetica-a-Romaniei-2016-2030.pdf

Ministerul Dezvoltării Regionale. Administraţiei Publice şi Fondurilor Europene - MDRAP (2017a, January). Strategia naţională a locuirii. Retrieved from http://www.mmediu.ro/app/webroot/uploads/files/2017-01-13_Strategia_Nationala_a_Locuirii_2016-2030.pdf

Ministerul Dezvoltării Regionale. Administraţiei Publice şi Fondurilor Europene - MDRAP (2017b, October). Strategia pentru mobilizarea investiţiilor în renovarea fondului de clădiri rezidenţiale şi comerciale, atât publice cât şi private, existente la nivel naţional. Retrieved from http://www.mdrap.ro/userfiles/Strategie_renovare_cladiri_2017%20final_23octombrie2017.pdf

Mishra, A. K., & Ramgopal, M. (2013). Field studies on human thermal comfort—An overview. *Building and Environment, 64*, 94–106. doi:10.1016/j.buildenv.2013.02.015

Moga, L. M., & Moga, I. (2017). *Punţi termice specifice planşeelor terasă, de pod, deasupra subsolului şi plăcilor pe sol la clădiri cu pereţi din zidărie*. Cluj-Napoca, Romania: U.T. Press.

Moga, L., & Bucur, A. (2018). Nano insulation materials for application in nZEB. *Procedia Manufacturing*, *22*, 309–316. doi:10.1016/j.promfg.2018.03.047

Molina, C., Anthony, C., Pickering, C., Valbjorn, O., & Bortoli, M. (1990). *Sick Building Syndrome: A Practical Guide*. Luxembourg: Commission of the European Communities. Retrieved from http://www.buildingecology.com/publications/ECA_Report4.pdf

Monostori, L., Kádár, B., Bauernhansl, T., Kondoh, S., Kumara, S., Reinhart, G., ... Ueda, K. (2016). Cyber-physical systems in manufacturing. *CIRP Annals*, *65*(2), 621–641. doi:10.1016/j.cirp.2016.06.005

Morelo, D. (2019). *What is Internet of Everything*. Retrieved from https://www.bporiver.com/how-to/%20technology/what-is-internet-of-everything

Mori, S., Suen, C. Y., & Yamamoto, K. (1992). Historical review of OCR research and development. *Proceedings of the IEEE*, *80*(7), 1029–1058. doi:10.1109/5.156468

Moschetti, R., & Brattebø, H. (2016). Sustainable business models for deep energy retrofitting of buildings: State-of-the-art and methodological approach. *Energy Procedia*, *96*, 435–445. doi:10.1016/j.egypro.2016.09.174

Mukherjee, S., & Chakraborty, D. (2013). Is environmental sustainability influenced by socioeconomic and sociopolitical factors? Cross-country empirical evidence. *Sustainable Development*, *21*(6), 353–371. doi:10.1002d.502

Musa, S. (2016). *Smart City Roadmap*. Retrieved from https://www.academia.edu/21181336/Smart_City_Roadmap

NASA, Atmospheric Science Data Center. (n.d.). *Surface meteorology and Solar Energy, RETScreen Report August 2015*. Retrieved from https://eosweb.larc.nasa.gov

Nejat, P., Jomehzadeh, F., Taheri, M., Gohari, M., & Majid, M. (2015). A global review of energy consumption, CO2 emissions and policy in the residential sector (with an overview of the top ten CO2 emitting countries). *Renewable & Sustainable Energy Reviews*, *43*, 843–862. doi:10.1016/j.rser.2014.11.066

Neugschwandtner, M., Neugschwandtner, G., & Kastner, W. (2007). Web services in building automation: Mapping knx to obix. In *Proceedings of the 2007 5th IEEE International Conference on Industrial Informatics* (Vol. 1, pp. 87–92). IEEE.

Neuhoff, K., Stelmakh, K., Amecke, H., Novikova, A., Deason, J., & Hobbs, A. (2012). Financial incentives for energy efficiency retrofits in buildings. In ACEEE Summer Study on Energy Efficiency in Buildings (pp. 236–248).

Newsham, G., Brand, J., Donnelly, C., Veitch, J., Aries, M., & Charles, K. (2009). Linking Indoor Environment Conditions to Job Satisfaction: A Field Study. *Building Research and Information*, *37*(2), 129–147. doi:10.1080/09613210802710298

NIBS. (2014). National Institute of Building Sciences (the website). Retrieved from https://web.archive.org/web/20141016190503/http://www.nationalbimstandard.org/faq.php#faq1

Nicolescu, O., & Verboncu, I. (2008). *Fundaments of organization management (Fundamentele managementului organizaţiei)*. Universitara Publishing House.

Norhidayah, A., Chia-Kuanga, L., Azharb, M. K., & Nurulwahidab, S. (2013). Indoor Air Quality and Sick Building Syndrome in Three Selected Buildings. *Procedia Engineering*, *53*, 93–98. doi:10.1016/j.proeng.2013.02.014

Nsiah, K. A., Schappacher, M., Rathfelder, C., Sikora, A., & Groza, V. (2018). An open-source toolkit for retrofit industry 4.0 sensing and monitoring applications. In *Proceedings of the 2018 IEEE International Instrumentation and Measurement Technology Conference (I2MTC)* (pp. 1–6). IEEE. 10.1109/I2MTC.2018.8409633

Oehler, S. (2018). Emissionsfreie Gebäude. Das Konzept der „Ganzheitlichen Sanierung" für die Gebäude der Zukunft.

Oehler, Stefan (2018). Emissionsfreie Gebäude. Das Konzept der „Ganzheitlichen Sanierung" für die Gebäude der Zukunft.

Olander, D. (2007). *General Thermodynamics*. Hoboken, NJ: CRC Press. doi:10.1201/9781420007558

OMA. (2017). Lightweight machine to machine technical specification. *Technical Specification OMA-TS-Lightweight-M2M-V1.0 approved.*

ONU Information Center for Romania. (in press). *Climate changes - A global problem.* Retrieved from http://www.onuinfo.ro/resurse/schimbari_climatice

OPC Foundation. (2017). *OPC UA Specification: Part 1 – Concepts, Core Specification Parts, Version 1.04.* OPC Foundation. Retrieved from https://opcfoundation.org

Østergård, T., Jensen, R. L., & Maagaard, S. (2016). Building simulations supporting decision making in early design - A review. *Renewable & Sustainable Energy Reviews*, *61*, 187–201. doi:10.1016/j.rser.2016.03.045

Osterwalder, A. (2004). *The Business Model Ontology: A proposition in a design science approach.* Universite de Lausanne, Ecole des Haute Etudes Commerciales.

Osterwalder, A., & Pigneur, Y. (2010). *Business model generation – A handbook for visionaries, game changers, and challengers*. Hoboken, NJ: John Wiley & Sons.

Paiho, S., Abdurafikov, R., Hoang, H., & Kuusisto, J. (2015). An analysis of different business models for energy efficient renovation of residential districts in Russian cold regions. *Sustainable Cities and Society*, *14*, 31–42. doi:10.1016/j.scs.2014.07.008

Panetta, K. (2017). Gartner top 10 strategic technology trends for 2018. *Gartner*. Retrieved from https://www.gartner.com/smarterwithgartner/gartner-top-10-strategic-technology-trends-for-2018/

Pardo-Castellote, G. (2003). Omg data-distribution service: Architectural overview. In *Proceedings. 23rd International Conference on Distributed Computing Systems Workshops, 2003.* (pp. 200–206). IEEE.

Passarelli, G. R. (2009). Sick Building Syndrome: An Overview to Raise Awareness. *Journal of Building Appraisal*, *5*(1), 55–66. doi:10.1057/jba.2009.20

Passer, A., Kreiner, H., & Maydl, P. (2012). Assessment of the environmental performance of buildings: A critical evaluation of the influence of technical building equipment on residential buildings. *The International Journal of Life Cycle Assessment*, *17*(9), 1116–1130.

Pätäri, S., & Sinkkonen, K. (2014). Energy service companies and energy performance contracting: Is there a need to renew the business model? Insights from a Delphi study. *Journal of Cleaner Production*, *66*, 264–271. doi:10.1016/j.jclepro.2013.10.017

Pelau, C., & Al Pop, N. (2018). Implications for the energy policy derived from the relation between the cultural dimensions of Hofstede's model and the consumption of renewable energies. *Energy Policy*, *118*, 160–168. doi:10.1016/j.enpol.2018.03.042

Pelau, C., & Chinie, A. C. (2018). Econometric Model for Measuring the Impact of the Education Level of the Population on the Recycling Rate in a Circular Economy. *Amfiteatru Economic*, *20*(48), 340–355.

Petäjäjärvi, J., Mikhaylov, K., Yasmin, R., Hämäläinen, M., & Iinatti, J. (2017). Evaluation of LoRa LPWAN technology for indoor remote health and wellbeing monitoring. *International Journal of Wireless Information Networks*, *24*(2), 153–165. doi:10.100710776-017-0341-8

Pfundstein, M., Rodolphi, A., Spitzner, M. H., & Gellert, R. (2012). Dämmstoffe. Grundlagen, Materialien, Anwendungen: DETAIL.

Piccardo, C., Dodoo, A., Gustavsson, L., & Tettey, U. Y. A. (2019a). Comparative Life-Cycle Analysis of Building Materials for the Thermal Upgrade of an Existing Building. *IOP Conf. Ser.: Earth Environ. Sci., 225*(1), 12044. 10.1088/1755-1315/225/1/012044

Piccardo, C., Dodoo, A., Gustavsson, L., & Tettey, U. Y. A. (2019b). Energy and carbon balance of materials used in a building envelope renovation. *IOP Conf. Ser.: Earth Environ. Sci., 225*(1), 12045. 10.1088/1755-1315/225/1/012045

Pickton, D., & Broderick, A. (2005). *Integrated Marketing Communications*. Prentice Hall.

Pombo, O., Rivela, B., & Neila, J. (2016). The challenge of sustainable building renovation: Assessment of current criteria and future outlook. *Journal of Cleaner Production, 123*, 88–100. doi:10.1016/j.jclepro.2015.06.137

Popa, E. S. (2011). The buildings automation. In *The 11th National Multidisciplinary Conference with International Participation "Professor Dorin Pavel - founder of the Romanian energy,"* Sebes, Romania.

Pressman, R. S. (2005). *Software Engineering: A Practitioner's Approach*. McGraw-Hill Companies Inc.

Rabiner, L. R., & Juang, B.-H. (1986). An introduction to hidden Markov models. *IEEE ASSP Magazine, 3*(1), 4–16.

Ratasuk, R., Vejlgaard, B., Mangalvedhe, N., & Ghosh, A. (2016). NB-IoT system for M2M communication. In *Proceedings of the 2016 IEEE Wireless Communications and Networking Conference (WCNC)*, (pp. 1–5). IEEE. doi:10.1109/WCNC.2016.7564708

Rehbein, D. A., & Pederson, A. (1996). Impact of the OLE for Process Control (OPC) standard on the process industry. In *Proceedings of the Industrial Computing Conference* (Vol. 6, pp. 285–293).

Rexroth, S., May, F., & Zink, U. (Eds.). (2014). *Wärmedämmung von Gebäuden. Zeitgemäß und wandlungsfähig. 1. Neuerscheinung*. Berlin, Germany: VDE VERLAG GMBH.

Reynders, B., & Pollin, S. (2016). Chirp spread spectrum as a modulation technique for long range communication. In *2016 Symposium on Communications and Vehicular Technologies (SCVT)* (pp. 1–5). IEEE. doi:10.1109/SCVT.2016.7797659

Richter, M. (2012). Utilities' business models for renewable energy: A review. *Renewable & Sustainable Energy Reviews, 16*(5), 2483–2493. doi:10.1016/j.rser.2012.01.072

Ricquebourg, V., Menga, D., Durand, D., Marhic, B., Delahoche, L., & Loge, C. (2007). The Smart Home Concept: our immediate future. In *2006 1ST IEEE International Conference on E-Learning in Industrial Electronics*. doi:10.1109/ICELIE.2006.347206

Riedy, C., Lederwasch, A., & Ison, N. (2011). *Defining zero emission buildings – review and recommendations: final report. Prepared for sustainability Victoria by the Institute for Sustainable Futures*. Sydney: University of Technology.

Riley, M., & Cotgrave, A. (2018). Construction Technology 1. House Construction (4th ed.). Oxford: Red Globe Press.

Ritchie, H., & Roser, M. (2018). CO_2 and other Greenhouse Gas Emissions. *OurWorldInData*. Retrieved from https://ourworldindata.org/co2-and-other-greenhouse-gas-emissions

Robichaud, L., & Anantatmula, V. (2010). Greening project management practices for sustainable construction. *Journal of Management Engineering, 27*(1), 48–57. doi:10.1061/(ASCE)ME.1943-5479.0000030

Rocchi, L., Kadzinski, M., Menconi, M. E., Grohmann, D., Miebs, G., Paolotti, L., & Boggia, A. (2018). Sustainability Evaluation of Retrofitting Solutions for Rural Buildings through Life Cycle Approach and Multi-Criteria Analysis. *Energy and Building*, *173*, 281–290. doi:10.1016/j.enbuild.2018.05.032

Rocky Mountain Institute. (2015) How to Calculate and Present Deep Retrofit Value - A Guide for Investors. Boulder, CO, Rocky Mountain Institute. Retrieved from http://www.rmi.org/retrofit_depot_deepretrofitvalue

Romanian Ministry of Transport, Construction, and Tourism. (2006). Calcul methodology of the energy performance of buildings. Indicative Mc 001/2006. Published in Official Monitor, Bucharest, Romania.

Rosca, E., Arnold, M., & Bendul, J. (2016). Business models for sustainable innovation - an empirical analysis of frugal products and services. *Journal of Cleaner Production*, 162, 133-145.

Ross, S.A., Westerfield, R. W., & Jordan (2002). *Fundamentals of Corporate Finance*. USA: The McGraw-Hill Companies.

Rostron, J. (2008). Sick Building Syndrome: A Review of Causes, Consequences and Remedies. *Journal of Retail & Leisure Property*, *7*(4), 291–303. doi:10.1057/rlp.2008.20

Roulet, C. A. (2018). *Well being and energy in buildings, an architectural issue*. France: École Polytechnique Fédérale de Lausanne.

Roven, A. (2019). *AI in Smart Cities*. Retrieved from https://medium.com/neuromation-blog/ai-in-smart-cities-dfe2fa7d2829

Sætnan, A. R., Schneider, I., & Green, N. (2018). *The Politics and Policies of Big Data: Big Data, Big Brother?* Routledge - Taylor & Francis. Retrieved from http://www.uniteforsight.org/global-health-university/quality-data

Sahlberg, B., Gunnbjörnsdottir, M., Soon, A., Jogi, R., Gislason, T., Wieslander, G., ... Norback, D. (2013). Airborne molds and bacteria, microbial volatile organic compounds (MVOC), plasticizers and formaldehyde in dwellings in three North European cities in relation to sick building syndrome (SBS). *The Science of the Total Environment*, *444*(1), 433–440. doi:10.1016/j.scitotenv.2012.10.114 PMID:23280302

Sanhudo, L., Ramos, N.M.M., Martins, J.P., Almeida, R.M.S.F., Barreira, E., Simões, M.L. & Cardoso, V. (2018), Building information modeling for energy retrofitting – A review. *Renewable and Sustainable Energy Reviews*, *89*, 249-260. doi:10.1016/j.rser.2018.03.064

Sardianou, E., & Genoudi, P. (2013). Which factors affect the willingness of consumers to adopt renewable energies? *Renewable Energy*, *57*, 1–4. doi:10.1016/j.renene.2013.01.031

Saxon, R. (2018). Getting the dimensions of BIM into focus. *BIMplus*. Retrieved from http://www.bimplus.co.uk/people/getting-dimensions-bim-focus/

Scheel, C. M., Rosing, W. C., & Farone, A. L. (2001). Possible sources of sick building syndrome in a Tennessee middle school. *Archives of Environmental Health*, *56*(5), 413–417. doi:10.1080/00039890109604476 PMID:11777022

Schlechtendahl, J., Keinert, M., Kretschmer, F., Lechler, A., & Verl, A. (2015). Making existing production systems Industry 4.0-ready. *Production Engineering*, *9*(1), 143–148. doi:10.100711740-014-0586-3

Schleipen, M. (2008). OPC UA supporting the automated engineering of production monitoring and control systems. In *Proceedings of the IEEE International Conference on Emerging Technologies and Factory Automation ETFA 2008* (pp. 640–647). IEEE.

Schlick, J. (2012). Cyber-physical systems in factory automation-Towards the 4th industrial revolution. In *2012 9th IEEE International Workshop on Factory Communication Systems (WFCS),* (pp. 55–55). IEEE.

Schmidt, B., Schmidt, D., & Venzmer, H. (Eds.). (2017). Energielieferant Altbau. Potenziale, Nachhaltigkeit, Objektbeispiele. Berlin: Beuth Verlag GmbH.

Schmidt, B., Schmidt, D., & Venzmer, H. (Eds.). (2017). Energielieferant Altbau. Potenziale, Nachhaltigkeit, Objektbeispiele. Beuth Verlag. Auflage. Berlin: Beuth Verlag GmbH.

Schmidt, T., & Mangold, D. (2010, November). *The multi-functional heat storage in Hamburg-Bramfeld - innovative extension of the oldest German solar energy housing estate.* Solites, IRES 2010, Berlin, Germany.

Schmidt, T., Mangold, D., & Muller-Steinhagen, H. (2003, June). *Seasonal thermal energy storage in Germany.* Solites, ISES Solar World Congress 2003, Goteborg, Schweden.

Schönwiese, C. (2019). *Klimawandel kompakt. Ein globales Problem wissenschaftlich erklärt.* Stuttgart: Borntraeger.

Schröpfer, T., & Menz, S. (2019). Dense and Green Building Typologies. In Research, Policy and Practice Perspectives. Singapore: Springer. doi:10.1007/978-981-13-0713-3

Schröpfer, T. (2016). *Dense + Green. Innovative Building Types for Sustainable Urban Architecture.* Berlin: Birkhäuser.

Scully, P. (2018). The Top 10 IoT Segments in 2018 – based on 1600 real IoT projects. *IoT Analytics.* Retrieved from https://iot-analytics.com/top-10-iot-segments-2018-real-iot-projects/

SDH. (n.d.). *Solar District Heating - The SDH Projects.* Retrieved from http://www.solar-district-heating.eu

SemtechCorp. (2015). *LoRa Modulation Basics, AN1200.22. LoRa Modulation Basics.* Semtech Corporation.

Shahin, H. S. M. (2019). Adaptive building envelopes of multistory buildings as an example of high performance building skins. *Alexandria Engineering Journal, 58*(1), 345–352. doi:10.1016/j.aej.2018.11.013

Shan, X., Zhou, J., Chang, V. W. C., & Yang, E. H. (2016). Comparing Mixing and Displacement Ventilation in Tutorial Rooms: Students' Thermal Comfort, Sick Building Syndromes, and Short-Term Performance. *Building and Environment, 102*, 128–137. doi:10.1016/j.buildenv.2016.03.025

Sheikhi, A., Ranjbar, A. M., & Safe, F. (2011). *Optimal dispatch of a multiple energy carrier system equipped with a CCHP.* Tehran: Department of Electrical Engineering Sharif University of Technology. doi:10.24084/repqj09.675

Shelby, Z., Hartke, K., & Bormann, C. (2014). *RFC7252: The constrained application protocol (CoAP). Internet Engineering Task Force.* IETF.

Simader, G. (2014). *Implementation of the EPBD Energy Performance of Buildings Directive.* EPBD.

Sinha, A., Gupta, R., & Kutnar, A. (2012). *Sustainable Development and Green Buildings.* Retrieved from www.researchgate.net

Sinha, R. S., Wei, Y., & Hwang, S.-H. (2017). A survey on LPWA technology: LoRa and NB-IoT. *Ict Express, 3*(1), 14–21. doi:10.1016/j.icte.2017.03.004

Skagestad, B., & Mildenstein, P. (in press). District Heating and Cooling Connection Handbook. International Energy Agency IEA, District Heating and Cooling DHC. *Combined Heat & Power Association.*

Smith, R. (2007). An overview of the Tesseract OCR engine. In *Ninth International Conference on Document Analysis and Recognition ICDAR 2007* (Vol. 2, pp. 629-633). IEEE.

Sobek, W. (2015). *Residentials by Werner Sobek.* Stuttgart: Avedition.

Social Media Marketing Industry Report. (2018). Socialmediaexaminer. Retrieved from https://www.socialmediaexaminer.com/social-media-marketing-industry-report-2018/

Şoimoşan, T.-M. (2015), *Contributions regarding the optimization of the heating systems in urban localities.* Unpublished doctoral dissertation, Technical University of Cluj-Napoca, Romania.

Şoimoşan, T.-M. (2015). *Contributions regarding the optimization of the heating systems in urban localities* (Unpublished doctoral dissertation). Technical University of Cluj-Napoca, Romania.

Şoimoşan, T.-M., Danku, G., & Felseghi, R. A. (2017). *Termotehnica construcţiilor.* Cluj-Napoca, Romania: U.T. Press.

Şoimoşan, T.-M., Felseghi, R.-A., & Corsiuc, G. D. (2015). Influence of the Operating Regime on the Performances of Thermal- Solar Systems Integrated in Heating Networks. *Applied Mechanics and Materials*, *772*, 531–535. doi:10.4028/www.scientific.net/AMM.772.531

Soon, J. J., & Ahmad, S. A. (2015). Willingly or grudgingly? A meta-analysis on the willingness-to-pay for renewable energy use. *Renewable & Sustainable Energy Reviews*, *44*, 877–887. doi:10.1016/j.rser.2015.01.041

Sornin, N., Luis, M., Eirich, T., Kramp, T., & Hersent, O. (2015). *LoRaWAN specification, available from the LoRa Alliance (technical specification).* LoRa Alliance.

SR EN 1907-1. (2014). *Romanian Standard regarding Heating installations. Heating demand calculation.* Matrix Rom Publisher.

Sřrensen, P. A., Nielsen, J. E., & Battisti, R. (2012). Solar district heating. Collection of fact sheets WP3 - D3.1 & D3.2. Inteligent Energy Europe, IEE.

Stanford Encyclopedia of Philosophy. (2005). *Well-being* (R. Crisp, Ed.). Stanford University; Retrieved from http://plato.stanford.edu/entries

Staniaszek, D. (2013). *A Guide to Developing Strategies for Building Energy Renovation.* Buildings Performance Institute Europe (BPIE). Retrieved from www.bpie.eu

Statista. (2018). Number of monthly active Instagram users from January 2013 to September 2017. Retrieved from www.statista.com/statistics/253577/number-of-monthly-active-instagram-users/

Steiner, T., Eßmann, F., & Dornigg, I. (Eds.). (2018). *Ökologie und Ökonomie des Dämmens. Analyse und Bewertung von Dämmmaßnahmen in der Altbausanierung. Fraunhofer IRB-Verlag.* Stuttgart: Fraunhofer IRB Verlag.

Stocker, E., Tschurtschenthaler, M., & Schrott, L. (2015). Cost-optimal renovation and energy performance: Evidence from existing school buildings in the Alps. *Energy and Building*, *100*, 20–26. doi:10.1016/j.enbuild.2015.04.005

Stop CO2. (n.d.). Nivelul de CO2 pentru producerea unui kwh de electricitate, in scadere. Retrieved from http://www.stopco2.ro/2017/10/23/nivelul-de-co2-pentru-producerea-unui-kwh-de-electricitate-in-scadere/

Stoppok, M., Jess, A., Freitag, R., & Alber, E. (2018). Of culture, consumption and cost: A comparative analysis of household energy consumption in Kenya, Germany and Spain. *Energy Research & Social Science*, *40*, 127–139. doi:10.1016/j.erss.2017.12.004

Sundin, J. (2012). *The experience of living with Sick Building Syndrome.* Unpublished Thesis, NOVIA University of Applied Sciences, Vaasa, Finland.

Swan, W. (Ed.). (2013). *Retrofitting the built environment.* Hoboken, NJ: Wiley Blackwell. doi:10.1002/9781118273463

Takigawa, T., Saijo, Y., Morimoto, K., Nakayama, K., Shibata, E., Tanaka, M., ... Kishi, R. (2012). A Longitudinal Study of Aldehydes and Volatile Organic Compounds Associated with Subjective Symptoms Related to Sick Building Syndrome in new Dwellings in Japan. *The Science of the Total Environment*, *417–418*, 61–67. doi:10.1016/j.scitotenv.2011.12.060 PMID:22265600

Tan, B., Yavuz, Y., Otay, E. N., & Çamlibel, E. (2016). Optimal selection of energy efficiency measures for energy sustainability of existing buildings. *Computers & Operations Research*, *66*, 258–271. doi:10.1016/j.cor.2015.01.013

Tantau, A., & Staiger, R. (2018). Business Models for Renewable Energy Initiatives: Emerging Research and Opportunities. Hershey, PA: IGI Global. doi:10.4018/978-1-5225-2688-9

Tantau, A., & Fratila, L. (2018). *Entrepreneurship and Business Development in the Renewable Energy Field*. Hershey, PA: IGI Global. doi:10.4018/978-1-5225-3625-3

Tantau, A., Pelau, C., & Pop, M. (2018). Fake news in the energy industry. In *Fourth BASIQ International Conference on New Trends in Sustainable Business and Consumption (BASIQ)*, Editura ASE (pp. 204-210).

Tantau, A., & Staiger, R. (2017). *Business Models for Renewable Energy Initiatives*. Hershey, PA: IGI Global.

The American Heritage Dictionary of the English Language. (n.d.). Retrieved from https://ahdictionary.com/

The Energy Audit. (n.d.). Retrieved from www.the-energyaudit.com

theNBS. (n.d.). Retrieved from https://www.theNBS.com

Tolkamp, J., Huijben, J., Mouric, R., Verbong, G., & Bouwknegt, R. (2018). User centered sustainable business model design. The case of energy efficiency in the Netherlands. *Journal of Cleaner Production*, *182*, 755–764. doi:10.1016/j.jclepro.2018.02.032

Torregrossa, M. (2015). *Energy-efficiency investment with special regard to the retrofitting of buildings in Europe. In B. Galgoczi (Ed.), Europe's energy transformation in the austerity trap. European Trade Union Institute*. Brussels: ETUI.

Trotta, G. (2005). The determinants of energy efficient retrofit investments in the English residential sector. *Energy Policy*, *120*, 175–182. doi:10.1016/j.enpol.2018.05.024

U.S. Green Building Council's Leadership in Energy and Environmental Design. (n.d.). Retrieved from https://www.usgbc.org/resources

UN-Habitat. (2013). State of the World's Cities 2012/2013: Prosperity of Cities. Routledge, New York, and UN-Habitat.

United Nations Environment Program -UNEP. (2007). *Buildings and Climate Change: Status, Challenges and Opportunities*.

United Nations. (2004). *World urbanization prospects: the 2003 revision Data tables and highlights*. New York: United Nations.

United Nations. (2018). *World Urbanization Prospects: The 2018 Revision – key facts*. Retrieved from https://population.un.org/wup/Publications/Files/WUP2018-KeyFacts.pdf

Upward, A., & Jones, P. (2016). An ontology for strongly sustainable business models defining an enterprise framework compatible with natural and social science. *Organization & Environment*, *29*(1), 97–123. doi:10.1177/1086026615592933

US Energy Information Administration. (2018). *Residential energy consumption survey (RECS). 2015 RECS survey data*. Retrieved from https://www.eia.gov/consumption/residential/data/2015/index.php?view=characteristics#house

van Hoof, J. (2010). Thermal comfort: Research and practice. *Frontiers in Bioscience*, *15*(1), 765–788. doi:10.2741/3645 PMID:20036845

Van Kranenburg, R., Anzelmo, E., Bassi, A., Caprio, D., Dodson, S., & Ratto, M. (2011). The Internet of Things. In *1st Berlin Symposium on Internet and Society: Exploring the Digital Future.* Retrieved from https://www.researchgate.net/profile/Matt_Ratto/publication/228360933_The_Internet_of_Things/links/0912f513755ebd1e87000000.pdf

VDMA. (2018). *OPC UA Companion Specification Robotics – Part 1: Condition monitoring, asset management, predictive maintenance, vertical integration.* VDMA.

Vejlgaard, B., Lauridsen, M., Nguyen, H., Kovács, I. Z., Mogensen, P., & Sorensen, M. (2017). Coverage and capacity analysis of sigfox, lora, gprs, and nb-iot. In *Proceedings of the 2017 IEEE 85th Vehicular Technology Conference (VTC Spring), Sydney, Australia* (pp. 4–7). 10.1109/VTCSpring.2017.8108666

Vinod Kumar, T. M., & Dahiya, B. (2017). Smart Economy in Smart Cities. In T. Vinod Kumar (Ed.), *Smart Economy in Smart Cities. Advances in 21st Century Human Settlements.* Singapore: Springer; doi:10.1007/978-981-10-1610-3_1

Volland, B. (2016). The role of risk and trust attitudes in explaining residential energy demand: Evidence from the United Kingdom. *Ecological Economics, 132*, 14–30. doi:10.1016/j.ecolecon.2016.10.002

Von Neumann, J. (1974). *The Computer and the Brain.* New Haven: Yale University Press.

Walters, J. (2018, *March 20*). Plight of Phoenix: how long can the world's 'least sustainable' city survive? *The Guardian.* Retrieved from https://www.theguardian.com/cities/2018/mar/20/phoenix-least-sustainable-city-survive-water

Wanga, C., Yong, K., & Chob, Y. K. (2015). Performance Evaluation of Automatically Generated BIM from Laser Scanner Data for Sustainability Analyses. In *International Conference on Sustainable Design, Engineering and Construction.* Elsevier Ltd. doi: Retrieved from https://www.sciencedirect.com/science/article/pii/S1877705815021864 doi:10.1016/j.proeng.2015.08.531

Wang, B., Xia, X., & Zhang, J. (2014). A multi-objective optimization model for the life-cycle cost analysis and retrofitting planning of buildings. *Energy and Building, 77*, 227–235. doi:10.1016/j.enbuild.2014.03.025

Wang, J. (2012). *Modern Thermodynamics. Based on the Extended Carnot Theorem. 2011.* Berlin, Germany: Springer.

Westermann, R., Meyer, Ü., Beobachter, D. S., Zeugin, K., Bolliger, B., & Faller, K. (2013). *Mein Haus energetisch sanieren. Komfort verbessern, Kosten senken.* Klima schützen: Beobachter.

Westermann, R.; Meyer, Ü.; Beobachter, D. S.; Zeugin, K.; Bolliger, B.; Faller, K. (2010). Mein Haus energetisch sanieren. Komfort verbessern, Kosten senken, Klima schützen: Beobachter-Edition.

Wiegman, J. W. E., & van der Kolk, E. (2012). Building integrated thin film luminescent solar concentrators: Detailed efficiency characterization and light transport modelling. *Solar Energy Materials and Solar Cells, 103*, 41–47. doi:10.1016/j.solmat.2012.04.016

Wiener, N. (1966). *God and Golem, Inc* (pp. 20–22). Cambridge, MA, USA: MIT Press.

Wikipedia. (n.d.). Retrieved from https://en.wikipedia.org/wiki/

Wilkinson S., (2012). Analysing sustainable retrofit potential in premium office buildings, *Struct Survey/ International Journal of Building Pathology and Adaptation, 30*(5).

Wilkinson, S.J. & Dixon, Tim (2016). Green roof retrofit. Building urban resilience.

Wilkinson, S. (2012). Analysing sustainable retrofit potential in premium office buildings. *Structural Survey, 30*(5), 398–410. doi:10.1108/02630801211288189

Wilson, L. (2011). Average household electricity use around the world. *Shrink That Footprint*. Retrieved from http://shrinkthatfootprint.com/average-household-electricity-consumption

Wilson, J. (2013). *The greened house effect. Renovating your home with a deep energy retrofit*. White River Junction, VT: Chelsea Green Publishing.

Wind G. (2017). Graue Energie.

Wind, G. (2017). Graue Energie. Retrieved from http://www.ibwind.at/download/GraueEnergie081121.pdf

Wirtz, B., Pistoia, A., Ulrich, S., & Gottel, V. (2016). Business Models: Origin, development and future research perspectives. *Long Range Planning*, *49*(1), 36–54. doi:10.1016/j.lrp.2015.04.001

Wolf, W. H. (2009). Cyber-physical systems. *IEEE Computer*, *42*(3), 88–89. doi:10.1109/MC.2009.81

Woods, J. E., Drewry, G. M., & Morey, P. (1987). Office worker perceptions of indoor air quality effects on discomfort and performance. In *Proceedings of the 4th International Conference on Indoor Air and Climate*, Berlin, Germany (pp. 464-408).

World Bank. (2010). Cities and Climate Change: An Urgent Agenda. Retrieved from https://openknowledge.worldbank.org/handle/10986/17381

World Bank. (2015). Competitive Cities for Jobs and Growth: What, Who, and How. Retrieved from https://openknowledge.worldbank.org/handle/10986/23227

World Bank. (2016). The Cost of Air Pollution. Strengthening the Economic Case for Action. Retrieved from http://documents.worldbank.org/curated/en/781521473177013155/pdf/108141-REVISED-Cost-of-PollutionWebCORRECTEDfile.pdf

World Health Organization - WHO. (1983). World Health Organization. Indoor Air Pollutants: Exposure and Health Effects. *EURO Reports and Studies*, *78*, 1–42.

World Health Organization - WHO. (1986). Indoor Air Quality Research – EURO Reports and Studies. WHO Regional Office for Europe, Copenhagen.

World Health Organization - WHO. (1999). *Strategic approaches to indoor air policy-making*. Bilthoven: WHO European Center for Environment and Health, Retrieved from http://www.paho.org/tierra/index.php?option=com_multicategories&view=article&id=75%3Astrategic-approaches-to-indoor-airpolicymaking&catid=9%3A19962000&Itemid=136& lang=en

Würtenberger, T., Bleyl, J., Menkveld, M., Vethman, P., & Tilburg, X. (2012). Business models for renewable energy in the built environment. IEA. Retrieved from http://iea-retd.org/wp-content/uploads/2012/04/RE-BIZZ-final-report.pdf

Wu, Z., Xia, X., & Wang, B. (2015). Improving building energy efficiency by multiobjective neighborhood field optimization. *Energy and Building*, *87*, 45–56. doi:10.1016/j.enbuild.2014.10.079

Xu, P., Chan, E., & Qian, Q. (2011). Success factors of energy performance contracting (EPC) for sustainable building energy efficiency retrofit (BEER) of hotel buildings in China. *Energy Policy*, *39*(11), 7389–7398. doi:10.1016/j.enpol.2011.09.001

Yang J. (2017). Influence of envelope insulation materials on building energy consumption.

Yang, C. H., & Heinsohn, P. A. (2007). Sampling and analysis of indoor microorganism (1st ed.). Wiley-Interscience. doi:10.1002/0470112433

Yoo, S. H., & Kwak, S. Y. (2009). Willingness to pay for green electricity in Korea: A contingent valuation study. *Energy Policy*, *37*, 5408–5416. doi:10.1016/j.enpol.2009.07.062

Zhao, X., Hwang, B. G., & Lu, Q. (2018). Typology of business model innovations for delivering zero carbon buildings. Journal of cleaner production, 196, 1213-1226.

Zuo, J., & Zhao, Z. (2014). Green building research-current status and future agenda: A review. *Renewable & Sustainable Energy Reviews*, *30*, 271–281. doi:10.1016/j.rser.2013.10.021

About the Contributors

Adrian Tantau is Professor at ASE Bucharest, UNESCO Chair for Business Administration, where he has been coordinating since 2014 the MBA Entrepreneurship and Business Administration in Energy. Between 2012 and 2016, he was Dean of the Faculty of Business Administration. Since 2012, he has been President at "Society for Business Excellence". Between 2009 and 2013, he was a member of the Board of the Company Hidroserv Portile de Fier S.A., Drobeta Turnu Severin. Mr. Adrian Dumitru Tantau has a PhD in Electrical Engineering (Politechnical University Bucharest) and a PhD in Management (Bucharest University of Economic Studies). Also, he graduated from the Faculty of Electrotechnic (UPB) and the Faculty of Management (ASE). His research activity is focused on business models in energy, especially renewable energy, energy security, and energy efficiency. At present, he is coordinating his PhD thesis in this field. He is a member at the European Foundation for Entrepreneurship Research, since 2006, member at the Commission of Economical Science and Business Administration CNATDCU since 2012, member at the Scientific Committee at International Conference on Business Excellence, Brasov 2009-2014, member at the Scientific Committee at International Conference Business Excellence in energy, Bucharest 2016. Also, he is an honorary member at the Romanian National Committee at Global Energy Council, in 2014.

* * *

Raluca Caplescu is an Assistant Professor at The Bucharest University of Economic Studies, Department of Statistics and Econometrics. Besides teaching, she also has experience in data analysis for finance and market research sectors.

Cătălin Curmei is lecturer in the UNESCO Department for Business Administration at the Bucharest University of Economic Studies. His interest fields are the marketing in small and medium size enterprises and the relational marketing in energy sector.

Andreea Curmei-Semenescu is an associate professor in the Finance Department of the Bucharest University of Economic Studies, she has as major interest fields corporate finance, capital budgeting and company valuation.

Raluca Andreea Felseghi is a Civil Engineer and Independent Researcher, she graduated the Technical University of Cluj-Napoca, Faculty of Building Services. Starting in 2011, she attended the postgraduate work (PhD) within Technical University of Cluj-Napoca and obtained the Doctoral Diploma in a fundamental field: Engineering Sciences, Civil Engineering. Her main fields of interests are energy efficiency of buildings; stationary application of renewable energies; fuel cell hydrogen technology; hybrid power generation systems; the impact of alternative energies on the environment and climate change. She is a member of the Romanian Association of Building Services Engineering (AIIR) - Transylvania Branch and member of Romanian Association for Hydrogen Energy (AEHR) - Cluj-Napoca Branch.

Constantin Filote is Full Professor with Computer Science, Electronics, and Automation Department of Stefan cel Mare University of Suceava. He received both the BSc (1988) and the PhD (1997) in Electronics and Telecommunications from Technical University of Iasi, Romania. His main research interests are in the fields of Cloud Computing, Digital Signal Processing, Process Modelling and Simulation, Engineering Education. Dr Filote is author and co-author of more than 120 technical papers in scientific conference proceedings and journals (67 ISI Web of Science), 4 national patent, 7 application patents, as well as 8 books and book chapters. He has worked as project manager (FP7-ICT-10-2013.1.6-610576-NUBOMEDIA) and researcher/financial responsible in several national and international research grants (EEA-Mommypreneurs, EEA-PAZEWAIA, LIFE+-ECOREG, EASME EYE-BYE-8, CIVIL JUSTICE-EIRENE, CBC-Think Green, SICRAMAS, FILOLET, ECODIN, SMiLE-EV). In 2012, Dr. Filote was awarded IT/Electronics/Electrical Invention Order of Merit at WIAF 2012 by KINEWS, Seoul, Korea.

Laurentiu Fratila is Assoc. Prof. at Bucharest University of Economic Studies in Romania. He holds a PhD in Applied Information Systems. He published numerous studies and books in the fields of interest: applied computer science, management information systems in energy and management information system in finance. He has participated in national and international research projects as project leader or member of research teams.

Giani Gradinaru is a Professor of the Statistics and Econometrics Department, The Bucharest University of Economics Studies and reasearcher at Institute of National Economy, Roamian Academy. He held didactic activity both undergraduate and post graduate level. He received the title of doctor in cybernetics and economic statistics in 2004 and habilitation in 2015. Postdoctoral scientific research aimed environmental statistics field, objectifying in books, articles and scientific workpapers. He is an expert evaluator, research project manager and PhD advisor.

Michael Krutwig is founder and managing director of krumedia GmbH, a German software company specializing in energy efficiency. His areas of expertise are software tools and SaaS for energy management and energy audits, Smart City, Smart Production and security aspects. He holds a Master's degree in Computer Science and is currently a PhD student at the Bucharest University of Economics (ASE).

Maria Alexandra Maassen is a lecturer, The Bucharest University of Economic Studies, Faculty of Business Administration in foreign languages (2019-present); Assistant prof. dr. The Bucharest University of Economic Studies, Faculty of Business Administration in foreign languages (2016-2019); PhD in Economics, specialization in Business Administration, The Bucharest University of Economic Studies (2012-2015); Member of Society for Business Excellence (since 2016); Member of organizing committee of the International Conference on Business Excellence (since 2016); Senior Procurement process associate for the German market, Genpact (2015-2019); Visiting researcher (POSDRU), The

University of Wageningen, Netherlands (2014-2015); Scientific secretary of the Research Center for Intellectual Capital and Entrepreneurship (since 2016).

Eugen Mitrică is Associate Professor in the Finance Department of the Bucharest University of Economics, where he has been teaching different finance topics, based on financial modeling, since 1998. He earned his PhD in Finance from the Bucharest University of Economics in 2007. He also has an extended practical experience in the finance field, working as: Auditor, Financial Controller, Chief Financial Officer and Financial Consultant for investments. He was involved for a while in the study of efficiency of capital markets. During the last years, his research was focused on public investment decisions and especially on the economic analysis, in the cost benefit analysis for investment projects regarding the nature preservation and mitigating the pollution effects. Also he was involved in a multi-disciplinary project related to climate change impact on the water supply and demand.

Alina Moise is an architecture student at Ion Mincu University of Architecture and Urbanism in Bucharest, Romania ('19). She has a particular interest in sustainable design, and has worked as part of an interdisciplinary team designing energy-efficient solar houses for the International Solar Decathlon. Alina has international experience in the architectural field taking part in the Erasmus Exchange program studying in Gdańsk, Poland, and working as a junior designer in a renewed architecture office in Helsinki, Finland. In addition to her studies and work experiences, Alina enjoys discovering new places and cultures and photographing her travels.

Cristian Bogdan Onete was born in 1959 in Bucharest, Romania and is full professor at The Bucharest University of Economic Studies, Faculty of Business and Tourism. His actual position is Head of Department Business, Consumer Science & Quality Management, The Bucharest University of Economic Studies, Faculty of Business and Tourism since 2007. He is now member of The University Senate, member of The Council of Business and Tourism Faculty and member of Business Administration Doctoral School of The Bucharest University of economic Studies. He holds a B.Sc. degree in Commerce and a Ph.D. degree in field of Economics from The Bucharest University of Economic Studies. And member of The University Senate. Professor Cristian - Bogdan Onete is specialist in the fields of: Information Systems Management, E-commerce, Consumer Sciences and Quality, teaching at bachelor, master and Ph. D. programs. Professor Cristian - Bogdan ONETE has interests focus on information technology, green economy and sustainable development. He also developed researches in product quality through social media, electronic commerce models in relation to consumer, the use of social media in business organization and also researches regarding new consumer behavior.

Corina Pelau is Associate Professor Ph.D. at the UNESCO Department for Business Administration and Vice-Dean of the Faculty for Business Administration, in foreign languages, within the Bucharest University of Economic Studies, Romania. Her main field of interest is consumer research with focus on the impact of digitalization and artificial intelligence on consumer behavior, sustainable consumption and neuromarketing. She is the author or coauthor of 3 books and published more than 60 articles in scientific journals indexed in international databases or at international conferences. She has been a member in 4 research grants. She is member of the scientific committee of several international journals or conferences. Her research work has been rewarded with the Best Paper in an Economic Journal Award by AFER (Association of Economic Faculties in Romania) in 2018 and the FESTO Young Researchers Award, at DAAAM International Conference, Trnava, Slovakia (2008). She is an expert evaluator for ARACIS, Commission Economic Science I (since 2014), member of the Association for Consumer Research (since 2015) and member of the European Marketing Academy (since 2018).

Doru Alexandru Pleşea is full professor and Vice Dean at the Bucharest University of Economic Studies, Faculty of Commerce. He holds a B.Sc. degree in Physics from the University of Bucharest and a Ph.D. degree in field of Economics from the Bucharest University of Economic Studies. Professor Plesea is currently teaching Science of Industrial Commodities, Electronic Commerce at bachelor programmes and Business Intelligence at master programmes. His professional interests focus on e-commerce, domotics, smart cities, green economy and sustainable development. In addition he is member of International Association of Commodity Science, Innovation and Sustainability (IACSIS) and of the Association for Innovation and Quality in Sustainable Business (BASIQ). Between 2006 and 2010 he was member of the editorial board of the "Tovari i rinki" journal (Commodities and Markets, ISSN 1998-2666). Between 2001 and 2004 professor Plesea was honored with Georgescu Roegen award for his contribution to scientific research. He is author of 9 books and more than 65 scientific papers and studies in specialized journal and volumes of scientific conferences. Also he has a relevant contribution to 15 research grants.

Andra Pop has a Master Degree in Entrepreneurship and Business Administration, within the Bucharest University of Economic Studies, Romania. Her main fields of interest are CSR, consumer knowledge in the energy field and education. She is Campaign Manager at the Gandul Aprins NGO and is responsible with the development and implementation of cultural and social projects.

Mihai-Ionut Pop is a Ph. D. student at the Business Administration Doctoral School within the Bucharest University of Economic Studies, Romania. His main fields of interest are digitalization, communication in the online environment and behaviour patterns of online users. He is the author of several scientific articles and the Project Manager at Digitix Media, a digital marketing agency, being responsible with the development and implementation of online communication strategies and private online projects.

Maria Simona Raboaca is working as Researcher in to National Research and Development Institute for Cryogenics and Isotopic Technologies ICSI Rm. Valcea, Hydrogen and Fuel Cell Department. Her PhD is "Theoretical and practical Contribution regarding to sustain with hybrid energy a Passive House" in Faculty of Building Services Engineering in Technical University of Cluj-Napoca, Romania. Now, she is project manager at ICSI to project "Smart conductive charging station, fixed and mobile, for electric propulsion transportation (SMiLE-EV)" proposes the deployment of fixed and mobile EV & PHEV charging stations to meet the mobility needs of tomorrow's society and to prepare active/potential industrial partners for knowledge/technology transfer at the component or system level in prepare launching new products. She has been contributing to the field of Renewable Energy, Green Buildings, Passive House Concept, Hydrogen Energy and Stationary and Mobile applications. She is author and co-author of more technical papers in scientific conference proceedings and ISI journals.

Teodora Melania Şoimoşan is working as Head of Works in Faculty of Civil Engineering, in Technical University of Cluj-Napoca, Romania. She has been contributing to the field of Renewable Energy and Energy Systems by performing various roles such as Head of Works, Researcher, Author and Reviewer.

She is Doctor of Philosophy in Civil Engineering, she has Master Degree in Management and Performant Technologies in Buildings and Bachelor Degree in the field of Building Services Engineering.

Robert Staiger lives in south of Germany. After his apprenticeship as a light mechanic he did a technical degree in mechanical/electrical on the University of Applied Science in Furtwangen (Dipl. Ing. (FH)). He also holds various degrees in energy and emissions protection related topics. In 2013, he finished up his economic studies (MBA) on the University Lahr (WHL). In 2014, he started his postgraduate work (PhD) in the field of sustainability, renewable energy and hydrogen as a renewable energy carrier, on the Bucharest University of Economic studies (supervisor Prof. Dr. Dr. A Tantau).

Ion Zgură is assistant professor at the Bucharest University of Economic Studies, Faculty of Business and Tourism. He holds a B.Sc. degree in Unconventional Technology. Dr. Zgura has a Ph.D in Economics from the Bucharest University of Economic Studies with the Thesis ''Relational models in electronic commerce''. In 2009, he obtained a Master's degree from the Bucharest University of Economic Studies in Business Administration. He is a active member of Association for Innovation and Quality in Sustainable Business (BASIQ). Dr. Zgura is a researcher at the the Research Center for Business Administration where is involved in several projects. He has also acted as web developer and part of scientific committees and conference organizing committees internationals in Germany, Austria, Italy, Romania. Dr Zgura published as a author and co-author of articles and scientific publications in magazines indexed in international databases, ISI Web of Knowledge and ISI Proceedings. His research interests are the field of technology, architectural informatics systems and unconventional technology.

Index

www.ingramcontent.com/pod-product-compliance
Lightning Source LLC
LaVergne TN
LVHW062137300926
849161LV00010B/2557